SUSTAINABLE SUGARCANE PRODUCTION

SUSTAINABLE SUGARCANE PRODUCTION

Edited by
Priyanka Singh, PhD
Ajay Kumar Tiwari, PhD

CRC Press
Taylor & Francis Group
Boca Raton London New York

CRC Press is an imprint of the
Taylor & Francis Group, an **informa** business

First published 2018 by Apple Academic Press, Inc.

Published 2019 by CRC Press
Taylor & Francis Group
6000 Broken Sound Parkway NW, Suite 300
Boca Raton, FL 33487-2742

First issued in paperback 2021

CRC Press is an imprint of the Taylor & Francis Group, an informa business

No claim to original U.S. Government works

ISBN 13: 978-1-77463-397-7 (pbk)
ISBN 13: 978-1-77188-702-1 (hbk)

Library and Archives Canada Cataloguing in Publication

Sustainable sugarcane production / edited by Priyanka Singh, PhD, Ajay Kumar Tiwari, PhD.

Includes bibliographical references and index.
Issued in print and electronic formats.
ISBN 978-1-77188-702-1 (hardcover).--ISBN 978-1-351-04776-0 (PDF)
1. Sugarcane. 2. Sustainable agriculture. I. Tiwari, Ajay Kumar, editor II. Singh, Priyanka (Scientist), editor

| SB226.2.S87 2018 | 633.6'1 | C2018-900408-8 | C2018-900409-6 |

CIP data on file with US Library of Congress

Visit the Taylor Francis Web site at
http://www.taylorandfrancis.com

and the CRC Press Web site at
http://www.crcpress.com

CONTENTS

ABOUT THE EDITORS

Priyanka Singh, PhD
Scientific Officer, Uttar Pradesh Council of Sugarcane Research, Shahjahanpur, India

Priyanka Singh, PhD, is the Scientific Officer of the Uttar Pradesh Council of Sugarcane Research, (UPCSR) Shahjahanpur, India. She has worked at the Indian Institute of Sugarcane Research, Lucknow, India for nine years. She has 19 years of research experience with specialization in organophosphorus chemistry and cane quality/postharvest management of sugar losses with the help of chemicals as well as eco-friendly compounds. She has extensive experience in using electrolyzed water to preserve cane quality, and she was the first person to report on the immense potential of electrolyzed water to preserve postharvest sucrose losses.

Dr. Singh has synthesized and characterized 37 new organophosphorus compounds belonging to the chalcone series, of which two important chemicals (*Chalcone thiosemicarbazone* and *Chalcone dithiocarbazate*) were found to be highly fungitoxic to the sugarcane parasitic fungi *Colletotricum falcatum, Fusarium oxysporum*, and *Curvularia pallescene*. She has worked on the extraction of volatile constituents from higher plants and their biological activity against agricultural pests; the management of postharvest formation of nonsugar and polysaccharide compounds in sugarcane and the effect of bioproducts on growth, yield, and quality of sugarcane and soil health; and indicators of postharvest losses in sugarcane. She is also credited for identifying Mannitol as one of the most important indicators of postharvest losses in sugarcane.

Presently, she is working on a varietal spectrum of sugarcane for the selection of elite sugarcane varieties so as to recommend the proper varietal balance of sugarcane varieties in Uttar Pradesh (India). She is also modulating the activities of sucrose metabolizing enzymes through bio-active silicon (orthosilicic acid) for increased cane and sugar productivity, which will benefit farmers as well as the sugar industry. She is also lending her expertise to a project on "Assessment of postharvest quality deterioration in promising sugarcane varieties under sub-tropical condition," which is expected to reduce postharvest losses and increase sugar recovery; and undertaking a project on "Varietal screening for jaggery" production at UPCSR, Shahjahanpur, for the recommendation of elite sugarcane varieties for the commercial production of jaggery.

Dr. Singh is a research advisor for a dissertation on genetic diversity in sugarcane and a training advisor for MSc students. She has organized several short-term training programs on techniques in microbiology, biotechnology, and molecular biology.

She received an "Award of Excellence" from Sinai University, Al Arish, Egypt, in 2008. She is also serving as Managing Editor for the journal Sugar Tech and as Executive Editor for journal *Agarica*. In addition, she is a reviewer for several international journals. She is one of the editors of the *Proceeding of International Conference IS-2011*. A prolific author, Dr. Singh has authored a book on innovative healthy recipes with jaggery, edited three books on postharvest management of sugarcane, and written several annual reports and a paper on 100 years of sugarcane research. She has also published three book chapters and more than 50 research papers in various national and international journals and proceedings. She has attended several national and international conferences and workshops in China, Egypt, Thailand, and India, and has coordinated technical as well as plenary sessions in India, China, and Egypt.

Dr. Singh completed her PhD on "Efficacy of organophophorus derivatives against fungal pathogens of sugarcane" in 2000 from DDU, Gorakhpur University (Uttar Pradesh, India). She was awarded a postdoctoral fellowship from DST, New Delhi.

Ajay Kumar Tiwari, PhD
Scientific Officer, Uttar Pradesh Council of Sugarcane Research, Shahjahanpur, India

Ajay K. Tiwari, PhD, is a Scientific Officer at the UP Council of Sugarcane Research, Shahjahnapur, UP, India. He is a regular member of the British Society of Plant Pathology, Indian Phytopathological Society, Sugarcane Technologists Association of India, International Society of Sugarcane Technologists, Society of Sugarcane Research and Promotion, Prof H. S. Srivastava Foundation, and Society of Plant Research. He has published 75 research articles and 12 review articles in national and international journals. He has also published six book chapters in edited books and has also authored seven edited books. He has submitted more than 150 nucleotide sequences of plant pathogens to Genbank.

Dr. Tiwari is a regular reviewer and member of the editorial board for many international journals. He is managing editor of Sugar Tech journal and Chief Editor of Agrica journal. He has been awarded the Young Researcher Award in Italy 2011, and the Young Scientist Award by DST-SERB, and he was nominated for the Narshiman Award by the Indian Phytopathological Society. Very recently, he was awarded the Young Scientist Award by the Chief Minister of the State Government of UP for his outstanding contribution in the area of plant pathology. Dr. Tiwari is the recipient of many international travel awards given by DST, DBT, and CSIR from India; PATHOLUX from Luxembourg; and IOM from Brazil. He has visited China, Italy, Germany, and Thailand for conferences and workshops. He has been involved in research on molecular characterization and management of agricultural plant pathogens for the last nine years. Currently, he is working on molecular characterization of sugarcane phytoplasma and their secondary spread in nature.

Dr. Tiwari earned his PhD in 2011 on Cucurbit viruses from the Biotechnology Department of CCS University, Meerut, UP, India.

LIST OF CONTRIBUTORS

R. O. Bordonal
Brazilian Bioethanol Science and Technology Laboratory – CTBE, Brazilian Center for Research in Energy and Materials – CNPEM, Giuseppe Maximo Scolfaro Street 10000, Cidade Universitária, Campinas, 13083–100, Brazil, E-mail: henrique.franco@ctbe.cnpem.br

B. M. M. N. Borges
Brazilian Bioethanol Science and Technology Laboratory – CTBE, Brazilian Center for Research in Energy and Materials – CNPEM, Giuseppe Maximo Scolfaro Street 10000, Cidade Universitária, Campinas, 13083–100, Brazil, E-mail: henrique.franco@ctbe.cnpem.br

C. D. Borges
Brazilian Bioethanol Science and Technology Laboratory – CTBE, Brazilian Center for Research in Energy and Materials – CNPEM, Giuseppe Maximo Scolfaro Street 10000, Cidade Universitária, Campinas, 13083–100, Brazil, E-mail: henrique.franco@ctbe.cnpem.br

H. Cantarella
Agronomic Institute of Campinas, Soils and Environmental Resources Center, P.O. Box 28, 13001-970 Campinas, SP, Brazil

M. O. Cardoso
Universidade Federal Rural de Pernambuco, Departamento de Engenharia Agrícola, Rua Dom Manoel de Medeiros, S/N, Dois Irmãos, Recife, Pernambuco, 52.171-900, Brazil

J. B. Carmo
Department of Environmental Sciences, Federal University of Sao Carlos (UFSCar), Rod. João Leme dos Santos Km 110, 18052-780 Sorocaba, SP, Brazil

S. G. Q. Castro
Brazilian Bioethanol Science and Technology Laboratory – CTBE, Brazilian Center for Research in Energy and Materials – CNPEM, Giuseppe Maximo Scolfaro Street 10000, Cidade Universitária, Campinas, 13083–100, Brazil, E-mail: henrique.franco@ctbe.cnpem.br

Amaresh Chandra
Division of Plant Physiology and Biochemistry, ICAR-Indian Institute of Sugarcane Research, Lucknow–226002, India, E-mail: amaresh_chandra@rediffmail.com

C. A. C. Crusciol
São Paulo State University (UNESP), School of Agriculture, Department of Crop Science, P.O. Box: 237, 18610-307 Botucatu, State of São Paulo, Brazil

Anna Durai
ICAR- Sugarcane Breeding Institute, Coimbatore – 641007, Tamilnadu, India

H. C. J. Franco
Brazilian Bioethanol Science and Technology Laboratory – CTBE, Brazilian Center for Research in Energy and Materials – CNPEM, Giuseppe Maximo Scolfaro Street 10000, Cidade Universitária, Campinas, 13083–100, Brazil, E-mail: henrique.franco@ctbe.cnpem.br

R. Gomathi
Plant Physiology, ICAR-Sugarcane Breeding Institute, Coimbatore–641007, Tamil Nadu, India, E-mail: gomathi_sbi@yahoo.co.in

L. M. P. Guimarães
Universidade Federal Rural de Pernambuco, Departamento de Engenharia Agronomia, Rua Dom
Manoel de Medeiros, S/N, Dois Irmãos, Recife, Pernambuco, 52.171-900, Brazil

Yupa Hanboonsong
Entomology Division, Faculty of Agriculture, Khon Kaen University, Thailand, E-mail: yupa_han@
kku.ac.thyupa_han@yahoo.com

María La O. Hechavarría
Institute of Sugarcane Research (INICA), Carretera ISPJAE, Km 1, Boyeros–19390, Havana, Cuba

Yaquelin Puchades Izaguirre
Institute of Sugarcane Research (INICA), Carretera ISPJAE, Km 1, Boyeros–19390, Havana, Cuba

V. P. Jaiswal
ICAR – Indian Institute of Sugarcane Research, Lucknow–226002, Uttar Pradesh, India

O. T. Kölln
Brazilian Bioethanol Science and Technology Laboratory – CTBE, Brazilian Center for Research in
Energy and Materials – CNPEM, Giuseppe Maximo Scolfaro Street 10000, Cidade Universitária,
Campinas, 13083–100, Brazil, E-mail: henrique.franco@ctbe.cnpem.br

Pavan Kumar
Division of Crop Improvement, ICAR-Indian Institute of Sugarcane Research, Lucknow–226002, Uttar
Pradesh, India,
Department of Biotechnology, Bundelkhand University, Jhansi – 284128, Uttar Pradesh, India

Arvind Kumar
UPCSR, Sugarcane Research Institute, Shahanjanpur – 242001, UP, India

Menhi Lal
Division of Crop Production, ICAR-Indian Institute of Sugarcane Research, Lucknow–226002, India

Eida Rodríguez Lema
Institute of Sugarcane Research (INICA), Carretera ISPJAE, Km 1, Boyeros–19390, Havana, Cuba

Chang-Ning Li
Guangxi Key Laboratory of Sugarcane Genetic Improvement, Key Laboratory of Sugarcane
Biotechnology and Genetic Improvement (Guangxi), Ministry of Agriculture, Sugarcane Research
Center, Chinese Academy of Agricultural Sciences; Sugarcane Research Institute, Guangxi Academy of
Agricultural Sciences, Nanning 530007, China, E-mail: liyr@gxaas.net

Yang-Rui Li
Guangxi Key Laboratory of Sugarcane Genetic Improvement, Key Laboratory of Sugarcane
Biotechnology and Genetic Improvement (Guangxi), Ministry of Agriculture, Sugarcane Research
Center, Chinese Academy of Agricultural Sciences; Sugarcane Research Institute, Guangxi Academy of
Agricultural Sciences, Nanning 530007, China, E-mail: liyr@gxaas.net

Qiang Liang
Guangxi Key Laboratory of Sugarcane Genetic Improvement, Key Laboratory of Sugarcane
Biotechnology and Genetic Improvement (Guangxi), Ministry of Agriculture, Sugarcane Research
Center, Chinese Academy of Agricultural Sciences; Sugarcane Research Institute, Guangxi Academy of
Agricultural Sciences, Nanning 530007, China, E-mail: liyr@gxaas.net

Xi-Hui Liu
Guangxi Key Laboratory of Sugarcane Genetic Improvement, Key Laboratory of Sugarcane
Biotechnology and Genetic Improvement (Guangxi), Ministry of Agriculture, Sugarcane Research
Center, Chinese Academy of Agricultural Sciences; Sugarcane Research Institute, Guangxi Academy of
Agricultural Sciences, Nanning 530007, China, E-mail: liyr@gxaas.net

S. R. V. L. Maranhão
Universidade Federal Rural de Pernambuco, Departamento de Engenharia Agronomia, Rua Dom Manoel de Medeiros, S/N, Dois Irmãos, Recife, Pernambuco, 52.171-900, Brazil

C. A. C. Nascimento
São Paulo State University (UNESP), School of Agriculture, Department of Crop Science, P.O. Box: 237, 18610-307 Botucatu, State of São Paulo, Brazil

P. M. Pardalos
Department of Industrial and Systems Engineering, University of Florida, Center of Applied Optimization, Distinguished Professor, Paul and Heidi Brown Preeminent Professor, 401 Weil Hall, Gainesville, FL 32611-6595, USA, E-mail: pardalos@ise.ufl.edu

E. M. R. Pedrosa
Universidade Federal Rural de Pernambuco, Departamento de Engenharia Agrícola, Rua Dom Manoel de Medeiros, S/N, Dois Irmãos, Recife, Pernambuco, 52.171-900, Brazil

Andréa Chaves Fiuza Porto
Universidade Federal Rural de Pernambuco, Estação Experimental de Cana-de-açúcar de Carpina, Rua Angela Cristina C. P. de Luna, S/N. Bairro de Santa Terezinha, Carpina, Pernambuco, 55810-000, Brazil, E-mail: achavesfiuza@yahoo.com.br

Mérida Rodríguez Regal
Institute of Sugarcane Research (INICA), Carretera ISPJAE, Km 1, Boyeros–19390, Havana, Cuba

R. Rossetto
Agência Paulista deTecnologia do Agronegócio (APTA), Centro de Cana-de-Açúcar do IAC, Rodovia SP 127 km 30, 13400–970 Piracicaba, SP, Brazil, E-mail: raffaella@apta.sp.gov.br

G. M. Sanches
Brazilian Bioethanol Science and Technology Laboratory – CTBE, Brazilian Center for Research in Energy and Materials – CNPEM, Giuseppe Maximo Scolfaro Street 10000, Cidade Universitária, Campinas, 13083–100, Brazil, E-mail: henrique.franco@ctbe.cnpem.br

Gulzar S. Sanghera
PAU, Regional Research Station, Kapurthala, Punjab, 144601, India, E-mail: sangheragulzar@gmail.com

T. Rajula Shanthy
Extension, ICAR-Sugarcane Breeding Institute, Coimbatore–641007, Tamil Nadu, India, E-mail: rajula.sbi@gmail.com

Lalan Sharma
ICAR – Indian Institute of Sugarcane Research, Lucknow–226002, Uttar Pradesh, India

Allan T. Showler
Knipling-Bushland, U.S. Livestock Insects Research Laboratory, USDA-ARS, 2700 Fredericksburg Road, Kerrville, TX78028, USA, E-mail: allan.showler@ars.usda.gov

S. K. Shukla
ICAR – Indian Institute of Sugarcane Research, Lucknow–226002, Uttar Pradesh, India, E-mail: sudhirshukla151@gmail, sudhir.shukla@icar.gov.in

A. K. Singh
Division of Crop Production, ICAR-Indian Institute of Sugarcane Research, Lucknow–226002, India, E-mail: shantaanil@yahoo.com

Akhilesh Kumar Singh
Division of Agricultural Engineering, ICAR-Indian Institute of Sugarcane Research, Lucknow–226002, India, E-mail: aksingh8375@gmail.com

Ekta Singh
Division of Crop Production, ICAR-Indian Institute of Sugarcane Research, Lucknow–226002, India

Xiu-Peng Song
Guangxi Key Laboratory of Sugarcane Genetic Improvement, Key Laboratory of Sugarcane Biotechnology and Genetic Improvement (Guangxi), Ministry of Agrioulture, Sugarcane Research Center, Chinese Academy of Agricultural Sciences; Sugarcane Research Institute, Guangxi Academy of Agricultural Sciences, Nanning 530007, China, E-mail: liyr@gxaas.net

Sangeeta Srivastava
Division of Crop Improvement, ICAR-Indian Institute of Sugarcane Research, Lucknow–226002, Uttar Pradesh, India,
E-mail: Sangeeta.Srivastava@icar.gov.in, sangeeta_iisr@yahoo.co.in

Hong-Wei Tan
Guangxi Key Laboratory of Sugarcane Genetic Improvement, Key Laboratory of Sugarcane Biotechnology and Genetic Improvement (Guangxi), Ministry of Agriculture, Sugarcane Research Center, Chinese Academy of Agricultural Sciences; Sugarcane Research Institute, Guangxi Academy of Agricultural Sciences, Nanning 530007, China, E-mail: liyr@gxaas.net

K. S. Thind
PAU, Regional Research Station, Kapurthala, Punjab, 144601, India

G. Vlontzos
Department of Agriculture Crop Production and Rural Development, School of Agricultural Sciences, University of Thessaly, Fytoko, 38446 Volos, Greece

Wei-Zan Wang
Guangxi Key Laboratory of Sugarcane Genetic Improvement, Key Laboratory of Sugarcane Biotechnology and Genetic Improvement (Guangxi), Ministry of Agriculture, Sugarcane Research Center, Chinese Academy of Agricultural Sciences; Sugarcane Research Institute, Guangxi Academy of Agricultural Sciences, Nanning 530007, China, E-mail: liyr@gxaas.net

Jian-Ming Wu
Guangxi Key Laboratory of Sugarcane Genetic Improvement, Key Laboratory of Sugarcane Biotechnology and Genetic Improvement (Guangxi), Ministry of Agriculture, Sugarcane Research Center, Chinese Academy of Agricultural Sciences; Sugarcane Research Institute, Guangxi Academy of Agricultural Sciences, Nanning 530007, China, E-mail: liyr@gxaas.net

Li-Tao Yang
State Key Laboratory for Conservation and Utilization of Subtropical Agro-Bioresources, Agricultural College, Guangxi University, Nanning 530005, China, E-mail: liyr@gxu.edu.cn

LIST OF ABBREVIATIONS

ABA	abscisic acid
ACIAR	Australian Centre for International Agriculture Research
ADH	alcohol dehydrogenase
AFLP	amplified fragment length polymorphism
AMMI	additive main effects and multiplicative interaction
ANP's	anaerobic polypetides
BAC	bacterial artificial chromosome
BOD	biochemical oxygen demand
CEC	cation exchange capacity
COD	chemical oxygen demand
CSIATI	Chinese Sugarcane Industry Association for Technological Innovation
DANIDA	Danish International Development Agency
DBIA	dot-blot immunoassay
DCD	dicaynadiamide
DD	differential display
DGR	Directorate of Groundnut Research
DOC	dissolved organic carbon
ERF	ethylene responsive factor
FAO	Food and Agriculture Organization
FDS	fixed dissolved solids
FDV	Fiji disease virus
FIRB	furrow irrigated raised
FISH	fluorescence in situ hybridization
FNAAS	Fellow of National Academy of Agricultural Sciences
GEOBIA	geographic object-based image analysis
GFP	green fluorescent protein
GISH	genomic in situ hybridization
GSD	grassy shoot disease
IAPSIT	international association for professionals in sugar and integrated technologies
ICT	information and communication technology

IFC	International Finance Corporation
IIAST	Integral Institute of Agricultural Science and Technology
IISR	Indian Institute of Sugarcane Research
IM	isomaltulose
IMDH-RD	Integrated, Multi-Disciplinary and Holistic Approach to Rural Development
IPM	integrated pest management
ISA	Indian Society of Agronomy
ISPP	Indian Society of Plant Physiology
JA	jasmonic acid
LAI	leaf area index
LAR	leaf area ratio
MHAT	moist hot air treatment
MOP	muriate of potash
NAR	net assimilation rate
NGS	next generation sequencing
NI	neutral invertase
NRA	nitrate reductase activity
NUE	nitrogen use efficiency
OM	organic matter
PA	precision agriculture
PETS	photosynthetic electron transfer system
PGQP	plant germplasm quarantine program
PI	phosphoinositide
PMC	press-mud cake
PMI	phosphomannose isomerase
PPP	public private partnership
PR	pathogenesis related
RFLP	restriction fragment length polymorphism
RGAs	resistant gene analogues
RGR	relative growth rate
RMD	ratoon management device
ROS	reactive oxygen species
RRL	resistance response library
RSD	ratoon stunting disease
RWC	relative water content
SAI	soluble acid invertase
SAM	s-adenosyl methionine

SAP	society of agricultural professionals
SCAR	sequence characterized amplified region
SCBV	sugarcane bacilliform virus
SCMV	sugarcane mosaic virus
SCSMV	sugarcane streak mosaic virus
SCSV	sugarcane streak virus
SCWL	sugarcane white leaf
SCYLV	sugarcane yellow leaf virus
SEC	strategic extension campaign
SNP	single nucleotide polymorphism
SPM	sulphitation press-mud
SPS	sucrose phosphate synthase
SRA	Sugar Research Australia
SRAP	sequence related amplified polymorphism
SRL	susceptible response library
SSF	suspended solids fixed
SSI	sustainable sugarcane initiative
SSP	single superphosphate
SSRP	Society for Sugar Research and Promotion
SSRs	simple sequence repeats
STP	spaced transplanting technique
SUGESI	The Sugarcane Genome Sequencing Initiative
TBIA	tissue-blot immunoassay
TDFs	transcript-derived fragments
TFP	total factor productivity
TFs	transcriptional factors
TIBs	temporary immersion bioreactor system
TRAP	target region amplification polymorphism
TS	total solids
TSS	total suspended solids
UPAAS	Uttar Pradesh Academy of Agricultural Sciences
VDS	volatile dissolved solids
VSS	volatile suspended solids
WR	water requirement
YLD	yellow leaf disease

FOREWORD

Sugarcane, a climate friendly crop, earlier considered as a major source of sweetener, has now acquired a novel status in the international market owing to its tremendous agro-industrial value. The crop sequesters carbondioxide from the environment; possess high varietal resistance, and helps in preserving bio-diversity. Sugarcane production and productivity differ widely from country to country due to climatic diversity and management practices. Brazil has highest area under sugarcane while Australia has highest productivity per unit area. The largest producers are Brazil, India, and China, accounting for more than 50% of world production.

Besides sugar, sugarcane crop provides biofuel, fiber, organic fertilizer, and numerous by-products and co-products with ecological sustainability. Sugarcane is one of the most efficient photo-synthesizer C4 plant, which can convert up to 2% of the incident solar energy into biomass. The over-exploitation of natural resources and increased human activities have made our ecosphere prone to biotic stresses like drought, water logging, salinity, and multiple soil-related problems such as nutrient deficiencies. These have severely eroded the sustainability of many food and industrial crops, including sugarcane. The prevailing thrust on high production intensive agriculture has led to excessive use of inputs, viz. irrigation, fertilizers, and pesticides, which are proving to be a long-term threat to crop productivity, soil, environment, and human health.

Eco-friendly sustainable sugarcane and sugar production calls for judicious use of inputs and bio-intensive sugarcane agriculture, encompassing cultivation of improved varieties with long field stability. The crop and field management should focus on resource conservation methods integrating BMP and ERP for improving production efficiency. The current marketing and handling infrastructure in sugarcane and integrated industry requires major reforms to increase efficiency and reduce transit losses. In order to become economically viable and globally competitive, the industry and scientists must support and educate sugarcane farmers with innovative methods of production, protection, and crop management. Sugarcane agriculture could be sustained only if the profit ability is ensured by reducing the cost of cultivation and improving the productivity per unit area; improved varieties, innovative farming technologies, mechanization, and post-harvest handling of crop will hold the key.

The sugar industry not only serves millions of farmers but also plays an important role in socio-economic development. Over the years, many alternative products and processes have been developed, utilizing the valuable co-products of sugarcane processing. The crop is a major source of bioethanol, bioelectricity, animal feed, fodder, and many bio-based products; however, economic exploitation of these value-added products depends on cane-biomass productivity and sustainability.

The book titled *Sustainable Sugarcane Production*, is a comprehensive repository of all aspects of scientific cane management based on the practical experience and knowledge of the authors. It encompasses chapters on agro-techniques, irrigation management, nutrition, ratoon management, mechanization, molecular approaches for sugarcane genetic improvement, crop protection, sugarcane maturity, postharvest management, etc. The book also highlights issues related to environmental protection by the sugarcane industry.

This compendium on sugarcane will be extremely useful to all stakeholders connected with the sugarcane and sugar industry.

—**Dr. S. Solomon, PhD**
Vice-Chancellor
Chandra Shekhar Azad University of Agriculture & Technology
Kanpur–208002, Uttar Pradesh, India
Tel: +91-512-2534155 (Off.), Fax: +91-512-2533808,
E-mail: vc@csauk.ac.in, drsolomonsushi11952@gmail.com

PREFACE

Sugarcane occupies a commanding position as an agro-industrial crop and is commercially grown in about 115 tropical and subtropical countries of the world. A quantum of world sugar is produced from sugarcane; however, this crop faces a number of problems, such as low cane productivity, biotic and abiotic stresses, high cost of cultivation, unavailability of seed cane of newly released varieties, post-harvest losses, and low sugar recovery.

The world population is projected to reach 9.7 billion by 2050. To feed the increasing world population, global food security will remain a worldwide concern. It is imperative that agricultural development should fully exploit the potential of sugarcane by practicing modern and scientific method of farming. Judicious use of available and newly developed practices and principles need knowledge of appropriate methods of implication. In addition, effective source timing applications during crop growth cycle is another important measure to realize maximum production and reduce cost production to meet the new requirements. Sugarcane agriculture could be sustained only if profitability can be ensured by reducing the cost of cultivation and by improving the productivity per unit area. This is possible through new research innovations, mechanization, and technological interventions in sugarcane agriculture.

The book *Sustainable Sugarcane Production* includes all the aspects of sustainable sugarcane cultivation, development, and management, such as farming and biotechnology, entomology, pathology, breeding, physiology, agronomy, ratoon management, abiotic stresses, mechanization, etc. The book contains modern crop production methods in a comprehensive and easily understood manner.

In addition this book also covers the latest information from the literature at the international level to make it usable for most agroecological regions of the world. This book can be used as reference material by agronomists, breeders, plant physiologists, farmers, and students of agricultural sciences. The objective of this book is to provide a comprehensive account of all the major achievements based on worldwide scientists in sugarcane research. The book is a compilation of recent advancements made on sugarcane development and cultivation and on improvement in cane and sugar yields using conventional and biotechnological approaches by different agricultural scientists and researchers of the major sugarcane growing countries.

We hope this book will serve as an important reference for students, scientists, and industry professionals involved in sugarcane and related crops.

— **Priyanka Singh, PhD**
Ajay Kumar Tiwari, PhD

CHAPTER 1

FARMING TECHNOLOGIES FOR SUGARCANE PRODUCTION IN UPLAND FIELDS

YANG-RUI LI,[1] XIU-PENG SONG,[1] JIAN-MING WU,[1]
CHANG-NING LI,[1] QIANG LIANG,[1] XI-HUI LIU,[1] WEI-ZAN WANG,[1]
HONG-WEI TAN,[1] and LI-TAO YANG[2]

[1] *Guangxi Key Laboratory of Sugarcane Genetic Improvement, Key Laboratory of Sugarcane Biotechnology and Genetic Improvement (Guangxi), Ministry of Agriculture, Sugarcane Research Center, Chinese Academy of Agricultural Sciences; Sugarcane Research Institute, Guangxi Academy of Agricultural Sciences, Nanning 530007, China, E-mail: liyr@gxaas.net*

[2] *State Key Laboratory for Conservation and Utilization of Subtropical Agro-Bioresources, Agricultural College, Guangxi University, Nanning 530005, China, E-mail: liyr@gxu.edu.cn*

CONTENTS

ABSTRACT

China is the third biggest sugar producing country in the world just after Brazil and India. In the milling year 2007/08, the total sugar production in China reached 14.83 MT. However, more than 80% of sugarcane is grown in upland field. China has developed a series of unique farming technologies for commercial sugarcane production in the upland fields. These technologies include deep ploughing and fine preparation of soil, intelligent fertilization system, trash addition to field, water saving irrigation, use of pathogen free healthy seed cane, rational application of vinasse in sugarcane field, chemical control, mechanization for sugarcane management, and comprehensive control of diseases, pests, weeds, rats, etc. The exploitation and comprehensive application of the new sugarcane farming technologies have promoted the Chinese sugar industry to a new level in about every 5 years, and made China become the third biggest sugar producing country in the world. However, the sugar industry has been experiencing a very difficult time in the recent two milling years because of the worldwide low sugar price and the high production cost at the domestic level, which led to a substantial reduction in sugarcane growing areas and sugar production. Mechanization and the related sugarcane variety selection and farming technology development have become the bottleneck for sustaining the development of the sugar industry in China.

1.1 INTRODUCTION

More than 90% of the sugar production is contributed by sugarcane in China. Sugarcane is mainly grown in upland fields where irrigation is not available. Development of the farming technologies for commercial sugarcane production in the upland fields has made significant progress which has promoted the fast development of Chinese sugar industry, and made China the third biggest sugar producer in the world after Brazil and India (Li, 2010; Li and Wei, 2006; Li and Yang, 2009, 2015a). China produced 11.99 tons of sugar in the milling year 2006/07, which was 36.06% higher than that in 2005/06 and; 14.87 million tons in 2007/08, which was 24.04% higher than that in 2006/07. The cane sugar production reached 10.75 tons in 2006/07, which was 89.66% of the total sugar production and; 13.67 million tons in the milling year 2007/08, which was 92.2% of the total sugar production. The sugarcane productivity and sugar recovery in China have been improved significantly since 1980s. In China, the average cane productivity increased from about 55 tons/ha in 1990s to about 65 tons/ha in 2000s to about 75 tons/ha in recent years, and average sugar recovery increased from about 10% in 1990's to about 12% in recent years.

The improvement in sugarcane production and productivity is mainly attributed to the success in breeding drought resistant sugarcane varieties and development of appropriate farming technologies for rain fed upland sugarcane production (Li, 2010; Li and Solomon, 2004, 2006; Li and Wei, 2006; Li and Yang, 2015a; Li et al., 2008, 2011).

In recent years, a group of new elite GT varieties have been released, and the varieties GT29, GT32, GT40, GT42, GT43, GT44, GT46, GT47, and GT49 have performed well in upland fields. Compared with ROC22, they show higher productivity and better ratoon-ability (Table 1.1).

We strongly recommend the use of multiple sugarcane varieties by each sugar mill, at least 8 to 10 varieties in a proportion of 30:50:20% for early:intermediate:late maturing varieties, respectively.

Combined with new sugarcane variety breeding and popularization of the improved varieties, we have developed a series of updated farming technologies for commercial sugarcane production in upland areas. These technologies include deep ploughing and fine preparation of soil, plastic film mulching, prescription fertilization, trash retention in field, water saving irrigation, pathogen free healthy seed cane, rational application of vinasse as a

TABLE 1.1   The Economic Attributes of Newly Selected GT Varieties

Variety	Cane yield				Sucrose % cane		Sugar yield	
	Total mean (t/ha)	vs. ROC22 (±%)	Ratoon (t/ha)	vs. ROC22 (±%)	Mean	vs. ROC22 (±%)	Mean	vs. ROC22 (±%)
GT29	97.4	-8.3	94.0	+10.5	15.62	+0.60	15.00	+4.2
GT32	104.1	+2.4	98.8	-1.83	14.72	+0.50	15.31	+6.1
GT40	89.0	+0.8	97.7	+16.8	15.24	+0.64	13.59	+6.0
GT42	101.7	+9.3	103.5	+13.2	14.77	+0.66	15.03	+14.5
GT43	101.3	+9.3	103.5	+13.8	14.47	+0.36	14.66	+11.7
GT44	107.9	+5.2	114.4	+11.8	15.25	+0.79	16.35	+12.8
GT46	114.5	+21.1	108.7	+24.0	14.36	+0.16	16.49	+21.3
GT47	104.6	+10.6	100.4	+21.5	14.40	+0.20	15.09	+12.1
GT49	103.7	+8.9	91.9	+11.2	14.45	+0.53	15.21	+12.5

liquid fertilizer, chemical control, mechanization for sugarcane management and comprehensive control of diseases, pests, weeds, and rats, and so on (Li and Yang, 2015b).

1.2 BREEDING OF DROUGHT RESISTANT SUGARCANE VARIETIES

Before 2002, the combinations and seedling numbers were very limited in China, and the breeding efficiency was very low. To improve our sugarcane breeding, we decided to increase the combination number to 500–1000, and total seedling numbers to 100,000–300,000, since 2002. The highest seedling number reached 600,000 in 2012, but we decided to keep about 300,000 seedlings of about 1100 combinations each year after 2013. We have selected a group of elite sugarcane varieties from the seedlings of 2002 and later, for example, GT29 (GT02-761), GT40 (GT02-1156), GT42 (GT04-1001), GT43 (GT05-3084), GT44 (04-1545), and GT46 (GT06-244). These varieties are now developing fast in commercial sugarcane production in upland areas. To improve the combinations, we have paid high attention to the germplasm innovation, such as exchanging germplasm with other countries, utilizing local wild germplasm collections including *Saccharum spontaneum, Erianthus arendinaceus* and *Narenga porphyrocoma* which were crossed and backcrossed with commercial sugarcane varieties, and some promising clones have been selected from the BC1 and BC2 progeny.

1.3 DEEP PLOUGHING AND FINE PREPARATION OF SOIL

Deep ploughing and fine preparation (Figure 1.1) of soil began to be popularized since early 1990s, and more than 90% of the sugarcane fields are ploughed by tractors. Experiments (Liao et al., 2010; Ye et al., 1995) showed deep ploughing to 45–60 cm and fine preparation of soil increased soil moisture, which is good for germination and emergence, rooting, tillering and fast growing, and finally for production of more millable stalks and thicker and longer stalks. These also improved the lodging resistance and sugar accumulation in stalks, resulting in 20% or more increase in cane and sugar productivity of sugarcane in rain-fed upland field. Deep ploughing and fine preparation of soil are considered key technologies for upland sugarcane production in China.

1.4 PLASTIC FILM MULCHING

Plastic film mulching (Figure 1.2) is favorable to retain the soil moisture and nutrients. It increased the soil temperature when seed canes were planted in winter and spring, resulting in 10–20 days of earlier germination and emergence, improving emergence rate by 15.3–26.1% with 15000–30,000 plants/ha. An increase in number of healthy and uniform plants increased cane productivity by 14.96%, and improved sucrose content by 0.53% (Li, 2010). Plastic film mulching has become one of the major farming practices

FIGURE 1.1 Deep ploughing and fine preparation of soil.

FIGURE 1.2   Plastic film mulching.

in upland sugarcane field. This practice covered 221,000 ha in Guangxi in 2011, that is, 52.1% of the total plant cane area.

1.5   PRESCRIPTION FERTILIZATION

A prescription fertilization system has been established for sugarcane, based on years of related data, and has been popularized since 1990s. After application of this system, farmers have decreased fertilizer application rate and production cost, and improved cane productivity by 10–25%, sucrose content by 0.4–0.8%, and fertilizer use efficiency by 4.5–8.2% (Li, 2010). This technology has been popularized since 1990s, and has been developed as an intelligent expert system. In 2011, the prescription fertilization system was applied in 172,000 ha of sugarcane fields.

1.6   TRASH RETENTION IN FIELD

Sugarcane trash contains rich nutrients such as N, P, K, Ca, Mg and micronutrients and organic matters. Sugarcane trash retention in field can improve the soil structure and physico-chemical properties, increase the organic matter content, and improve the soil fertility. There are two ways for sugarcane trash incorporation in field (Figure 1.3), one is chopping the trash and mixing with soil (Figure 1.3A), and the other is collecting the trash and placing in every alternate row (Figure 1.3B). The former practice is better to increase the organic mater and nutrients in the soil but needs machine operation. The latter practice is simple and easy, and can be done by any field so it is more

FIGURE 1.3 Sugarcane trash: (A) chopping the trash and mixing with soil; (B) collecting the trash and placing in alternate rows.

popular in China. Experiments from 1987–2010 showed that trash retention in field has improved soil structure and physico-chemical properties. As a result, soil organic matter level was increased from 1.79 to 2.60%, total nitrogen from 0.09 to 0.14%, available phosphorus from 13.37 to 43.25 mg/kg, available potassium from 79.63 to 233.33 mg/kg, and the average cane productivity was improved by 7.7%. Trash incorporation to field significantly increased the microbe population, and the population of bacteria, fungi and actinomycetes were 2.38, 1.80 and 2.74 times more than that in the conventional control (Liao et al., 2014). Besides, trash blanketing is also good to retain soil moisture and to control weeds (Li, 2010). In recent years, sugarcane trash incorporation/mulching in field has been carried out in over 300,000 ha each year.

1.7 WATER SAVING IRRIGATION

Water saving irrigation practices include spray, micro-spray (Figure 1.4) and drip irrigations, which are developing fast in China in recent years. Application of these water saving irrigation technologies especially fertigation, saves water, fertilizer and labor, and improves fertilizer use efficiency. Experiments showed that fertigation improved cane productivity by 19.2 to 56.4%, and fertilizer use efficiency by 90%, and saved water by 30–60% (Li, 2010; Chen et al., 2012; Xu et al., 2010, 2011). In dry upland sugarcane areas, water saving irrigation practices are becoming increasingly popular since 2000s. In Chongzuo City, for example, the water saving irrigation techniques were applied in 18,867 ha in 2001.

FIGURE 1.4   Micro-spray (A) and spray irrigation (B) in sugarcane field.

1.8   HEALTHY SEED CANE PRODUCTION AND UTILIZATION

Pathogen free healthy seed cane production technology has been developed and put to use since 2000s in China. For micropropagation standard, the detected pathogen rates are zero in the plantlets derived from sugarcane stem tip tissue culture, and also pathogen free (100% pathogen removal) is required for mosaic, ratoon stunt, yellow leaf and smut, and pathogen detection and verification must to be done before the T1 plants are transplanted to field. Experimental results proved that use of pathogen free healthy seed cane improved cane productivity by 15.1 to 52.1% and sucrose content by 0.12 to 1.71% because of control of various diseases such as ratoon stunting disease, mosaic viruses, yellow leaf disease, etc. in the seed canes (Li, 2010). Using temporary immersion bioreactor system (TIBs) further improved the propagation rate was further improved up to 40 times compared with traditional tissue culture method (Yang et al., 2011a). However, in many places, the results were not satisfactory with respect of the variety ROC22, because long duration monoculture of variety ROC22 in large scale had already accumulated substantial pathogen load of smut and of pests in the fields and the surrounding environments. So the pathogen seed cane production program needs to be taken up for the newly released varieties so that these new varieties can be extended faster and grown in the field for a longer period.

1.9   VINASSE AS A LIQUID FERTILIZER

Vinasse is the waste liquid of molasses fermentation for alcohol production in the sugarcane mills. Its chemical components are mainly various soluble inorganic and organic nutrients, which are nontoxic, so vinasse should be a very good liquid complete fertilizer for sugarcane and other crops. But vinasse contains high content of proteins which are easily oxidized producing a foul smell, and it also contains high concentrations of permeable ions especially K^+ which could harm the crops if applied without appropriate treatment. For a long time, vinasse was a major pollutant in sugar producing areas, but now application in fields after appropriate dilution can solve this problem very well, needing no other additives. An environment-friendly technology system has been developed for application of vinasse in both plant and ratoon crops of sugarcane in China, based on research on the effects of vinasse application rates on soil properties, sugarcane growth under different soil types and farming conditions, and evaluation of the physico-chemical property of vinasse, the movements of different vinasse components in soil and soil adsorption characteristics on vinasse after application (Li, 2010; Li et al., 2008; Jiang et al., 2012; Su et al., 2012; Yang et al., 2012, 2013; Zhu et al., 2009). Experiments proved that spraying 75 t/ha vinasse of Brix 6–8 in plant cane, and 105 t/ha vinasse of Brix 8–9 in ratoon cane, and mulching with plastic film, could ensure good growth of sugarcane plants without additional application of any chemical fertilizer. This improved cane productivity by 10–30%, and sucrose content by 0.2–1.0 units compared with traditional fertilization (Figure 1.5). Spraying vinasse with high-pressure spray gun could make very good soil coverage on the newly planted seed canes (Figure 1.5B).

By continuous application of vinasse in subsequent years, the Changling Farm in Guangxi realized average cane yields over 120 t/ha for years since 2005. This technology not only solves the vinasse pollution problem completely, but also provides a superior liquid fertilizer for crop production, which improves soil fertility, yield and quality of cane, and reduces sugarcane production cost. Hence, this technology is fast becoming popular in recent years. As sugarcane vinasse also contains rich organic components including high concentration of proteins and many active substances such as enzymes and vitamins, recently it has been used to made biofertilizers and organic-inorganic complex fertilizers with other inorganic N, P, and K fertilizers.

1.10 CHEMICAL REGULATION OF PLANT GROWTH

Application of plant growth regulators is very effective for promoting plant growth and improving sugar productivity in sugarcane. Different

FIGURE 1.5   Vinasse as a liquid fertilizer applied in sugarcane. (**A**) pumping vinasse into tank; (**B**) applying vinasse in plane cane field; (**C**) applying vinasse in ratoon cane field.

concentrations of ethephon show different physiological regulatory effects on sugarcane, including promoting plant growth, improving drought and cold resistance, and improving cane and sugar productivity, so ethephon is a versatile growth regulator for sugar cane industry (Li, 2004, 2006, 2010; Li and Solomon, 2003, 2006; Jian et al., 2012). Chemical ripening with glyphosate-borate complex and high concentration of ethephon are very effective to promote sugar accumulation in sugarcane and improve sugar productivity (Li and Solomon, 2004, 2006). In the past, the biggest limitation for application of growth regulators in large scale was the difficulty in application as foliar spray. We first tried to spray the chemicals by human-operated airplane (Figure 1.6A) in large scale (Yang et al., 2011b), and then by unmanned aerial vehicle (Figure 1.6B). The latter is better, and is developing fast in recent years.

FIGURE 1.6   Foliar spray of chemicals by airplanes in sugarcane. (A) Man-operated airplane operation; (B) Unmanned aerial vehicle operation.

1.11   MECHANIZATION FOR FIELD MANAGEMENT

With the urbanization in China, labor is becoming scarce, and labor cost is increasing, thereby, favoring mechanization for field management in sugarcane production. Almost 100% mechanization has been attained in soil preparation and in most field operations such as planting, fertilizer application, mulching with trash and plastic film, weed and pest controls, etc., but very little mechanization is practiced for sugarcane harvesting (Li, 2010; Li and Yang, 2015a; Li et al., 2008, 2011).

1.12   COMPREHENSIVE CONTROL OF DISEASES, PESTS, WEEDS, AND RATS

Comprehensive control of diseases, pests, weeds, and rats included selecting resistant sugarcane varieties, using pest and pathogen free healthy seed canes, sterilization of seed canes, removing sources of diseases, pests and rats, and using techniques like mechanical trapping in the field (Figure 1.7A), biological control of borers using pheromone, *Trichogramma* and Cuban flies; termite control with *Metarhizium*, light trapping of borer, longhorn beetle and scarab (Figure 1.7B), pre- and post-emergent weed control with herbicide, rodent control in large scale, and so on. One sample is that in Nanning East Asia Sugar Corporation Ltd., a workshop for *Trichogramma* production has been setup, and the cards of *Trichogramma* are given to all the farmers who supplies produce the millable cane for the Corporation without any charge, and the technicians also predict borer occurrence using pheromone

FIGURE 1.7   Insect pests control by field trapping of longhorn beetle (A) and light trapping for many kinds of pests (B).

and guide the farmers to apply the *Trichogramma* cards into the sugarcane field. By large scale of application of *Trichogramma*, the borer occurrence has decreased for more than 30% and the sucrose cane percentage increased for 0.5% (absolute value) in recent years in this Corporation.

Based on these innovative technologies, improved packages of practices have been recommended for different sugarcane growing areas and farmers, which considerably improved the sugar productivity in China. However, different areas may have different alternatives because of the differences in the natural and social conditions. For example, vinasse is not available in many sugar mills, and lower quantity of vinasse should be applied in sand soils and fertile soils.

1.13   MAJOR PROBLEMS FOR FARMING TECHNOLOGIES IN SUGARCANE PRODUCTION IN UPLAND AREAS

The biggest challenge for sugarcane agriculture in China is mechanization of cane harvesting. The major reasons are the small size of farm holdings and the quality requirement of millable cane prescribed by sugar mills. For mechanizing the harvest operations, it is necessary to combine small farms into bigger farms. Sugar mills need to scale down the quality requirement standards for millable cane and improve the cane cleaning system. Obviously, it still needs some more time, but is promising.

For soil preparation, strong-power tractors more than 160 horse powers are still not popular in most sugarcane growing areas, so the plough depth is only about 30 cm, which results in weak soil water holding capacity and shallow root system, so the cane crop may get easily affected by drought or lodging, limiting sugarcane productivity and sucrose content.

Most farmers have a misconception that more fertilizer will give higher productivity, so overfertilization is a serious problem with respect to N, P, and K fertilizers. Based on our experiment, in general, the N, P, and K applications are 75, 50, and 30% more than that sugarcane needs, respectively. Overfertilization not only increases the production cost, but also leads to low fertilizer utilization efficiency.

The sugarcane farmers have been growing the variety ROC22 since a long time, after rejuvenation treatments. But this variety is susceptible to cold, is a poor ratooner, and is susceptible to borers and smut. Especially, after continuous cultivation of this variety in the same fields, the soils have accumulated considerable smut pathogens. That is why healthy seed cane application for ROC22 does not give expected good results in sugarcane production.

Finally, the sugarcane ratooning is limited to few ratoons in China. The sugarcane production cycle is 3 years mostly (one plant crop, two ratoon crops), or even shorter, which leads to high production cost compared to that in most other countries like Brazil, having 5 years or longer production cycle. The bad ratoon performance might be related to dry condition in winter and spring for the upland sugarcane growing area, which occupied about 80% of the total sugarcane growing area. So water scarcity is another big limitation for sugarcane productivity. Also, the ratoonability and ratoon performance of sugarcane varieties and field management practices needs to be improved.

1.14 FUTURISTIC APPROACH

Keeping in mind the high cost of production and the intense competition at the global level, it can be concluded that mechanization is the need of the hour in China. Consolidation of farms, use of suitable improved sugarcane varieties, modifications in farming techniques to suit mechanization etc., are important aspects to be taken care of before the introduction of mechanization in the sugarcane farms.

The biological nitrogen fixing ability of sugarcane, releasing phosphorus and potassium, growth promotion of soil microbes, etc., should be exploited to decrease the use of chemical fertilizers in sugarcane production, thereby, decreasing production cost and improving fertilizer use efficiency, while assuring high cane and sugar productivity.

Pathogen free healthy seed cane production should be combined with development of new elite sugarcane varieties to accelerate the spread of new sugarcane varieties and to improve the seed cane quality for commercial production. Deep plough and fine preparation of soil are important, and large tractors should be extended to plough soil depth to 50–70 cm. This is important for high and stable productivity and lodging resistance in rain-fed upland sugarcane growing areas.

Cultivation of sugarcane varieties with strong resistance and ratoon ability such as GT29, GT32, GT42 and GT44 and suitable improved agronomic practices should be further stressed for multi-ratooning (up to four ratoons per cycle, in place of two ratoons at present) to increase the production cycle from the present 3 years to 5 years, to decrease the cost and improve the efficiency of Chinese sugarcane production.

1.15 CONCLUDING REMARKS

Continuous exploitation and comprehensive application of the new sugarcane farming technologies have improved the sugarcane productivity in upland areas. New elite drought resistant sugarcane varieties, and simple and easily-operated farming practices, water saving irrigation especially fertigation and ratoon-prolonged practices will play more important role in the future.

KEYWORDS

- **China**
- **development**
- **farming technology**
- **sugar industry**
- **sugarcane**

REFERENCES

Chen, G. F., Tang, Q. Z., Li, Y. R., Huang, Y. Y., Liu, B., Xu, L., & Huang, H. R., (2012). Effects of sub-soil drip fertigation on sugarcane in field conditions. *Sugar Tech, 14*, 418–421.

Jiang, Z. P., Li, Y. R., Wei, G. P., Liao, Q., Su, T. M., Meng, Y. C., Zhang, H. Y., & Lu, C. Y., (2012). Effect of long-term vinasse application on physic-chemical properties of sugarcane field soils. *Sugar Tech, 14*, 412–417.

Li, Y. R., & Solomon, S., (2003). Ethephon: A versatile growth regulator for sugar cane industry. *Sugar Tech, 5*, 213–223.

Li, Y. R. (2004). Development of sugar industry in China. In: Li, Y. R., & Solomon, S., (eds.). *Sustainable Sugarcane and Sugar Production Technology*. Beijing: China Agriculture Press, 47–59.

Li, Y. R. (2006). Research and development strategies to improve sugar productivity in China. In: Li, Y. R., & Solomon, S., (eds). *Technologies to Improve Sugar Productivity in Developing Countries*. Beijing: China Agriculture Press, 7–14.

Li, Y. R., & Wei, Y. A., (2006). Sugar industry in China: R&D and policy initiative to meet sugar and biofuel demand of future. *Sugar Tech, 8*, 203–216.

Li, Y. R., & Yang, L. T., (2009). New developments in sugarcane industry and technologies in China since 1990's. *Southwest China Journal of Agricultural Sciences, 22*, 1469–1476.

Li, Y. R., & Yang, L. T., (2015a). Sugarcane agriculture and sugar industry in China. *Sugar Tech, 17*, 1–8.

Li, Y. R., & Yang, L. T., (2015b). Research and development priorities for sugar industry of China: recent research highlights. *Sugar Tech, 17*, 9–12.

Li, Y. R., (2004). China: an emerging sugar super power. *Sugar Tech, 6*, 213–228.

Li, Y. R., (2006). *Techniques and Principles for Applying Ethephon to Improve Cane Productivity and Sucrose Content in Sugarcane*. Beijing: China Agriculture Press, 1–9.

Li, Y. R., (2010). *Modern Sugarcane Science*. Beijing: China Agriculture Press, 12–492.

Li, Y. R. (2008). Status of sugar industry development in China. In: Li, Y. R., Nasr, M. I., Solomon, S., & Rao, G. P., (2008). *Meeting the Challenges of Sugar Crops & Integrated Industries in Developing Countries*. Cairo: Engineering House Press Co., Egypt, 759–764.

Li, Y. R., Srivastava, M. K., Rao, G. P., Singh, P., & Solomon, S., (2011). *Balancing Sugar and Energy Production in Developing Countries: Sustainable Technologies and Marketing Strategies*. Lucknow: Army Printing Press, India, 16–24, 100–114, 170–178, 280–302, 312–321, 353–376.

Li, Y. R., Zhu, Q. Z., & Wang, W. Z., (2008). Multiple location experiment of technique system for direct application of vinasse from cane mill in sugarcane fields. *Southwest China Journal of Agricultural Sciences, 21*, 749–755.

Li, Y. R., Zhu, Q. Z., Wang, W. Z., & Solomon, S., (2007). Pre-emergence application of vinasse on sugarcane growth and sugar productivity in China. *Sugar Tech, 9*, 160–165.

Liao, Q., Wei, G. P., Liu, B., Chen, G. F., Huang, D. L., & Li, Y. R., (2010). Effects of mechanized deep ploughing and scarification cultivation technology on growth and yield of sugarcane. *Guangxi Agricultural Sciences, 41*(6), 542–544.

Liao, Q., Wei, G. P., Chen, G. F., Liu, B., Huang, D. L., &. Li, Y. R., (2014). Effect of trash addition to the soil on microbial communities and physico-chemical properties of soils and growth of sugarcane plants. *Sugar Tech, 16*, 400–404.

Su, T. M., Li, Y. R., Wei, G. P., Jiang, Z. P., Liao, Q., & Zhu, S. B., (2012). Macronutrients absorption and surface runoff losses under different fertilizer treatments in sugarcane field. *Sugar Tech, 14*, 255–260.

Xu, L., Huang, H. R., Huang, Y. Y., Chen, G. F., Yang, L. T., & Li, Y. R., (2011). Spatial distribution of sugarcane root and soil available nutrients with subsurface drip irrigation in sugarcane field. *Guangdong Agricultural Sciences, 1*, 78–80.

Xu, L., Huang, H. R., Yang, L. T., & Li, Y. R., (2010). Combined application of NPK on yield and quality of sugarcane applied through SSDI. *Sugar Tech, 12,* 104–107.

Yang, L. T., Mo, F. L., Zhu, Q. Z., Li, N., Ou, Z. X., & Li, Y. R., (2012). Effect of vinasse application on sugarcane growth and yield. *Journal of Southern Agriculture, 43,* 18–21.

Yang, L. T., Zhang, B. Q., Zhu, Q. Z., Li, Z. G., Wang, W. Z., Chen, W. J., & Li, Y. R., (2011b). Effects of application of drought resistant sucrose-yield promoter sprayed by aircraft in large area of sugarcane. *Chinese Journal of Tropical Crops, 32,* 189–197.

Yang, L., Qin, G., Yang, L. T., Wu, J. M., Luo, R. H., Wei, Y. W., & Li, Y. R., (2011a). Optimization of sugarcane rapid propagation in temporary immersion bioreactors system. *Journal of South China Agricultural University, 32*(3), 36–41.

Yang, S. D., Liu, J. X., Wu, J., Tan, H. W., & Li, Y. R., (2013). Effects of vinasse and press mud application on the biological properties of soils and productivity of sugarcane. *Sugar Tech, 15,* 152–158.

Ye, Y. P., Yang, L. T., & Li, Y. R., (1995). Effects of deep ploughing and scarification on, N, P, and K absorption and cane yield and quality in sugarcane. *Sugarcane, 2*(1), 50–51.

Zhu, Q. Z., Li, Y. R., Wang, W. Z., & Liao, J., (2009). Effect of continual quantitative rational application of vinasse in sugarcane field. *Sugar Crops of China, 2,* 10–13.

CHAPTER 2

HEADWAYS IN AGRO-TECHNIQUES FOR HEIGHTENED YIELD OF SUGARCANE: INDIAN PERSPECTIVE

A. K. SINGH,[1] MENHI LAL,[2] and EKTA SINGH[3]

[1]*Division of Crop Production, ICAR-Indian Institute of Sugarcane Research, Lucknow–226002, India, E-mail: shantaanil@yahoo.com*

[2]*Division of Crop Production, ICAR-Indian Institute of Sugarcane Research, Lucknow–226002, India*

[3]*Division of Crop Production, ICAR-Indian Institute of Sugarcane Research, Lucknow–226002, India*

CONTENTS

2.1   INTRODUCTION

Sugarcane cultivation on the earth can be found between 36.7°N and 31.0°S latitude and from sea level to 1000 m of altitude. In India, it is grown from latitude 8°N to 33°N latitude. On global basis, the contribution of sugarcane to the total sugar production recorded a quantum jump from 56 to 80% in the last two decades. This kind of trend is a clear indication that the dependence on sugarcane as raw material for sugar manufacture has also increased in countries producing sugar from sugar beet. Brazil is the largest producer of sugarcane (728.13 million tonnes) in the world from 19.2 m ha acres with the productivity of 78.85 t/ha (About 25% of world sugar and ethanol producer). India has the second position in area and production of sugarcane. Other important sugarcane producing countries are China (123.46 mt), Thailand (96.50 mt), and Pakistan (58.49 mt). Area under the crop in countries like Australia, Cuba, Indonesia, Mexico, and South Africa ranges between 51.73 to 0.7 m ha in each country (Table 2.1).

Sugarcane plays a key role in Indian economy by contributing substantially to national agricultural gross domestic product (GDP) by way of contributing excise duty and payment to cane growers. With 453 sugar factories, constituting 252 mills from the co-operative sector and 134 mills from the private sector (up to 2016) located in rural areas, the Indian sugar industry is a prime catalyst in converting the potential agro-industrial rural sector into economic strength of the country. Over 40 million farmers are involved

TABLE 2.1   Country-Wise Sugarcane Area, Production, and Productivity (2015–16)

Country	Production (mt)
Brazil	728.13
India	349.56
China	123.46
Thailand	96.50
Pakistan	58.49
Maxico	51.73
Colombia	38.75
Phillipphines	32.90
United States	28.00
Indonesia	27.40

Source: FAO statistics (2015, 2016).

in sugarcane cultivation, harvesting, and ancillary activities. The industry employs over 0.5 million skilled and unskilled workers mainly from rural areas. Thus, over 8.8% of our rural population is directly or indirectly dependent on sugar industry.

The main product of sugarcane is sucrose, which accumulates in stalk (internodes). Sucrose extracted and purified in specialized mill factories is used as important food material in human diet, raw material in human food industries or is fermented to produce ethanol, a low pollution fuel. So the sugarcane products include table sugar, molasses, bagasse, and ethanol. During the past 50 years, this crop has metamorphosed from sugarcane to fiber cane to alcohol cane to energy cane. Sugarcane as biofuel crop has much expanded in the last decade yielding anhydrous ethanol (gasoline additive) and hydrated ethanol by fermentation and distillation of sugarcane juice and molasses (Gunkel et al., 2007). By-product vinasse, a liquid waste is sometimes used for fertigation purpose. Bagasse a by-product of both sugar and ethanol production is burnt to generate electricity to run the mill and it can also be used for production of biodegradable plastic. It provides most of the fuel for steam and electricity for sugar mills in Australia and Brazil (Hartemic, 2008). Few factories in India also have diversified into by products based industries and have invested to establish distilleries, organic chemical plants, paper and board factories as well as co-generation units.

In India, sugarcane (*Saccharum*spp. hybrid complex) is the only source of sugar, which holds a prominent position as a cash crop for peasants and planters. The crop occupies about 2.68% (4.99 m ha) of the gross cropped area in the country and its production has increased from 241.0 million tonnes during 1990–91 to 359.33 million tonnes in 2014. However, after a record sugar production of 28.4 million tonnes in 2006–07, the sugar year 2009–10 has indicated a shortfall of 33.3% with production of 18.9 million tonnes. Among states, Uttar Pradesh has the largest acreage (2.13 m ha) under sugarcane, which is 42.6% of the total acreage of the crop in the country. Maharashtra by commanding 0.77 million hectares under the crop holds the next spot. Tamil Nadu, Karnataka, Gujarat, Andhra Pradesh, Uttarakhand, Bihar, Haryana, Punjab, and Madhya Pradesh are the other important states for sugarcane cultivation. Subtropical northern region of the country occupies about 60% of total area under the crop; however, the productivity is higher in tropical region being highest in Tamil Nadu (104.3 t/ha) followed by Karnataka, Maharashtra, and Andhra Pradesh (Table 2.2).

TABLE 2.2   Area, Production, and Productivity of Sugarcane in Different States of India (2015–16)

State	Area ('000/ha)	Production ('000 tonnes)	Yield (t/ha)
Andhra Pradesh	157	12460	79.4
Assam	29	1048	36.1
Bihar	258	14240	55.2
Chhattisgarh	29	79	2.7
Gujarat	185	13040	70.5
Haryana	113	8588	76.0
Jharkhand	10	705	69.1
Karnataka	400	34200	85.5
Kerala	2	137	91.5
Madhya Pradesh	73	3343	45.8
Maharashtra	987	75087	76.1
Orissa	13	880	67.7
Punjab	99	7131	72.0
Rajasthan	4	282	67.4
Tamil Nadu	263	27615	105.0
Uttar Pradesh	2160	133203	61.7
Uttarakhand	98	6047	61.7
West Bengal	20	2300	115.0
Others	18	1040	57.5
All India	**4918**	**341425**	**69.4**

Source: NFCSF (2016). National Federation of Cooperative Sugar Factories Limited: Cooperative Sugar, *42*(7), 57–98.

2.2   CLIMATE AND SOIL

Sugarcane is a warmth and bright weather-loving crop that can keep growing and store sugar under favorable conditions for even two years. It however, needs very mild, cold weather for sugar accumulation. It can tolerate intermittent droughts in its vegetative phase and requires a mild dry soil during maturity. It is, therefore, a crop of the tropics and the near tropics or subtropics. Various components of the climate which affect sugarcane crop are temperature, rainfall, humidity, sunshine, and wind velocity. The minimum threshold temperature for cane growth is 16°C. Its vegetative growth almost ceases when the mean daily air temperature drops below

21°C. Rising temperatures in the range of 21°–32°C has been found to be in close correlation with growth of the crop. No stoppage of cane growth due to higher temperatures under adequate moisture supply has been reported. Under deficient moisture, temperature above 32°C may exert a retarding effect over growth (Burr et al., 1957). In India, sugarcane is cultivated from 8°N to 33°N latitude wherein areas receiving average annual rainfall from 500 mm to 2500 mm fall. The crop water requirement varies from 1200 mm to 2500 mm depending upon the length of growing season and atmospheric demand at various growth stages of the crop. Being a crop of 12 months or even longer duration, sugarcane encounters all the three seasons, i.e., summer, monsoon, and winter in its life cycle particularly under subtropics. It requires high temperature (above 30°C), low relative humidity (below 50%), sunny weather with minor showers for its germination and tillering phase, moderate temperature with high relative humidity (above 70%) and evenly distributed rainfall for grand growth phase and low temperature, and moderately dry and sunny period for ripening phase. The temperature below 20°C reduces rate of germination, high light intensity, and its long duration promote tillering whereas continuous rainfall, cloudy weather, and short days affect it adversely. In both tropics and subtropics, grand growth phase commences with the onset of southwest monsoon when the temperature and humidity both are high. During this phase high wind velocity (above 60 km/hr) is harmful and standing cane become more prone to lodging.

The sugarcane growing areas of India are classified broadly in five major agro-climatic zones (Table 2.3).

Deep, well-drained, and medium textured soils having sufficient water holding capacity with near neutral reaction are most suitable for sugarcane cultivation. However, sugarcane can be grown in any kind of soil varying from very heavy to very light in texture except problem soils such as saline, sodic, and acidic soils. In India sugarcane is grown in three major soil groups, i.e., alluvial, black, and red soils. In Uttar Pradesh and Uttarakhand, sugarcane is mainly grown on Gangetic alluvial soils. Bihar consists of two physiographic units' viz., northern plains and southern plateau of Chhota Nagpur. North Bihar is the white sugar belt of state and its soils are alluvial deposits laid down by the Ganges and Brahmputra. In Punjab and West Bengal sugarcane is grown on alluvial soils. The alluvial soils of Punjab have poor structure with hard or *kankar* pan while west Bengal alluvials are developed as riverine low lands, and categorized as Gangetic riverine, Vindhya riverine and Buxa riverine. The main groups of soils in Andhra Pradesh are black,

TABLE 2.3   Sugarcane Agro-Climatic Zones of India

Sl. No.	Name of zone	Area coverage	Particulars
Subtropical			
1.	North western zone	Uttar Pradesh (western and central), Punjab, Haryana, and parts of Rajasthan	Area is characterized by extreme climate especially temperature. Crop suffers with extreme moisture deficit during summer. Elongation phase is short and growth ceases because of low temperature with the commencement of winter.
2.	North central zone	Eastern U.P., Bihar, and West Bengal	The crop suffers with flood and gets inundated for long periods.
3.	North eastern zone	Assam and other northeastern states	Sugarcane is grown in patches and area under cultivation is very less.
Tropical			
4.	East coast zone	Coastal Odisha, coastal Andhra Pradesh, and coastal Tamil Nadu	Sugarcane productivity is very high and most suitable climatic conditions for sugarcane growth.
5.	Peninsular zone	Maharashtra, Gujarat, Karnataka, Kerala, Madhya Pradesh, and parts of Andhra Pradesh and Tamil Nadu	Supports sugarcane crop of longer duration. (*Adsalicrop* of 18 months). The most favorable climatic conditions for sugar accumulation. Long hours of sun-shine, and low temperature at night adds to high sugar recovery in this zone.

Source: AICRP (2011).

alluvial, red and lateritic, and sugarcane is cultivated on almost all type of soils in the state. In Tamil Nadu sugarcane is grown in red and lateritic black and alluvial soils, which are dominant in the state. In Karnataka, this crop is grown on black, red and lateritic soils. In Madhya Pradesh and Maharashtra sugarcane is grown in medium to heavy black soils (Table 2.4).

2.3 FITS MORE AS CROP ECOLOGY

Sugarcane is a perennial plant generally cultivated for different durations to fit in a location specific cropping pattern determined by climatic, edaphic, and socio-economic factors. In India sugarcane growing belts fall in two broad zones: tropics (south of 23°N, which is best suited for sugarcane) and subtropics, north of 23°N. It is historical anomaly that a crop, which is native of tropics is being cultivated in subtropics at such a large scale. Early thin cane varieties are more suitable for North Indian conditions better availability of industrial infrastructure like rail and road created during British period

TABLE 2.4   Sugarcane Growing Soils of India

Sl. No.	Sugarcane growing states	Soil type
1.	Andhra Pradesh	Black, alluvial, laterite
2.	Assam	Alluvial, Tilla lands
3.	Bihar	Alluvial
4.	Gujarat	Medium and deep black, alluvial
5.	Haryana	Alluvial
6.	Karnataka	Medium and deep black, red soils
7.	Madhya Pradesh	Medium to heavy black
8.	Maharashtra	Medium to heavy black
9.	Odisha	Red, black, laterite
10.	Punjab	Alluvial
11.	Rajasthan	Deep and medium black
12.	Tamil Nadu	Black and alluvial
13.	Uttar Pradesh	Alluvial
14.	Uttarakhand	Alluvial
15.	West Bengal	Alluvial

Source: AICRP (2011).

in North India are the factors responsible for higher sugarcane acreage in subtropics than in tropics (Table 2.5).

The growing season or crop cycle of sugarcane is generally restricted to 40–52 weeks in subtropics compared to 52–72 weeks (depending on planting time) in tropics. Lower temperature (less than 20°C) onwards mid-November to February restricts the planting time in subtropics whereas planting may continue round the year in tropics. The tillering span is restricted due to unimodal rainfall pattern (mid June to mid September) in northern India. The climate is generally favorable in elongation phase, but erratic nature of monsoon invariably affects the crop adversely. Cane growing tracts in Andhra Pradesh and Tamil Nadu enjoy warmer conditions and in the absence of growth arresting temperature (>20°C), the duration for crop elongation is more than double as compared to subtropics. However, the susceptibility to cyclonic weather and less hours of bright sun shine in eastern coastal belts normally offset the crop performance in comparison to Maharashtra, where weather remains uniform with low relative humidity during ripening phase and less diurnal temperature variations rarely exceeding 15°C. In fact, highest sugar recovery is recorded in this belt (20–22° latitude) in both the hemispheres. The best growing conditions for sugarcane crop in India is available in Deccan region on account of freedom from cyclonic weather, receipt of high solar radiation and occurrence of sufficiently low and appropriate ambient temperatures during winter. On the contrary, the active crop season is restricted to April through September in Punjab and March to October in other locations in North India due to unfavorable soil and ambient temperature. The shortening of growing period of sugarcane in Uttar Pradesh may be attributed to sharp decline in temperature, relative humidity and rainfall during winters. Problem of low recovery in coastal belts in tropics is related to humid and warm climate, being more conducive to vegetative growth of the crop than sucrose accumulation.

Being a long duration crop, sugarcane undergoes a series of interdependent but fairly distinct growth phases during its life cycle. These phases are germination (initial 30–45 days), tillering (45–135 days after planting), elongation or grand growth (135–240 days after planting), and ripening or maturity (240 days after planting onwards). Sugarcane requires warmer weather with better soil moisture status to ensure proper germination and emergence. Hot and dry weather is congenial for tillering where a short and humid season is required for grand growth. Crop duration and time of planting of sugarcane vary widely for tropical and subtropical parts of the country (Table 2.6).

TABLE 2.5 Varieties of Sugarcane Released and Notified From 2000–2010

Name of variety	Year of release	State (s) for which recommended	Key characteristics			Resistant to
			Maturity	Cane yield (t/ha)	Sucrose(%)	
Co 85004	2000	Gujarat, Maharashtra, Karnataka, Kerala, Interior of Tamil Nadu & A.P., Madhya Pradesh & Chhattisgarh	Early	90.5	19.5	Smut
Co 86032	2000	-do-	Midlate	102.0	20.1	Smut
Co 87025	2000	-do-	Midlate	98.2	18.3	Smut
Co 87044	2000	-do-	Midlate	101.0	18.3	Smut
Co 8371	2000	-do-	Midlate	117.7	18.6	Smut
CoM 88121	2000	-do-	Midlate	88.7	18.6	-
Co 91010	2000	-do-	Midlate	116.0	19.1	Smut
Co 94008	2004	-do-	Early	119.8	18.3	Red rot
Co 99004	2007	-do-	Midlate	116.7	18.8	Red rot
Co 2001–13	2009	-do-	Midlate	108.6	19.03	Red rot, Smut, Wilt
Co 2001–15	2009	-do-	Midlate	113.0	19.37	Red rot, Smut
Co 0218	2010	-do-	Midlate	103.77	20.79	Red rot
Co 86249	2000	Coastal Tamil Nadu and Andhra Pradesh, and Orissa	Midlate	104.2	18.7	Red rot, Smut
CoC 01061	2006	-do-	Early	110.8	17.4	Red rot
CoS 91230	2000	Punjab, Haryana, Rajasthan, Central and Western U.P., and Uttarakhand	Midlate	68.2	18.8	Red rot
CoPant 90223	2001	-do-	Midlate	73.3	18.5	Red rot, Smut

TABLE 2.5 (Continued)

Name of variety	Year of release	State (s) for which recommended	Key characteristics			
			Maturity	Cane yield (t/ha)	Sucrose(%)	Resistant to
CoH 92201	2001	-do-	Early	70.0	18.2	Red rot
CoS 95255	2004	-do-	Early	70.5	17.5	Red rot
CoS 94270	2005	-do-	Midlate	81.5	17.1	Red rot
CoH 119	2005	-do-	Midlate	82.8	17.5	Red rot
Co 98014	2007	-do-	Early	76.3	17.6	Red rot
CoS 96268	2007	-do-	Early	69.8	17.9	Red rot
CoPant 97222	2007	-do-	Midlate	88.2	18.2	Red rot
CoJ 20193	2007	-do-	Midlate	75.9	17.9	Red rot
CoS 96275	2007	-do-	Midlate	80.8	17.3	Red rot
Co 0118	2009	-do-	Early	78.2	18.45	Red rot, Smut, Wilt
Co 0238	2009	-do-	Early	81.08	17.99	Red rot, Wilt
Co 0124	2010	-do-	Midlate	75.71	18.22	Red rot
Co 0239	2010	-do-	Early	79.23	18.58	Red rot
Co 87263	2000	Eastern Uttar Pradesh, Bihar, West Bengal, and Jharkhand	Early	66.3	17.4	Red rot, Smut
Co 87268	2000	-do-	Early	78.9	17.5	Red rot, Smut
Co 89029	2001	-do-	Early	70.6	16.3	Red rot
BO 128	2001	-do-	Midlate	69.2	17.6	Red rot, Smut
CoSe 95422	2001	-do-	Early	67.8	17.7	Red rot

Name of variety	Year of release	State (s) for which recommended	Key characteristics			
			Maturity	Cane yield (t/ha)	Sucrose(%)	Resistant to
CoSe 92423	2001	-do-	Midlate	70.1	17.5	Red rot
CoSe 96234	2004	-do-	Early	64.1	17.9	Red rot
CoSe 96436	2004	-do-	Midlate	67.1	17.7	Red rot
CoLk 94184	2008	-do-	Early	76.0	18.0	Red rot
Co 0232	2009	-do-	Early	67.82	16.51	Red rot, Smut, Wilt
Co 0233	2009	-do-	Midlate	67.77	17.54	Red rot, Smut, Wilt
Co 0232	2009	Assam	Early	67.82	16.51	Red rot, Smut, Wilt
Co 0233	2009	-do-	Midlate	67.77	17.54	Red rot, Smut, Wilt

Source: AICRP (2011).

TABLE 2.6   Sugarcane Planting Time and Crop Duration in Different States of India

State	Planting time	Crop duration (months)
Subtropical India		
Uttar Pradesh, Uttarakhand, Punjab and Haryana	September–October (Autumn)	16
	February–March (Spring)	12
	April–May (Summer)	10
Bihar	October–November (Autumn)	13–15
	February–March (Spring)	10–12
Rajasthan	October (Autumn)	14-15
	February–March (Spring)	10-12
Madhya Pradesh	October–November (Autumn)	14
	January–February	12
Assam	January–March	10–12
West Bengal	October–April	12–15
Tropical India		
Maharashtra	July–August (*Adsali*)	18
	October–November (Pre-seasonal)	15
	January–February (*Suru*)	12
Andhra Pradesh	August–September (*Adsali*)	18
	December–January (Early varieties)	10
	January–February (Mid varieties)	11
	February–March (Late varieties)	12–14
Tamil Nadu	July–September (Special season)	13
	December–May (Main season)	12
	January–March (Ideal main season)	12
Karnataka	July–August (*Adsali*)	16–18
	October–November (Autumn)	14
	January–February (Spring)	10–12
Gujarat	October–November (Pre seasonal)	14
	January–February (*Eksali*)	12
Odisha	October–March	10–14
	January February (Most ideal)	
Kerala	October–December (Usual season)	10–12
	August (in hilly rainfed tract)	10–12

Source: AICRP (2011).

2.3.1   TILLAGE

Sugarcane requires a vary clean preparation of field with good tillage. Thorough ploughing specially in heavy soils helps the root system penetrating deep in to the soil and prevents crop lodging. In northern India, where sugarcane is mostly grown in alluvial loam soils, one deep ploughing with soil turning plough followed by two cross harrowings is sufficient. In peninsular India, where sugarcane is grown in clay and clay loam soils, more number of ploughings are required. Under the conditions of sub-soil hardpan, it was found that cross sub-soiling is beneficial to ensure proper root proliferation (IISR, 2011). Sugarcane is such a crop, which responds favorably to repeated inter-row tillage. This is why it is said that tillage is manure for both plant and ratoon sugarcane.

2.3.2   SEED CANE

Sugarcane is a vegetatively propagated crop and seed used in commercial plantings is sett of the stem carrying one, two or three buds. A good seed cane sett should carry healthy viable buds and have adequate sett moisture, high reducing sugars, high nitrogen content, and be free from insect pests and diseases.

Seed cane should be genetically pure, derived from a plant crop (8–10 months crop in tropical and 10–12 months crop in subtropical areas) unaffected from biotic (diseases, pests and abiotic, drought, waterlogging) stress seed crop should have good germinability of buds and viability of tillers commensurating with the genetic potential of the variety. In matured cane, top half of the cane can be used as seed leaving the bottom half; as the germination of setts from the bottom half is seldom satisfactory. The lower internodes are more mature with less moisture and of course sucrose content is higher than that of reducing sugars. Upper half of the cane is succulent and contains younger undamaged buds without any covering of scales. Because of high reducing sugar content young buds germinate better and faster. A crop raised exclusively for seed purpose is called a short crop (6 to 8 months). The short crop is fertilized about 6 weeks prior to harvest to improve the quality. The entire cane from a short crop is fit for planting. For good germination, the seed cane should have about 80% viable buds, more than 60% moisture and high reducing sugar content (Singh et al., 2001).

The cane grown for seed purpose should be harvested carefully and trash or other leaves should be removed by hands without any damage to the buds. It is preferable to plant the setts in furrows/pits/trenches immediately after cutting. In case the placement of setts in the field is likely to be delayed, it should be stored in shade or covered with sugarcane trash and sprinkled with water at proper intervals. Normally for planting one hectare of sugarcane field it requires 40,000 setts. Farmers use hand held choppers (*gandasa*) for cutting sugarcane setts however, Indian Institute of Sugarcane Research has developed a sett-cutting machine, which cut 12,000 (2 or 3 bud) setts/hr. In order to avoid spread of sett borne diseases the sett-cutting implement should be frequently dipped in fungicide solution.

Treatment of sugarcane setts for preventing diseases like smut is very important. Dipping of setts in 0.2% solution of systemic fungicide Carbendazim for 15 minutes before planting is recommended for this. In order to control sett borne diseases like red rot, smut, grassy shoot disease and wilt moist, hot air treatment (MHAT) should be adopted which involves placement of seed cane under 54°C temperature and 99% RH for 2–2.5 hours in MHAT plant designed by IISR, Lucknow. Raising of seed nurseries from heat treated canes offers the best scope for eliminating the primary infection of ratoon stunting and grassy shoot diseases. This also provides the basis for multiplication of disease free seed material.

In general, three bud setts are used for planting in the country except in Punjab, Haryana, and Western U.P. where two bud setts are still preferred to ensure better stand of the crop. Healthy and good cane setts should be selected by rejecting the cane setts, which have damaged buds, reddening at cut ends and infested by pests. The intact cane length should never be used as seed material due to prevalence of apical dominance in such canes that hinders the germination of lower buds. Also seed from ratoon crop should never be used as it is prone to many seed borne diseases like grassy shoot disease (GSD), ratoon stunting disease (RSD), and smut. Under late planting conditions, pre-germinated setts are preferred.

2.3.3 SEED RATE

Seed rate of sugarcane varies with the climatic conditions, quality of seed cane, planting season and the method of planting. Optimum plant popula-tion per unit area is essential to get profitable yield. About 38,000 to 40,000 three-bud setts are needed to plant one hectare during autumn in subtropical north, which can be obtained from 5–6 tonnes of cane. In spring season 48,000 to 50,000 setts and under late planting (summer) about 58,000 to 60,000 setts are needed from 6 to 8 tonnes of sugarcane. In tropical south, seed requirement is comparatively less as it ranges from 25,000 to 40000 three/two bud setts per ha. This is mainly due to wider row spacing of sugarcane planting adopted in tropical region. High seed rates are used in Punjab, Haryana and Tamil Nadu while low seed rates are common in west Bengal, Madhya Pradesh, Maharashtra, Andhra Pradesh, Karnataka and Gujarat. In U.P., Bihar, Rajasthan, Assam, Odisha and Kerala normal seed rates are used. In Tamil Nadu, due to pineapple disease, the germination is poor whereas in Punjab and Haryana the germination is poor because of the incidence of frost. In either case high seed rate is required.

2.3.4 PLANTING METHODS

Planting of sugarcane is done manually or mechanically. In manual planting several options are available depending on the agro-ecological and socio-economic conditions. Method of planting also differs on the basis of planting material used and nature of soil manipulation/earth work done. In general, flat method is very common in North India while in tropical south ridge furrow method is extensively followed. Adoption of planting methods also depends on objective of the grower and resource availability. For seed economy spaced transplanting technique (STP), bud chip method and poly bag raised settlings are adopted. Under adverse climatic and soil conditions like drought or salinity trench method may be followed. Ring-pit method of planting may be adopted for undulating topography, problem soils and yield maximization. Partha method is suitable for the situations where soil is excessively moist at planting.

Different methods of sugarcane planting are described in the following subsections.

2.3.4.1 Flat Method of Planting

In this method of planting land is prepared in a fine tilth by one or two deep ploughings followed by harrowing and planking. Shallow furrows of 10–15 cm depth at a distance of 90 cm in autumn and 75 cm in spring season are opened and setts are placed in furrows following either head to head

or bud-to-bud alignment and compacted with heavy wooden plank to conserve soil moisture. This method is very common in North India. Sugarcane
cutter-planter developed by IISR, Lucknow effectively reduces the time
and labor required for flat sugarcane planting. In place of 30–40 man-days,
only 5 man-days are required for planting with cutter planter. This machine
performs all planting related operations including fertilizer placement and
insecticide/fungicide spray in a single pass. In a day, about 2 ha of sugarcane
planting can be done with this planter.

2.3.4.2   Trench Method

Trench planting is adopted to save irrigation water. After the land preparation
25 cm deep and 30 cm wide trenches are made. Bottom of the trench is tilled
to pulverize the soil and incorporated with FYM. Cane setts are placed with
last bud overlapping at the bottom in two rows (on either side of the trench
base) and then covered with thin soil layer followed by irrigation. This system is recommended for the soils with high level of fertility and availability
of organic matter and fertilizer. Trench planting leads to better germination
because of higher moisture content and thin soil layer over the setts. Deep
trench planting provides better anchorage to sugarcane crop. Comparatively
high yields are observed under trench planting. Ratoon yields are also higher
from trench planted cane.

2.3.4.3   IISR 8626 Method of Planting

A planting technique, known as IISR 86206 (later IISR 8626) was developed by Panje et al. (1968). This technique was intended to substantially
exploit the sub-soil moisture and native soil fertility of Indo-Gangetic plains.
In this method, about 2 months before planting seed stalks are topped and
trash and leaves are removed. This facilitates the sprouting of lateral buds.
The main field is prepared by forming trenches of 30 cm depth and 20 cm
width at spacing of 90 cm from center. A 'tailed' or 'long' rayungan (cut
seed from topped and sprouted cane) are planted at a distance of 50–75 cm
vertically in trenches filled with the mixture of pulverized soil and fertilizer
to the brim followed by irrigation. Collectively the technique is referred to
as 'CAEGUS' system, which stands for cessation of auxin action, extension
of growth and unhindered utilization of soil. The yields are high and inputs

are efficiently utilized in this system but commercial adoption has been poor and hence the technique remained only of academic interest.

2.3.4.4   Spaced Transplanting (STP) Method

In conventional method of planting, 10–15% of cane produced is utilized for seed material, genetic potential of tillering is not well expressed and solar radiation is not harvested by the sugarcane at its optimum capacity. Considering these factors, an amelioration technology known as STP was developed at the IISR (Srivastava et al., 1981). In this technique single bud setts are used and settlings are developed from a raised seed bed. Approximately 50 m^2 land area and about 20 quintals of seed cane are needed to raise the settlings for transplanting in one hectare. Single bud setts are prepared by cutting just above the growth ring. Setts are planted vertically and adequate watering is done by rose can. The vertically planted setts are covered with cane trash or paddy straw. About 4–6 weeks old settlings are transplanted following 45 or 60 cm plant-to-plant spacing in the main field. Irrigation is given just after transplanting. This technique is well suited for achieving a reduction in seed rate by one-third. It increases the multiplication ratio of the buds, reduces lodging of cane and gives an yield rise of 20–25% in tropics and about 40–50% in subtropics over the conventional sett planting. This method is especially suited for seed cane multiplication of new varieties.

2.3.4.5   Ring-Pit Method of Planting

Ring-pit method of planting developed at IISR by Singh et al. (1984) encourages synchronous development of mother shoots and suppresses excessive tillering. In this method circular pit of 75 or 90 cm diameter and of 45 cm depth are dug at a center-to-center distance of 120 cm during autumn and 105 cm during spring planting season. The field is marked at regular intervals of 120 or 105 cm with the help of long rope, straight lines are drawn both ways and at the point of intersection a ring of corresponding diameter is made. A space of 30 cm is left between the two rings. Each pit receives 5–8 kg FYM or compost, 40 g urea and 10 g each of single superphosphate (SSP) and muriate of potash (MOP). Healthy setts with two or three eye buds, preferably obtained from heat-treated cane are chosen. Twenty setts are arranged horizontally in a circular fashion in each pit. Thus, each pit has 60 eye buds and about 25–35 millable canes are formed in each pit. A tractor mounted pit digger for ring planting has been developed at ICAR-IISR, Lucknow (Sharma and Singh, 1988) and this can make 500 pits a day in a well prepared field. This method though requires higher seed rate and initial investment in planting, the cane yield obtained is substantially higher than that from flat method of planting.

2.3.4.6 Partha Method

This method was developed by S.V. Parthasarathy for planting of sugarcane under excessive soil moisture conditions or swampy soils. It is more suited for heavy rainfall area of coastal Andhra Pradesh, Tamil Nadu, and Karnataka. In this method, three budded setts are planted in slanting position, 60° to vertical. One eye bud is thrust into the ground about 2.5 cm deep and two top buds sprout up. When the monsoon recedes, the slanting sett is pressed into soil to horizontal position and soil from the sides is added up to enable the shoots to strike roots.

2.3.4.7 Seblang Method

In Seblang method, seed crop is grown in well fertilized light soils to promote profuse tillering. As the tiller develops, it is separated from the mother shoot along with the roots and planted. The Seblang method is considered ideal for gap filling in ratoon as well as plant crop. Seblang method gives the highest rate of sett propagation under adverse edaphic and climatic conditions (Van Dillewijn, 1952).

2.3.4.8 Bud Chip Technique of Planting

Bud chips are the planting material in which bud along with a portion of nodal region is chipped off using bud chipping machine. The 'bud chip' is treated with fungicide and planted in a nursery bed or poly-bags filled with the mixture of organic matter (FYM or press mud), soil, and sand in 1:1:1 ratio. About 6–8 weeks old settling are transplanted in the main field. The objective of this technique is also the saving of seed material and the cane stalks after removal of bud chips can be sent to mills.

2.3.4.9   Poly-Bag Settling Transplanting

This is the replacement of labor-intensive STP method of planting. Single buds are planted vertically in perforated plastic bags filled with FYM, soil and sand in 1:1:1 proportion, 4–6 weeks old seedling are transplanted in main field. In this method higher rate of survival over that of STP method can be ensured. The polybag seedlings can be used for gap filling material in plant as well as ratoon. This method is also used for seed multiplication of newly developed varieties at higher rates.

2.3.5   SPACING, PLANT POPULATION AND PLANTING GEOMETRY

Number of millable canes per unit area is the single most important variable that contributes about 40% to the cane yield. Number of millable stalk and its length can be altered by manipulating the micro-environment and providing optimum condition to plant. A review on historical and theoretical aspects of spacing in sugarcane by Irvine and Benda (1980) reported that during early part of the eighteenth century, the spacing adopted for commercial crop varied from 45 to 60 cm in the United States and South America. Even now inter row spacings of 60–90 cm are adopted in many parts of Asia and Africa where man and animal powers are used. Closer rows resulted in higher cane yield particularly in temperate and subtropical areas. Increasing the seed rate without altering row spacing was found to be of little consequence, but the increased density coupled with narrower spacing holds promise for enhancing stalk population and yield (Singh et al., 1972).

2.4   NUTRIENT MANAGEMENT FOR SUSTAINABLE SUGARCANE PRODUCTION

Sugarcane owing to its long duration and huge biomass production removes substantial amount of plant nutrients from the soil, as reported from IISR a crop of 100 t/ha exhausts 208 kg N, 53 kg P and 280 kg K, besides 3.4 kg Fe, 1.2 kg Mn, 0.6 kg Zn, 0.2 kg Cu and 30 kg S. On the other hand, Indian soils are universally deficient in N except in some parts of north-eastern region. Further, nearly 50% soils are deficient in P and 20% in K. Of late, sulphur availability has also become critical in low organic matter coarse textured soils under S exhausting oilseed based cropping systems. Consequently, there exists a huge regional disparity in fertilizer (N, P, and K) use and the consumption of plant nutrients, hence the recommendations for nutrient application varies from state to state (Table 2.7). Presently, fertilizer application in some of the sugarcane growing states is far below the national average of 135.25 kg NPK/ha apart from wide ratio in their use. It is, therefore, imperative to adopt location specific judicious and balanced nutrient management practices in sugarcane for enhancing cane yield, improving produce quality and maintaining system sustainability. Nutrient management practices have one of the major roles in crop growth, modification of juice composition and accumulation of sugar in cane stalk.

High sugarcane yield with improved quality traits needs higher amount of plant nutrients. Glaring nutritional imbalances are being increasingly observed with the advent of new high intensity sugarcane based cropping

TABLE 2.7   State-Wise Recommendation for Fertilizer Use in Sugarcane

State	Recommendation (kg/ha)		
Assam	136	70	60
Andhra Pradesh	112	100	120
Bihar	150	85	60
Chhattisgarh	250	125	125
Gujarat	250	125	125
Haryana	150	150	0
Kerala	165	82	82
Karnataka	250	75	190
Madhya Pradesh	250	125	125
Maharashtra	250	115	115
Odisha	112	100	120
Punjab	150	60	60
Rajasthan	200	60	60
Coastal Tamil Nadu	275	63	113
Tamil Nadu	225	60	120
Uttar Pradesh (west)	150	60	60
Uttar Pradesh (east)	180	60	40
Uttarakhand	120	80	40
West Bengal	150	85	60

Source: AICRP (2011).

systems aimed at achieving high productivity. More so, the estimated magnitude of fertilizer use in sugarcane clearly shows the under nutrition of the crop as the current (2008–09) consumption (568.81 thousand tonnes) of N, P, and K is almost one third of total fertilizer requirement (1713.52 thousand tonnes) as well as the total nutrient removal (1542.61 thousand tonnes) by the crop (Table 2.8). Hence, it is advocated to use all possible sources of nutrients including manures, bio-fertilizers and crop residues (Singh et al., 2010).

Productivity and quality of sugarcane crop solely depend upon the quantity and quality of millable canes, which to a major extent is determined by the supply and uptake of nutrients at tillering stage. An optimum N concentration of 1.95–2.0% in plant at tillering has been estimated to be essential for maximum cane yield (Yadav, 2000). As protein synthesis is the highest during this phase, nitrogen application helps in it. Nitrogen content in index tissues has been related to the number of tillers produced and also the yield (Singh, 1978). Studies have also recorded a direct contribution of 40% of the number of millable canes to the agronomic yield of sugarcane crop followed by the length (27%), girth (3%), and weight (30%) of stalk. Therefore, management of plant nutrients plays a key role in influencing the number of tillers, height, girth and weight of cane. The role and need of different plant nutrients, which form the basis of fertilizer scheduling are elaborated herewith along with the fertilizer use efficiency.

2.4.1 NITROGEN

Nitrogen nutrition is of paramount importance for higher sugarcane productivity. Nitrogen increases the number of millable cane as well as the total weight of stalk. An estimated quantity of 1.2 kg N is removed from soil to produce 1.0 tonne cane. Based upon the field fertilizer trials, the rate of nitrogen application to sugarcane varies from 120 to 200 kg/ha in subtropical and 250 to 400 kg/ha in tropical belt depending upon the duration of crop whether it is annual or *adhsali* one. With an application of 67.3 kg N/ha, the average response to per kg additional N on cane yield was 106.2, 121.2, 105.8, 57.3, 115.4, and 81 quintals/ha in Punjab, Uttar Pradesh, Rajasthan, Delhi, Bihar, and parts of Haryana, respectively (Bhendia et al., 1964). Owing to intensive cropping and poor soil health management the soil fertility and productivity declined substantially and the recommendation for subtropics has gone

TABLE 2.8 State-Wise Status of Fertilizer Use in Sugarcane (2008–2009)

State	Area (mha)	Production (mt)	Productivity (t/ha)	*Nutrient removal ('000t) by sugarcane				**Fertilizer requirement ('000 t) by sugarcane				***Fertilizer consumption by sugarcane ('000t)			
				N	P	K	Total	N	P	K	Total	N	P	K	Total
Assam	0.03	1.1	37.9	2.29	0.58	3.08	5.95	4.08	2.10	1.80	7.98	0.97	0.41	0.48	1.86
Andhra Pradesh	0.20	15.38	78.50	31.99	8.15	43.06	83.21	22.40	20.00	24.00	66.40	26.87	13.30	7.73	47.90
Bihar	0.11	4.96	44.30	10.32	2.63	13.89	26.83	16.50	9.35	6.60	32.45	13.61	3.67	2.40	19.69
Gujarat	0.22	15.51	70.20	32.26	8.22	43.43	83.91	55.00	27.50	27.50	110.00	19.27	8.39	3.30	30.96
Haryana	0.09	5.13	57.00	10.67	2.72	14.36	27.75	13.50	13.50	0.00	27.00	13.32	4.41	0.41	18.14
Kerala	0.01	0.28	125.00	0.58	0.15	0.78	1.51	1.65	0.83	0.83	3.30	0.38	0.19	0.32	0.89
Karnataka	0.28	23.33	83.00	48.53	12.36	65.32	126.22	70.00	21.00	53.20	144.20	19.45	12.58	9.21	41.24
Madhya Pradesh	0.08	2.98	42.20	6.20	1.58	8.34	16.12	20.00	10.00	10.00	40.00	3.20	2.11	0.36	5.66
Maharashtra	1.09	60.65	78.90	126.15	32.14	169.82	328.12	272.50	125.35	125.35	523.20	64.76	36.10	23.08	123.93
Orissa	0.02	0.65	59.80	1.35	0.34	1.82	3.52	2.24	2.00	2.40	6.64	0.69	0.34	0.21	1.23
Punjab	0.11	4.67	57.60	9.712	2.48	13.08	25.26	16.50	6.60	6.60	29.70	18.27	5.23	0.78	24.28

Rajasthan	0.01	0.39	59.70	0.81	0.21	1.09	2.11	2.00	0.60	0.60	3.200	0.33	0.15	0.01	0.49
Tamil Nadu	0.31	32.80	99.909	68.22	17.38	91.844	177.45	69.75	18.60	37.20	125.55	34.31	13.53	19.29	67.13
Uttar Pradesh	2.08	109.05	52.30	226.82	57.80	305.34	589.96	312.000	124.80	124.80	561.60	232.36	72.59	20.20	325.15
Uttarakhand	0.11	5.59	52.20	11.63	2.96	15.65	30.24	13.20	8.80	4.40	26.40	9.805	2.64	1.11	13.55
West Bengal	0.02	1.64	70.00	3.41	0.87	4.59	8.87	3.00	1.70	1.20	5.90	1.45	0.86	0.84	3.15
Others	0.03	0.78	22.70	1.62	0.41	2.18	4.22	0.00	0.00	0.00	0.00	0.00	0.00	0.00	0.00
All India	4.42	285.03	67.90	592.86	151.07	798.08	1542.01	894.32	392.73	426.48	1713.52	344.32	148.91	75.58	568.81

Source: AICRP (2011); Anonymous, (2010); Singh et al., (2006).

*Nutrient removal has been estimated through sugarcane production and considering 100 t of sugarcane removes 208 kg N, 53 kg P and 280 kg K.

**Fertilizer requirement has been calculated on the basis of site-specific recommendations.

***Fertilizer consumption by sugarcane has been estimated through sugarcane area and average fertilizer consumption for the year 2006–2007.

to the extent of 150 kg N/ha for plant crop and 225 kg/ha for ratoon. The recent research data evince that there is need for upward revision of fertilizer recommendation particularly in north central, peninsular and coastal regions as positive response upto 125% of recommended N dose were recorded at most of the AICRP on sugarcane centers during 2010–11 (AICRP Annual Report 2011). Application of nitrogen should be tailored in such a way that its adequate availability is ensured during tillering. In Uttar Pradesh and Bihar whole of N is recommended to be applied below cane setts in furrows at planting or within 90 days in 2–3 splits. For Punjab and Haryana half of the recommended N is placed below the cane setts at planting and remaining half is top-dressed or drilled along the cane rows during April to June. In case of autumn planted cane in North India, one-third dose of nitrogen is applied at planting time and remaining dose equally in March, April and May. Under rainfed conditions, application of 75 kg N/ha at planting proves to be an optimum dose. Under tropical conditions nitrogen is applied in 2 to 4 splits of different proportions to get an optimum N use efficiency.

Comparative efficiency of ammonical and nitrate sources of N under controlled conditions has revealed better efficiency of ammonical sources in sugarcane. Amongst different sources urea showed a favorable effect on the level of total nitrogen, NO_3-N and C:N ratio in the soils (Zende and Kibe, 1981). Maintenance of supply of NO_3-N in the soil at about 20–40 ppm throughout the crop growth and its absorption by cane crop as reflected by N index in the leaf tissues ensures higher cane yields. The availability of nutrients in the soil is directly linked up with the monthly soil temperature and it increases in hot and decreases in cold months. However, under field conditions all the sources of fertilizer nitrogen were equally efficient in cane production. The choice therefore, rests on the relative cost and availability of fertilizer.

In order to enhance the efficiency of nitrogenous fertilizers by checking de-nitrification losses, various nitrification inhibitors such as N-serve, AM, thio-urea, dicaynadiamide (DCD), potassium azide and tetrazole have been evaluated and their effect as nitrification inhibitors was recorded in the order: N-serve > Telodrin > DCD (Singh, 1978).

2.4.2 PHOSPHORUS

Phosphate compounds in plant are called 'energy currency.' Deficiency of P severely reduces inorganic P concentration, the photosynthetic rate and carbohydrate translocation in plants. In order to produce one tonne of sugarcane 0.50 kg P_2O_5 is removed from the soil. The majority of the experiments have shown that phosphorus has no direct effect on maturity of cane rather it counteracts the adverse effect of excessive dose of nitrogen. However, at IISR, Lucknow soil as well as foliar application of P @ 30 kg P_2O_5/ha showed an improvement in plant-ratoon cane yield. In other multi-location trials, the response to P was observed up to 60 kg P_2O_5/ha (Table 2.9). Phosphorus is recommended at 50–80, 30–90, 120–180 and 60–100 kg P_2O_5/ha in northern, southern, western and eastern part of the country, respectively. It is best advisable to apply P based on its soil test basis as sugarcane is able to utilize only 10–20% of the applied P. The crop absorbs about 20–25% P up to tillering and remaining 75–80% is absorbed during elongation phase. Length, girth, population and yield of millable canes are known to be influenced by P application and the response becomes more pronounced at higher levels of nitrogen (Jafri, 1973). The relative efficiency of phosphatic fertilizers has been found to be in order of DAP > SSP > MRP. Di-calcium phosphate as a P carrier has a special advantage over super phosphate due to its higher P content, absence of free acidity and availability for a relatively longer period in soil, which show high P fixation. Best method of P application in sugarcane fields is to apply at half way down the ridge in light to medium deep soils. The availability index of Olsen's P was found to be positively correlated with total P in leaf sheath. The phosphate content in juice is helpful for eliminating its colloids during processing and 300 ppm P_2O_5 concentration in juice is rated as a critical limit for better crystallization and color of finished product 'sugar.'

TABLE 2.9 Effect of Phosphorus Application on Cane Yield, Sugar Content, Millable Canes, and Cane Length

P. doses (kg/ha)	Cane yield (t/ha)	Sugar content (%)	No. of millable cane (thousand/ha)	Cane length (cm)
0	92.5	16.3	167.2	183.7
30	92.0	16.1	153.0	171.2
60	104.2	17.3	166.6	193.2
CD (P = 0.05)	8.0	NS	NS	NS

Source: Dhillon et al., (1993).

2.4.3 POTASSIUM

Potassium is fundamental to sugarcane for the synthesis and translocation of proteins as well as carbohydrates. It is also involved in the process of sucrose accumulation in plants. One tonne of sugarcane removes 3.4 kg K_2O from the soil. However, Indian soils being rich in potassium, its response is inconsistent from state to state and therefore soil test is necessary. The role of K in plant- water relations is well recognized and its top dressing at late stage is advocated to minimize the adverse effect of severe soil-moisture stress during drought. The experimental results given in Table 2.10 showed that sugarcane crop responded significantly to K application up to 120 kg/ha in terms of cane yield and sucrose percentage (Tiwari et al., 1998). Potassium is recommended at 40–90, 75–190, 120–150 and 60–120 kg K_2O/ha in northern, southern, western and eastern parts of the country. Sulphate ion accompanying K is preferable to chloride ion in view of its better effect on tillering ratio, primary index, cane yield and sugar accumulation. Application of K at planting is as good as applied in splits. High potassium concentration in plant helps in maintaining higher cell moisture for longer period even under drought conditions, which ensures higher juice extraction. However, very high level of potassium in juice adversely affects the crystallization of sugar and leads to higher sugar loss in molasses.

2.4.4 SECONDARY NUTRIENTS

Among the secondary nutrients, the role of Ca and S has been found crucial in sugarcane based cropping systems involving pulses and oilseeds,

TABLE 2.10   Effect of Potassium Application in Sugarcane

Potassium dose (kg/ha)	Cane yield (t/ha)	Sucrose in cane juice (%)	K uptake (kg/ha)	Available soil K status (kg/ha)
0	73.3	15.82	191.6	402.0
60	91.7	16.08	216.8	450.0
120	93.3	16.18	236.1	482.0
CD (P = 0.05)	4.30	0.05	21.0	46.0

Source: Tiwari et al. (1998).

respectively. A tonne of cane removes 0.5 to 1.0 kg Ca and 0.91 kg S from the soil. Calcium maintains structural and functional entity of plant cell membranes and takes part in the formation of Ca pectate. Due to its deficiency, the rind becomes soft which has adverse effect on juice quality. Sulphur is directly connected with N utilization possibly by increasing nitrate reductase activity and in marginally deficient soils, application of 30 kg S/ha has shown good response in sugarcane. Of the different sulphur sources ammonium sulfate proved the best followed by iron pyrite and elemental sulphur.

2.4.5   MICRONUTRIENTS

Under field conditions deficiency symptoms of Mn in sugarcane exhibiting chlorosis has been reported from central Uttar Pradesh. With passage of time the deficiencies of more and more micronutrients viz., Zn, Fe and Cu have been observed. Foliar application of micronutrients has been found effective in increasing cane yield in Uttar Pradesh and Punjab. In Bihar, application of Zn, Mn and Fe each @ 25 kg/ha and Borax @ 5 kg/ha in combination with FYM in calcareous soils improved the cane yield (Sen et al., 1984).

2.4.6   FERTILIZER APPLICATION SCHEDULE IN SUGARCANE

Fertilizer application schedule in sugarcane is guided by soil texture, planting season and irrigation management. It is widely conceded that for a 10–12 months crop, the time of nitrogen application may be restricted to the first 45–90 days, i.e., the period of tillering. Fertilizer nitrogen should be applied in 2–3 splits and completed before the onset of monsoon. Reduction in the number of fertilizer nitrogen applications has also been suggested for 18 months *adhsali* crop where it can be done in 3 split doses, 10% at planting, 40% after 8 weeks and 50% at the time of earthing-up without loss in the yield as compared to 4 split applications. Phosphorus and potassium are applied as basal. Where chances of leaching are more as in coastal Tamil Nadu, 2 split application of potassium (one at planting and the other at six months) is recommended.

2.4.7   *ORGANICS TO SUSTAIN HIGH SUGARCANE YIELD*

The analysis of sugarcane productivity trends during the recent years reveals that the cane yield throughout the country has either plateaued or even declined in some states. Besides socio-economic constraints, the most significant reason for the yield decline is an intensive cultivation coupled with inadequate nutrient supply. Consequently an integrated approach involving chemical, organic and biological sources of nutrients will have to be applied in an intensive cropping system for long-term maintenance of soil fertility.

The use of organics in association with chemical fertilizers has proved superior over its individual component. Moreover, organic sources increase the nutrient use efficiency and bring about economy in fertilizer use. The fertilizer and manorial schedules in the system help in correcting the emerging deficiencies of nutrients other than, N, P and K particularly the micro-nutrients. Bio-manorial nutrient management strategies in sugarcane include:

(i)   integration of organics with fertilizers;

(ii) nutrient management through legumes including green manuring;

(iii) crop residues recycling with an accent on input economy;

(iv) bio-manuring integrating bio-fertilizers with organics;

(v)   utilizing sugar factory by-products/wastes to supply plant nutrients.

2.4.7.1   Integration of Organics With Fertilizers

Addition of bulky organic manures improves the physical properties of the soil and creates ideal rhizospheric environment. This provides congenial soil-water relations for better nutrient release and availability. Besides supplying major plant nutrients, the organic manures play a key role in meeting the requirement of micronutrients.

Application of FYM and/or green manure in sugarcane (Table 2.11) established its beneficial effect in improving the production efficiency of fertilizer N and more so at its optimal level. The response of sugarcane to FYM @ 25 t/ha was found to be 8.0 tonnes cane/ha averaged over 258 experiments conducted at 19 centers (Gaur, 1992).

TABLE 2.11 Cane Production Efficiency of N (kg cane/kg N applied) Under Different Organic Manuring Treatments

Organic manuring	N rates (kg/ha)	Row spacing		Mean
		90 cm	60 cm	
Control	150	192.7	246.0	219.3
	300	131.7	168.3	150.0
Farmyard manure-FYM	150	250.0	305.5	277.0
	300	145.3	193.3	169.3
Green manure-GM	150	272.6	329.3	301.0
	300	159.6	179.	169.3
FYM + GM	150	298.0	418.0	358.0
	300	169.0	218.3	193.7
Mean for N	150	253.3	324.7	289.0
	300	151.4	189.7	170.6

Source: Lal and Singh (2002).

2.4.7.2 Nutrient Management Through Legumes Including Green Manuring

Nutrients applied to crop are often partially utilized and enough residual and cumulative effects are carried over to second or third crops in the sequence. The current transfer of nitrogen from legume to non-legume during the same season and the carry over effects to the subsequent crop has occupied a prime place in integrated nutrient management strategies. Legumes in sugarcane based cropping systems are accommodated as dual purpose intercropped grain legumes with incorporation of green plants in soil or as an intercropped green manure to supplement chemical fertilizers.

Under National Agricultural Technology Project at IISR, Lucknow, the *in situ* incorporation of intercropped dual purpose cowpea (PusaKomal) in spring planted sugarcane after picking the green vegetable pods added 70.61 kg N/ha followed by *Sesbania* green manuring (45.90 kg N/ha). In winter-initiated ratoon inter-cropped berseem after cuttings for fodder left the highest amount of available nitrogen in soil (249.33 kg/ha in 0–15 cm and 240.50 kg/ha in 15–30 cm soil layer). At the harvest of sugarcane, the bulk density of soil under sugarcane ratoon + berseem/shaftal/menthi was lower

(1.28 g/cm^3) as compared to that under sugarcane ratoon sole (1.38 g/cm^3). Infiltration rate was more (5.5 mm/hours) in plots under sugarcane ratoon + shaftal due to addition of huge root biomass in top layer of soil.

2.4.7.3   Crop Residue Recycling With an Accent on Input Economy

Crop residues are renewable and readily available but are scattered organic resources. Intensive sugarcane based production system besides adding huge quantities of biomass (13.32 million tonnes stubble, 37.59 million tonnes root and 35.52 million tonnes trash/year) of sugarcane *perse,* has the enormous potential residue from cereals, pulses and oilseeds crops grown in succession and/or association. Since nutrients absorbed by cane plants from soil do not form the constituents of its marketable commercial product 'sugar,' there is good opportunity of organic recycling in this crop. The recycling of roots/trash directly in the soil through vermi-culture, green tops/molasses through ruminants in the form of cattle dung/urine, press-mud from juice as soil amendments/sulphur source and spent wash from distilleries as irrigation resource after dilution can return multi-nutrients to soil from sugarcane crop itself.

2.4.7.4   Integrating Bio-Fertilizers With Organics

Bio-fertilizers have an important role to play in improving the nutrient supplies through their availability to crop under upland conditions. Bio-fertilizers make up a judicious combination with chemical fertilizers and organic manures. The cane and sugar yield could be increased up to 9.39 and 1.34 tonnes/ha, respectively by inoculation of *Azotobacto*r and *Azospirillum* under graded levels of nitrogen. *Gluconacetobacter diazotrophicus*, a new class of bio-fertilizer has shown an excellent property of biological nitrogen fixation in sugarcane with non-inhibitory effect of NO_3-N under micro-aerophilic conditions. An endophyte, this bacterium is found in sugarcane root, stalk, leaf and trash and is capable of fixing sizeable amount of nitrogen (Suman, 2001).

2.4.7.5   Sugar Factory by-Products/Wastes as Source of Nutrients

The sugar industry by-products like press-mud from sugar factory and spent wash from distillery continue to be of economic importance. In the sulphitation factories press mud amounts to about 3% and in carbonation factories about 7%. Press-mud cake (PMC) has a great potential to supply plant nutrients (1–2% N, 2–4% P_2O_5 and 0.5–1.5% K_2O), besides having beneficial effects on physico-chemical and biological properties of soil. These inturn influence the availability and uptake of nutrients, cane yield and juice quality. An increase in sucrose content in juice was noticed by application of 12.5 t/ha of PMC over the recommended dose of NPK through fertilizer at Padegaon.

In Bihar, the utilization of sulphitation press-mud (SPM) cake along with inorganic fertilizer gave significantly higher yield of sugarcane as compared to farmyard manure and cane trash compost. It recorded higher cane and sugar yield and proved superior over inorganic amendments like pyrites on calcareous saline-sodic soils (Singh, 1993).

Distillery effluent (spent wash) is another important organic waste that contains appreciable amount of plant nutrients. The use of liquid and semi-solid distillery effluent as an organic manure increases cane yield significantly. Application of sugar mills effluent upto 800 m³/ha did not show any adverse effect on germination, tillering and growth of cane. It increased tiller production by 7–10% and cane yield by 10–15%. There was no adverse effect on juice quality parameters of cane (Singh, 2000).

2.5 WATER MANAGEMENT FOR SUSTAINABLE SUGARCANE PRODUCTION

Sugarcane requires large volume of water for remunerative cultivation. On an average 20 megaliters of water/ha is required by the crop (Shrivastva et al., 2011) to fulfill its consumptive use (metabolic activities), evapo-transpirational needs and losses during the course of irrigation and thereafter. Plant and ratoon crops differ in their efficiency to use water. Normally, 88 kg water/kg cane is required for a plant crop whereas, ratoon requires 118 kg water/kg cane produced. In India sugarcane is, by and large, an irrigated crop and acreage of sugarcane in most of the sugarcane growing states is 95 to 100% irrigated. However, irrigated area under sugarcane in Assam, Bihar, and West Bengal is limited to just 2.2, 23.7, and 44.6% of the total area under the crop. The average cane yield in irrigated areas are 67 t/ha, while that in partially irrigated or totally unirrigated areas is 41 t/ha. Thus nearly 43% of the cane production comes from about 25% of the area, which is fully irrigated and the remaining 57% is produced from 75% of the total hectarage, which does not receive optimum irrigation. Since sugarcane claims 6% of the total water resources of the country, limited availability of water to cane crop is the major factor responsible for low average cane yield in the country.

Research done to work out the water requirement of sugarcane at Padegaon (Maharashtra), Shahjahanpur, Lucknow (Uttar Pradesh), Anakapalle (Andhra Pradesh) and Coimbatore (Tamil Nadu) revealed that it ranges between 1400 mm to 2500 mm, being the lowest in Bihar and the highest in Maharashtra. Annual precipitation is the major contributor to the total water requirement of the crop as the highest water requiring grand growth or elongation phase wherein rate of dry matter accumulation is the highest coincides with the rainy season both under tropical and subtropical regions. However, water application through irrigation is highest during emergence and tillering phases, which pass through dry and desiccating atmospheric conditions during March to June. Factors like soil texture, planting season and weather, hence determine the total water requirement in different sugarcane growing states (Table 2.12). Volume of water required to be applied through irrigations varies from 60 to 200 cm in various states (Singh and Pannu, 1998). In the states of subtropical region where main season of planting is February–March, 60–80 cm water

TABLE 2.12   Water Requirement (WR) in Various Sugarcane Growing States of India

State	WR (ha-cm)
Subtropical India	
Bihar	140
Uttar Pradesh	160–180
Punjab	170–180
Tropical India	
Andhra Pradesh	160–170
Tamil Nadu	180
Karnataka	200–240
Maharashtra	
Plant cane (seasonal)	250
Plant cane (pre-seasonal)	300
Plant cane (Adsali)	350
Ratoon	300
Madhya Pradesh	270

Source: Srivastava and Johari (1979); Verma (2004).

is applied through 6–8 irrigations; 4–6 prior to the onset of monsoon during tillering phase and remaining two irrigations after withdrawal of monsoon for maturity and ripening. Irrigation is stopped one month before harvesting in subtropical region to improve sucrose accumulation. Comparatively higher amount of water (120–200 cm) and more frequent irrigations (20–36) are needed in tropical states like Tamil Nadu, Andhra Pradesh, Karnataka and Maharashtra. *Adhsali* crop, grown in Maharashtra needs irrigations at weekly interval; hence, irrigation water requirement goes up to 250–270 cm. Yadav and Prasad (1988) have reported that sugarcane yields are higher when more water is made available to the crop during the tillering phase.

2.5.1 IRRIGATION SCHEDULING

In order to get better growth of sugarcane crop irrigation has to be given at a time when soil does not cope up freely with the plant water requirements. Several approaches have been adopted to decipher the interval of irrigation and the quantity of water required to be applied in each irrigation based on all or some of the factors like crop stage, climate, soil texture and depth, effective root zone, and infiltration rate. In subtropical north irrigation at critical growth stages of sugarcane has been worked out and the tillering phase of crop that coincides with dry pre-monsoon (March–June) period has been found to be most critical. Normally 7.5 to 10 cm water is applied at each irrigation. Experimental results have indicated that irrigations at the entire first, second and third orders of tillering ensure highest cane yield (Singh et al., 1984). However, under the conditions of limited water availability if water is available for one irrigation only, it should be given at 3rd order of tillering and if it is sufficient for two irrigations, 2nd and 3rd order of tillering are the most responsive (Table 2.13). In tropical India, usually one or two irrigations are given at an interval of 3 or 4 days after planting to help the setts germinate and the seedling to establish well. Thereafter, in the absence of rains, cane is irrigated every 10 to 12 days during its growing period. In dry areas and in sandy loam soils, irrigation may be needed at an interval as short as 8 days. Towards the time of harvesting, irrigation frequency is reduced, and just before harvest, irrigation is withheld for about a month. Recently, irrigating the crop based on IW/CPE ratio has been found to result in the highest cane yield with limited use of water. The optimum ratio varies from 0.6 to 0.8 in different sugarcane growing regions. Irrigation

TABLE 2.13   Effect of Irrigation at Different Stages of Crop Growth on the Yield of Sugarcane

No. of irrigations	Time of irrigation	Yield (tonnes/ha)
One	I (emergence)	48.81
	II (first-order tillering)	45.38
	III (second-order tillering)	47.05
	IV (third-order tillering)	56.59
Two	I + II	50.97
	I + III	49.78
	I + IV	60.00
	II + III	56.51
	II + IV	60.29
	III + IV	60.01
Three	I + II + III	64.20
	I + II + IV	69.75
	I + II + IV	59.52
	II + III + IV	64.66
Four	I + II + III + IV	66.31
CD 5%		**11.12**

Source: IISR (1973).

at 1.0 IW/CPE ratio in Haryana recorded significantly higher yield (70.4 t/ha) over 0.75 (66.4 t/ha) and 0.5 (61.9 t/ha) ratios. Water application based on available soil moisture in active root zone has also been found effective in water conservation and irrigating the field at 50% and 75% depletion of soil moisture during tillering and maturity phases recorded higher cane yield under tropical conditions. In subtropics irrigations scheduled at 25% soil moisture depletion during first 120 days (germination and tillering phase), 50% soil moisture depletion during grand growth and sugar accumulation stage and 75% depletion at ripening has been found beneficial.

2.5.2   IRRIGATION METHODS

Since sugarcane is cultivated as a row crop planted widely through flat method surface irrigation is widely adopted. Farmers of subtropical region adopt flood irrigation because of convenience however a lot of water goes unused in this method owing to uneven distribution in the field and consequently water use efficiency is low. To make water use more efficient furrow irrigation is advocated wherein water is applied in furrows between two crop rows. The length of the furrow ranges from 10 to 30 m or more depending upon the soil type and slope of the land. The serpentine method of furrow irrigation is practiced for 18-month crop (*adhsali*) in many parts of central and tropical India. Water use efficiency was found to increase significantly at Lucknow by adopting skip furrow method of irrigation wherein instead of irrigating all the rows and inter-row spaces, one row is skipped and therefore alternate furrows are irrigated. This could save water to the extent of 36.5% over that of regular furrow irrigation and water use efficiency was enhanced by 64% with increase in yield of 2.2 t/ha (Srivastava and Johari, 1979).

In view of fast depletion of ground water in north and north western parts of the country covering major sugarcane growing and sugar producing regions having more than 80% ground water irrigation through deep tube well pumping it become imperative to adopt irrigation methods which are low water requiring, precise in application and where water use efficiency is highest. Drip method of irrigation has been found to address these concerns as with this method soil moisture is uniformly maintained at or near field capacity in the rhizosphere, which enhances nutrient uptake and eliminate unproductive losses of water (deep percolation, runoff, etc.). In Maharashtra drip irrigation is catching up and nearly 400 ha of sugarcane area have been brought under drip irrigation. Drip irrigation required 940 mm of water/ha as against 2150 mm in conventional flood method of irrigation. The cane yield observed under drip method was 170 t/ha compared to 128 t/ha with conventional flooding. Results from VSI, Pune indicated that surface and subsurface drip with daily irrigation under paired row planting saved 45–50% water, increased water use efficiency by 2.5 times, increased cane yield by 20–30% and improved sugar recovery by 0.2–0.6 unit. Under tropical conditions it was observed that subsurface drip at paired (40:140 cm) spacing produced significantly higher number of millable canes, cane length, and single cane weight compared with furrow irrigation (Yaduvanshi and Yadav, 1990).

Sprinkler system may be employed for soils of topographic conditions, which are undulating and not suited to surface irrigation. High-pressure nozzles (500–700 Kpa) are commonly used but recent advances include the development of low-pressure nozzles (100 Kpa or less), which are tailored as per the requirements of the soil and cultivation practices.

Covering of soil surface with trash mulch is an effective method of moisture conservation. In sugarcane it is done by uniform (10 cm thick) spreading of trash in inter-row spaces at completion of germination (45 days after planting). In addition to avoiding moisture loss through evaporation it maintains the soil moisture at higher level for a relatively longer time compared to uncovered soil surface besides, weed growth is also suppressed effectively. Trash mulching in sugarcane crop irrigated at two moisture regimes (25 and 50% ASM) caused an average increase in cane yield by 26%. Trash mulching brought about higher water use efficiency under both the moisture regimes, by economizing irrigation water. It saved 33.3 and 40.0% irrigation water at 25 and 50% ASM, respectively as compared to no trash mulching (Motiwale and Singh, 1973). Further, substantial amount of water can be saved by adopting improved methods of irrigation (Srivastava et al., 2011) as evident from the data of demonstrations carried out at farmers' fields near Lucknow (Table 2.14).

2.5.3 MOISTURE STRESS CONDITIONS

Water deficit in sugarcane hastens flowering and maturity of tissues, reduces nitrogen uptake and its utilization and increases cell wall thickness. Under soil moisture stress, sugarcane shows higher Brix and purity much earlier and maintains it for a stretch. In severe cases the moisture deficit increases fiber and bagasse percentage and also the contents of various intermediate products of metabolism through protein and carbohydrate hydrolysis in the sugarcane juice making clarification process difficult. The best way to cope under such situations is to select drought tolerant varieties viz., Co 94008, Co 8371, Co 87025, Co 87044, Co 2001–13 and Co 2001–15 in peninsular zone; CoC 01061 in east coast zone; Co 87268 and CoLk 94184 in north central zone and Co 98014, CoPant 97222 and CoH 119 in north west zone. Such varieties are characterized by deep and extensive root system and are able to tap water from lower soil horizons.

Drip irrigation proves to be a boon under water stress conditions in improving productivity and increasing water use efficiency. Spray application of potassium either alone or in combination with urea at very low concentration produces considerably higher yield under moisture stress conditions.

Headways in Agro-Techniques for Heightened Yield

TABLE 2.14 Effect of Irrigation Methods on Sugarcane Yield, Water-Saving and Water-Use Efficiency

Agronomic measures	No. of demonstrations	Cane yield (t/ha)			Water applied (ha-cm)			IWUE (kg/ha cm)		
		D	FP	Increase (%)	D	FP	Saving (%)	D	FP	Increase (%)
TM	28	80.2	63.8	25.7	47.7	65.4	37.2	1682.2	975.9	72.4
SF	32	88.5	63.8	38.8	53.7	65.4	21.7	1648.2	975.9	68.9
ICGS	24	81.8	63.8	28.2	42.3	65.4	44.5	1807.1	975.9	85.2

D, demonstration; FP, farmer's practice; IWUE, irrigation water use efficiency; TM, trash mulching; SF, skip-furrow method of irrigation; ICGS, irrigation at critical growth stages.

Source: Srivastava et al. (2011).

Potassium also improves the recoverable quality of juice. Application of 60 kg K_2O/ha at 240 days (before last irrigation) with trash mulching has improved the yield and juice quality. Soaking of seed material in saturated lime solution (80 kg lime in 400 liters of water for 2 hrs) has been found advantageous. Cane yield can also be improved by spray application of Kaoline (6%) during drought period.

2.5.4 MANAGEMENT OF WATER UNDER WATERLOGGED CONDITIONS

Though sugarcane can tolerate waterlogging and flooding with partial submergence for quite long periods, the prolonged submergence and swampy conditions of soil have adverse effects on the growth, yield and quality of cane. In order to avoid these conditions it is necessary that adequate drainage facilities be created for ample supply of oxygen, extensive root development, adequate nutrient supplies and enhanced microbial activities to support the vigor of above ground shoots.

In such situations, early planting is beneficial. Crop planted in the first week of February yields higher than the crop planted in March and April. Autumn planting is more beneficial over spring planting as by the time flooding occurs autumn crop attains sufficient vigor and height. Trench planting proves better over flat planting as roots penetrate deeper into the soil, which prevents lodging of cane (Srivastava and Dey, 2000). There was 59% increase in yield due to trench system over conventional flat planting. Seed rate should be increased to one and half times to compensate the effect of tiller mortality. Two or three split applications of nitrogen compensate for nitrate leaching. Spray application of urea during the period of waterlogging increases the cane yield. Similarly potassium and phosphorus application along with urea spray causes greater root proliferation and stiffness of cane. It brings about tolerance against insect pests damage and lodging. Support against lodging during rainy season may be provided by early earthing-up in the months of May and June. Sugarcane matures earlier under waterlogged conditions hence the crop affected by waterlogging must be harvested as early as possible after water has receded, it helps maintain the quality of cane for better sugar recovery. Processing of such canes at the earliest further reduces the chances of sugar loss.

2.6 WEED MANAGEMENT FOR SUSTAINABLE SUGARCANE PRODUCTION

Sugarcane is one among the most severely affected crops due to weeds. The extent of loss to cane yield due to weeds varies from 10% to total crop failure (Srivastava et al., 2005) depending upon the composition and diversity of weed flora under a particular climatic and edaphic condition. Large number of weeds flourishes in sugarcane fields due to a variety of factors. The slow germination as well as less initial growth of the crop provides ample opportunity to weeds to germinate and establish long before the crop plants can offer any competition. Availability of warm and humid micro-climate during the larger part of growing season in the north and throughout the growing season in south also favors weed growth, particularly of grasses in sugarcane fields. Cultural practices such as wide spacing between cane rows, frequent and heavy irrigations and application of heavy doses of manures and fertilizers enable the weeds to grow profusely and take away the lion's share of moisture and nutrients during initial stages. Weeds besides reducing the cane yield through direct interference in growth and development also prove harmful to the crop by providing shelter to insect-pests and disease causing pathogens.

2.6.1 WEED FLORA

In India sugarcane is grown under a variety of agro-climatic conditions, which are responsible for perceptible change in the composition of weed flora at different locations. Under subtropical climate of North India *Amaranthus spinosus*, *A. viridis*, *Asphodelus tenuifolius*, *Anagallis arvensis*, *Chenopodium album*, *Convolvulus arvensis*, *Cynodon dactylon*, *Cyperus* spp., *Commelina benghalensis*, *Digera arvensis*, *Euphorbia* spp., *Ipomoea* spp., *Phyllanthus niruri*, *Portulaca oleracea*, *Sorghum halepense*, and *Trianthema portulacastrum* predominantly infest the sugarcane fields.

2.6.2 WEED CONTROL METHODS

Cultural and mechanical methods of weed control have been found useful for the control of annual weeds infesting sugarcane fields. In subtropical India, critical period of crop weed competition coincides with the emergence of annual as well as dicot weeds. Hence, cultural and mechanical methods successfully check the loss in cane yield. In addition to the removal of weeds, mechanical methods also contribute towards maintaining the good physical condition of soil, which is very conducive for the tillering and growth of sugarcane. Among mechanical methods hand weeding, digging by spades and inter-culture with bullock drawn implements are commonly adopted. Removal of weeds by hand from inter- and intra-row spaces has been found best among the entire weed control methods at all the locations. Manual or mechanical hoeing at 30, 60, and 90 days stage of the crop has been found to effectively minimize the yield loss. However, hand weeding has not been found much effective against broad-leaved weeds and *Cyperus rotundus*. As far as cultural methods are concerned trash mulching and inter cropping has been found quite effective against weeds in sugarcane. Trash mulching in inter-row spaces suppresses the weeds and also excessive tillering resulting in reduction of weed dry matter production by 60% and increase in cane yield by 28.4% over that of no weeding (Srivastava, 2002). At many places trash mulching has been found to be more effective against weeds than the hand weeding and spray of chemicals. In case of ratoon crops trash mulching (10 t/ha) just after ratoon initiation has been found to provide best weed control.

Inter-cropping in between sugarcane rows with crops like chickpea and peas in autumn planted, and green gram, black gram and cowpea in spring planted sugarcane has been reported to reduce the weed competition and yield loss caused by weeds. For the control of *Striga* infestation in sugarcane fields sowing of trap crops like *Cholam (Andropogon sorghum)* and *Paspalum* sp. has been found effective. These crops induce the dormant seed of parasite to germinate, which could be killed easily by mechanical methods.

In order to avoid the problems associated with the cultural and mechanical methods, use of chemicals have been found effective. Pre-emergence application of atrazine @ 2 kg a.i./ha followed by 2, 4-D spray @ 1.0 kg/ha 60 days after planting have been found to be most effective in containing the weed population below economic threshold level (Srivastava et al., 2005). Binding weeds such as *Coccinia* spp., *Ipomoea* spp., and *Convolvulus arvensis* pose serious threat to sugarcane yield in coastal areas, north Bihar and eastern Uttar Pradesh for their effective control pre-emergence spray of metribuzin @ 1.25 kg/ha have been recommended.

Integration of various methods of weed control provides better management than any single method as one method fulfills the shortfalls of the other. For instance pre-emergence spray of atrazine followed by 2, 4-D spray 60 days after planting added with one hoeing at 90 days after planting gives better weed control in comparison to the use of herbicides alone. Continued use of a single herbicide leads to a shift in the weed flora and results in a preponderance of single weed species. Hence, a combination of synergistic herbicides is needed to control broader spectrum of weed flora. Srivastava (2003) while working on the efficacy of some new herbicides advocated the pre-emergence use of sulfentrazone 0.5 kg + atrazine 2.0 kg/ha in combination for better results.

2.7   EARTHING UP AND TYING

In most of the sugarcane growing regions the crop needs to be supported to remain erect in the field till harvest. It is essential not only for efficient trapping of solar radiation but also to avoid losses caused due to lodging of cane. A lodged sugarcane crop provides ample chance for emergence of late tillers that parasitize on full-grown canes and seldom turn into millable canes. Besides, there is significant loss in sucrose accumulation and content once a crop is lodged. Lodged crop also gets damaged by porcupines, jackals and rats. Hence, earthing up the sugarcane rows at the closing-in of crop canopy has been found beneficial for reaping higher cane yield. Under subtropical conditions two earthings are recommended, first in the month of June after receiving the pre-monsoon showers and the second during July to guard against high velocity winds and heavy rainfall. Earthing up in July also provides drainage channels for disposing of the excess water from sugarcane fields. In Maharashtra a local modern implement *dufan* is used for interculture, at 4½ month stage or when cane has formed 2–3 internodes earthing up is done with the help of a ridge maker. After earthing up furrows are converted into ridges and ridges into furrows. Earthing up operation gives support to the shoots and crop does not lodge later on.

At 6 to 8 month age crop attains a good height and becomes prone to lodging when high velocity winds are blown. In subtropical belt the crop may be bound by taking leaves from two adjacent clumps in a row twisting them into a rope and then tying the two clumps together. This operation is repeated after a month or so. At second time already tied clumps from two adjacent rows are brought together and tied by twisting their leaves. In tropical belt tying and propping of the crop is done when the crop attains a height of 130 to 150 cm. Another operation is carried out when the crop grows to a height of 200 cm or more.

2.8   RATOON MANAGEMENT

Ratooning is an integral component of sugarcane production system. It is an age old method of propagation of sugarcane wherein subterranean buds on left over stubble after harvest of preceding crop, emerge and give rise to a new crop stand termed as ratoon. Ratoon cropping is the pivot of sugarcane based production system around which cane productivity and economic profitability at farm level as well as sugar recovery at factory level revolve. By and large sugarcane crop in tropics and subtropics is ratooned at least once. In Mauritius, as many as 5 ratoons are taken, 85% of the cane milled each year is ratoon one. Planting once and harvesting thrice has always been considered a profitable venture in sugarcane culture. Ratooning in sugarcane economizes the cost of cultivation and increases the margin of profit in following ways:

- Saves the cost of seed cane.
- The land preparation and planting operations requiring 32 and 19% energy respectively are eliminated. This in turn results in saving of 51% of energy and expenditure thereon.
- The question of delay in sugarcane planting does not arise due to timely initiation of ratoon crop. Thus, the loss in yield due to delayed planting is automatically avoided.
- Supply of ripen ratoon cane to mills ensures high sugar recovery early in the crushing season.
- Early vacation of fields from ratoon cane enables the farmers to sow wheat crop timely in wheat-sugarcane plant-ratoon-wheat system in north-west zone.

At national level the productivity of ratoon cane stands at 58 t/ha against 85 t/ha for plant cane. It has also been observed that crop yield declines continuously with each successive ratoon crop particularly under subtropical conditions (Table 2.15). In view of 50% cane area occupied by ratoon crop annually, the national average of cane productivity goes down by 13.5 t/ha. Translated into practice, this means that around 54 million tones of sugarcane is lost every year by keeping half of the sugarcane acreage under ratoon crop with lower productivity. Potentially ratoons can yield as high as planted crops provided the recommended package of practices is followed earnestly. Through careful management of the crop factors responsible for poor ratoon yield viz., shorter growth period, loss of tilth in root zone, piece-meal harvesting, build up of insect-pests and diseases, sparse or gappy crop stand, soil sickness due to

TABLE 2.15 Yield of Subsequent Ratoon Crops from Single Planting at Lucknow

Years	Crop description	Yield (t/ha)
1973–74	Plant crop	78.5
1974–75	First ratoon	55.3
1975–76	Second ratoon	76.6
1976–77	Third ratoon	74.1
1977–78	Fourth ratoon	64.1
1978–79	Fifth ratoon	59.9
	Farmyard manure @ 15 t/ha was added after this)	
1979–80	Sixth ratoon	67.0
1980–81	Seventh ratoon	62.1
1981–82	Eighth ratoon	56.7
1982–83	Ninth ratoon	56.2
	(i) With P and K @ 100 kg/ha each	51.8
	(ii) Without P and K	54.0
	(iii) Average	
1983–84	Tenth ratoon	51.4

Source: Yadav (1992).

TABLE 2.16 Harvesting Schedule of Sugarcane for High Sugar Recovery

October	November	December	January	February	March	April
For subtropical region						
Ratoon II of APC	Plant of APC	Ratoon II of SPE	Ratoon I of SPE	Ratoon I of SML	Plant of SPM	Plant of SPM
Ratoon I of APC	Ratoon II of SPE	Ratoon I of SPE	Ratoon II of SPM	Plant of SPE	Plant of SPE	Plant of SPM
For tropical region						
Ratoon of October	Plant of October	Plant of November	Ratoon of December	Ratoon of January	Ratoon of February	-
Plant of October	Ratoon of November	Ratoon of December	Plant of December	Plant of January	Plant of February	-
				Plant of December	Plant of January	Plant of February

APC, autumn planted cane; SPE, spring planted early; SPM, spring planted mid-late.
Source: Modified from Shahi (1998).

mono-cropping, etc., can be effectively alleviated. Thus, there is a considerable scope to enhance the productivity of sugarcane ratoon by adopting improved agro-techniques. The management issues related to multiple ratooning are:

1. varietal choice and crop planning;
2. adequate crop stand and crop vigor;
3. weed perpetuation and inter-culture operations;
4. nutrient and water management;
5. harvesting schedule and cutting management.

2.9 SUGARCANE BASED CROPPING SYSTEMS

Sugarcane based cropping system has a distinct advantage since sugarcane leaves large crop residues. The plant crop of sugarcane is invariably followed by a ratoon crop. Mostly one to two ratoons are taken. However, the crop preceding sugarcane and succeeding ratoon crop varies in different agro-climatic and socio-economic situations. Some of the important sugarcane based cropping systems prevalent in the country are listed in Table 2.17a and b.

2.9.1　CROP DIVERSIFICATION IN SUGARCANE

In view of changing market scenario, consumers' preferences and global competitions, new income generating opportunities have emerged through crop diversification in sugarcane 'produce to product chain.' This would help in increasing the land utilization efficiency, reducing the production cost, economizing the use of market purchased costly inputs and making plant-ratoon system sustainable. This in turn raises the socio-economic status of small and marginal resource constrained farmers and generates employment especially for rural women and youths. Intercropping of different crops in between sugarcane rows offers all these advantages. For sugarcane based intercropping systems the recommended optimum plant population of base crop is suitably combined with an appropriate additional plant density of the associated crop and there is crop intensification in both space and time dimensions.

　　The crops and their combinations in intercropping systems are largely determined climatically. Inter-regional differences in the extent of

TABLE 2.17A Sugarcane Based Cropping System (Rotations) Commonly Practiced in Different States of India

Sl. No.	Sugarcane based cropping systems (rotations)	State
1.	Early rice -pea-sugarcane plant-ratoon-wheat	Eastern U.P.
2.	Early rice-autumn sugarcane-ratoon-moong/wheat	-do-
3.	Early rice-toria-sugarcane plant-ratoon-wheat	-do-
4.	Rice-early potato-sugarcane-ratoon-wheat	Western and Central U.P.
5.	Rice-wheat/mustard-sugarcane-ratoon-wheat	-do-
6.	Green manure-toria-sugarcane-ratoon-wheat	-do-
7.	Green manure-potato-sugarcane-ratoon-wheat	-do-
8.	Green manure-potato-sugarcane (autumn) + potato-ratoon-wheat	-do-
9.	Sorghum-early potato-sugarcane-ratoon-wheat	-do-
10.	Maize (early)-potato-sugarcane-ratoon-wheat	-do-
11.	Cotton + pea-sugarcane-ratoon-wheat	-do-
12.	Maize-wheat-sugarcane-ratoon-wheat	-do-
13.	Rice-pea-sugarcane-ratoon-wheat	-do-
14.	Rice/maize-sugarcane-ratoon	-do-
15.	Rice-green manure-sugarcane-ratoon-wheat	-do-
16.	Fallow-wheat-green manure-sugarcane-ratoon-wheat	-do-
17.	Maize-wheat-sugarcane-first ratoon-wheat	Punjab and western U.P.
18.	Fodder (*kharif*)-berseem-sugarcane-ratoon-wheat	Punjab
19.	Groundnut-wheat-sugarcane-ratoon	Gujarat
20.	Cotton-sugarcane-ratoon-sorghum	Maharashtra
21.	Sugarcane-ratoon-wheat	-do-
22.	Sugarcane-ratoon-cotton-gram	-do-
23.	Rice-sugarcane-ratoon	South India
24.	Ragi-sugarcane-ratoon	-do-
25.	Rice-groundnut-sorghum-finger millet-sunnhemp-sugarcane	Karnataka under canal irrigation-fixed 3 years rotation
26.	Sugarcane-fodder sorghum-groundnut-tobacco-cotton-green manure	Maharashtra canal irrigated areas with block system of irrigation
27.	Sugarcane-plant-ratoon-*kharif*-rice-winter rice-sunnhemp	A.P., Telangana region

Source: AICRP, Sugarcane (2011).

TABLE 2.17B   Crop Yield and Economics of Different Sugarcane Based Intercropping Systems (Pooled Data of Two Years)

Intercropping system	Yield (t ha−1)	Equivalent cane yield (t ha −1)	Average cost of cultivation (Rs. ha−1)	Average gross returns (Rs. ha−1)*	Average returns (Rs. ha−1)	B–C ratio
Autumn Sugarcane (Co.J.-64)						
Three-bud sugarcane alone	94.2	-	90.386.0	1,88.400.0	98.014.0	1.08
Single-bud sugarcane alone	120.8	-	88.292.5	2,41.600.0	1,53,307.5	1.73
Single-bud sugarcane + Gram	125.6+0.9	137.1	66,667.5	2,74,710.0	1,78,042.5	1.84
Single-bud sugarcane + Lentil	123.8+0.4	130.0	93,458.2	2,60,000.0	1,66,541.8	1.78
Spring Sugarcane (Co.H.-119)						
Three-bud sugarcane alone	91.4	-	88,654.5	1,82,800.0	94,145.5	1.06
Single-bud sugarcane alone	117.8	-	85,458.5	2,35,600.0	1,50,141.5	1.76
Single-bud sugarcane + Summer Moong	121.6+0.8	135.4	93,192.5	2,70,800.4	1,77,607.9	1.91
Single-bud sugarcane + Summer Mash	120.4+0.7	131.1	93,667.5	2,62,200.5	1,68,533.0	1.80

Source: Saini, S. P. et al. (2012)

intercropping depend on factors such as irrigation facilities, type of crops and soil management.

Intercropping permits more intensive cropping in a crop like sugar-cane, which is traditionally grown in 'single cropping system.' As sugar-cane plants have large lateral spread when fully grown, their optimum row spacing is usually 90 cm or more. The vacant space in between the rows can quickly be covered by weeds, which when allowed to grow unchecked eventually offer severe competition to the main crop. Instead of such a waste of available spaces, intercropping offers opportunity for profitable utili-zation of such space while cutting down on the direct cost of cultivation of the main crop by reducing the expenditure on weed control. The initial growth of sugarcane is very slow. It takes about 30–35 days to germinate in spring (February–March) and even more in autumn (October–November). Thereafter, further 100 days are taken to develop full canopy to cover entire inter-row spaces. In these 130–135 days another crop of short duration like short duration pulses or potato could easily be grown in this space without affecting the main crop.

There is tremendous scope of increasing area under pulses through inter-cropping in sugarcane under Indian tropics and subtropics (Table 2.18).

2.9.1.1 Intercropping With Autumn Planted Sugarcane

Sugarcane in autumn is planted at wider spacing of 90 cm or more, the bud sprouting is late and initial growth rate of crop is slow. Autumn planted sug-arcane provides 15–20% higher cane yield and 0.5 units more sugar recovery than spring planted cane but the area remains limited. However, companion cropping of winter pulses may promote autumn planting of sugarcane on account of higher returns and better resource use efficiency.

Intercropping of sugarcane + potato produces higher yield of compo-nent crops over potato – cane sequential system. Intercropping of two rows French bean (PDR-14) shows distinct positive effect on sugarcane growth similar to potato in terms of shoot count at grand growth stage. The system appeared to be quite profitable as it yielded 1664 kg/ha French bean yield. Autumn sugarcane intercropped with two rows of lentil (DPL 15) receiving 150 kg N/ha in combination with *Azospirillum* produced highest sugarcane equivalent yield, which was however, comparable to sub-optimal dose of 112.5 kg N/ha. These observations indicate that intercropping sugarcane

TABLE 2.18 Crops Suitable for Intercropping in Sugarcane

A. Tropical region	
Green gram (*Vigna radiata*)	Groundnut (*Arachis hypogea*)
Black gram (*Vigna mungo*)	Sesame (*Sesamum indicum*)
Cowpea (*Vigna unguiculata*)	Maize (*Zea mays*)
Sunflower (*Helianthus annuus*)	Finger millet (*Eleusine coracana*)
Soybean (*Glycine max)*	Radish (*Raphanus sativus*)
Coriander (*Coriandrum sativam*)	Onion (*Allium cepa*)
Okra (*Abelmoschus esculentus*)	

B. Subtropical region (autumn planting)	
Potato (*Solanum tuberosum*)	Onion (*Allium cepa*)
Radish (*Raphanus sativus)*	Garlic *(Allium sativum)*
Mustard (*Brassica campestris.*)	Coriander *(Coriandrum sativam)*
Carrot (*Daucus carota*)	Sugar beet (*Beta vulgaris*)
Lentil (*Lens esculentus*)	Wheat (*Triticum aestivum*)
Linseed (*Linum ustitatissimum*)	French bean (*Phaseolus vulgaris*)
Peas (*Pisum sativum*)	Maize (*Zea mays)*
Turnip (*Brassica sp.)*	Toria (*Brassica rapa*)
Grain amaranth (*Amaranthus* spp.)	

C. Subtropical region (spring planting)	
Green gram (*Vigna radiata*)	Tomato (*Lycopersicum esculentum*)
Black gram (*Vigna mungo)*	Cowpea (*Vigna unguiculata*)
Brinjal (*Solanum melongena*)	Dhaincha (*Sesbania sesban*)

D. Subtropical region (ratoon cane)
Berseem-Egyptian clover *(Trifolium alexandrium)*
Shaftal-Persian clover (*Trifolium resupinatum*)
Senji-Indian clover (*Melilotus parviflora*)
Lucerne (*Medicago sativa*)
Menthi-Fenugreek (*Trigonella foenum graecum*)
Oats *(Avena sativa)*

Source: Lal and Singh (2004); Verma et al., (1984).

with two rows of lentil could effect a saving of 37.5 kg N/ha. The compatibility of pulses as intercrop in sugarcane for enhancing system productivity has also been documented (Singh et al., 2006).

Among oilseed crops, intercropping of mustard (*Varuna*) in autumn sugarcane (1:1 ratio) enhances net return and cane equivalent yield. On an average, the system yields 80 tonnes sugarcane and 1500 kg mustard/ha. On the other hand, two rows of mustard *(Pusa Jaikisan)* intercropped with autumn sugarcane produced 2150 kg seed/ha. In recent years, the trials on sugarcane + linseed *(Garima)* at IISR, Lucknow gave very promising results producing 85 tonnes cane and 1600 kg linseed/ha.

2.9.1.2 Intercropping in Spring Planted Sugarcane

It is estimated that about one million hectare additional area can be brought under pulses by intercropping green gram and black gram in spring planted sugarcane especially in UP, North Bihar, Punjab and Haryana. Experiments conducted at Lucknow reveal that out of 5 genotypes of green gram minimum loss of cane yield is due to genotype PDM-11 and PDM-84-139 (4–5%). Among dual-purpose legumes, the highest net monetary return (Rs. 1208) works out for sugarcane + green gram (K- 851 for grains) system followed by sugarcane + cowpea (*Pusa Komal* for green pods), i.e., Rs. 1134/ha. After picking green pods for vegetable and/or mature pods for grains, the legume plants with longer leaf area duration are incorporated in the soil between the inter-row spaces of sugarcane as green manure. These systems affect nitrogen economy in sugarcane to the extent of 35–40 kg/ha besides producing bonus yield of pulses (Singh et al., 2007). The compatibility of intercrops varies with the row arrangements and genotypes selected as intercrop in the system.

Sugarcane based intercropping system involving *Sesbania sesban* at high density in spring planted cane effectively controls weed population and its *in situ* turning exhibits allelopathic effects on germination of *Cyperus rotundus* nuts, besides correcting nutritional imbalances.

2.9.1.3 Intercropping of Grain Cereals in Sugarcane

The crop compatibility especially with grain cereals must be adjusted as per the time of planting of sugarcane and the time available for intercrops. Accordingly for intercropping in sugarcane, such varieties of maize should be selected which could be harvested before the commencement of very hot weather, e.g., *Surya* and *Sweta*. However, for autumn planted sugarcane long duration maize like Azad Uttam could find place. Intercropping of maize in February planted cane as well as initiated ratoon offers great potential for midseason income generation in maize growing areas of subtropical belt. In such systems, the planting geometry of sugarcane and nitrogen nutrition of component crops are the key management issues. Staggered planting of sugarcane (CoPt 90223) at 45:105:45 cm accommodating two rows of maize *(Surya)* produces 83 t/ha sugarcane equivalent yield and generates sizeable amount of mid-season income from maize cobs.

2.9.1.4   Sugarcane Ratoon Based Intercropping Systems

Failure of subterranean bud sprouting in winter harvested plant cane is the major constraint in rationing in north-western zone. The problem is more acute in high sugar early maturing varieties. The management strategy lies in keeping the metabolic activities of stubble buds alive till the commencement of favorable temperature regime. Introducing early bulking high-density intercrops (*berseem, shaftal,* and *senji*) may help mitigate the problems asso-ciated with winter-initiated ratoon. Moreover, intercropped forage legumes serve as live much, regulate the rhizospheric thermal regime through root respiration, protect stubble buds from frost damage and encourage the sprouting of subterranean buds during spring.

2.9.1.5   Intercropping of Sugarcane + Wheat

Wheat–Sugarcane–Ratoon–Wheat is the most important cropping system, which has attracted the attention of the scientists and development workers in recent years. It is estimated that 0.3 million hectares of sugarcane area in India is under this system. The importance of the system lies in the fact that more than 60% of the sugarcane area in western U.P. and 10–14% in other states is covered by this system. A drastic reduction (30–50%) in sugarcane yield is a common feature when the sugarcane is planted late (summer- April end to May end) after the harvest of the wheat crop. High to very high tem-perature at planting and germination time, low humidity, over mature seed cane, little time for tillering, heavy infestation of weeds and insect-pests are the major factors responsible for poor cane yield. The recent approaches for the management of this system include simultaneous planting of sugarcane and wheat (taking wheat as reference crop) and Wheat + Sugarcane under Furrow Irrigated Raised Beds-FIRB.

2.9.1.6   Sugarcane Planting With Wheat Under FIRB System

In order to enhance the productivity of sugarcane in wheat-sugarcane system, an innovative, wheat + sugarcane intercropping under Furrow Irrigated Raised (FIRB) system has been developed at Indian Institute of Sugarcane Research, Lucknow. This technology is most suited for sandy loam and loamy soils. In this system, wheat is sown on raised beds and sugarcane is planted in furrows. In view of the fact that both the crops are heavy feeder for nutrients and compete for moisture and light. In this system, nutrients are applied in the rhizosphere of the respective component crops. Accordingly, nutrients for wheat are placed on the raised beds and for sugarcane in the furrows after wheat harvest. Since the irrigation water is applied only in the furrows, it creates variable moisture environment and dynamics of solute movement in the modified bed configuration. The root proliferation of the component crops is also in different depths. Thus, it provides ample opportunity for exploiting *spatial* and *temporal* complementarities enhances input use efficiency and increases the profitability and sustainability of the system. Economics of various intercropping systems are presented in Table 2.19.

Wheat, the reference crop of the system is sown on raised beds in the month of November/December. After field preparation and leveling, half of the nitrogen and full doses of phosphorus and potassium for wheat are applied by broadcast method followed by light harrowing. Remaining nitrogen for wheat is applied in two split doses on the raised beds at crown root initiation and tillering stage. Under this system, wheat requires 75% of NPK recommended for flat method. Wheat rows are sown at 17 cm row spacing on each raised bed (48–50 cm top width) keeping the seed rate of 75–80 kg/ha by tractor drawn Raised Bed Maker-cum-Seeder that makes raised beds and furrows and drills wheat seeds on raised beds concurrently. Immediately after sowing of wheat a light irrigation is given in the furrows (2/3 height) for proper germination of wheat. Subsequent irrigations are also applied in the furrows to maintain proper soil moisture environment on beds for wheat growth. Overall, six irrigations are required in wheat

TABLE 2.19   Economic Evaluation of Sugarcane Based Intercropping Systems

Intercropping system	Sugarcane yield (t/ha)	Intercrop yield (t/ha)	CEY* (t/ha)	Net returns (Rs/ha)	B:C ratio
Autumn Sugarcane					
Sugarcane sole	85.2	–	85.2	50199	1.63
Sugarcane + rajmash	86.8	1.94	132.8	89884	2.54
Sugarcane + lentil	76.5	1.16	99.0	59629	1.73
Sugarcane + mustard	70.7	1.44	94.9	55474	1.59
Sugarcane + maize (green cobs)	78.6	82,412**	125.9	83815	2.34
Sugarcane + potato	90.6	28.9	179.4	106736	1.67
Sugarcane + cabbage	103.00	3.47	166.10	98560	2.52
Sugarcane + onion	104.00	8.69	121.00	69462	2.79
Spring Sugarcane					
Sugarcane sole	77.3	–	77.3	42696	1.38
Sugarcane + cowpea (green pods)	75.2	2.90	90.4	51261	1.48
Sugarcane +green gram	76.6	0.57	91.6	52765	1.54
Winter Initiated Ratoon					
Sugarcane ratoon sole	73.2	–	73.2	42440	1.40
Sugarcane ratoon + berseem	79.4	56.8	109.3	73542	2.43
Sugarcane ratoon + shaftal	77.9	54.7	106.7	71072	2.35
Sugarcane ratoon + lucerne	72.6	41.2	94.3	59292	1.96

** No. of green cobs, Cane Equivalent Yield.

Source: Lal and Singh (2004).

crop. Germination, tillering and growth of wheat are much better due to improved soil tilth on the raised beds on one hand and the border effect on the other.

Sugarcane is planted manually in 80 cm spaced furrows (30–32 cm top width and 20–22 cm depth) by wet planting method in the month of November just after sowing of wheat that coincides with first irrigation. Sugarcane setts are placed in the irrigated furrows in the overlapping fashion using 50,000 three-bud setts and pressed down in soil by feet. Where wheat is sown in the month of December, sugarcane is planted in month of February in the standing wheat crop. In this case, sugarcane planting coincides with irrigation at boot leaf stage in wheat. For planting sugarcane in standing wheat, irrigation is applied in furrows preferably in the evening and next day, sugarcane setts are placed in furrows when the soil is in muddy condition and pressed into the soil manually. In order to chisel the furrows, a wheel hoe has also been designed to loosen the soil at the bottom of the furrow prior to irrigation for better placement of setts. After wheat harvest, the furrows are used for irrigating sugarcane till earthing-up operation. Sugarcane needs irrigation immediately after harvest of wheat. Fertilizer application is completed before onset of monsoon, i.e., in the last week of June. Thus under FIRB systems, sugarcane planting is advanced and gets optimum time span for tillering and grand growth phase and yields 35–40% higher cane as compared to wheat-sugarcane sequential system.

2.9.1.7 Harvesting

In order to ensure maximum yield of sugarcane and ultimately sugar from the crop, it would be necessary to have selective harvesting. In this way it is possible to have higher sugar recovery, which naturally increases the sugar production per unit area. The detailed harvesting schedule for plant and ratoon crops is given in Table 2.16.

In subtropical India, cane starts ripening from December onwards. Besides various external factors, the genetic and biochemical constitution of genotype, stage of maturity and cane quality at harvest affect the rate of post harvest deterioration of canes, staling of cane affect the sugar recovery and extraction efficiency. Delay in milling of harvested sugarcane starts after 24–36 h of harvesting, which results in loss of cane weight and sugar due to drying and inversion resulting in accumulation of reducing sugars. Therefore, harvested cane should be immediately sent to mill for crushing and 'kill to mill' period should be minimized.

Research to overcome post harvest bio deterioration in cane suggested that harvested cane should be stored in a shade, covered with trash and sprinkled with water. Spraying of formalin or polycide on harvested cane or dipping them in Actin ID solution (0.015%) or Leukokil (1.0%) was effective in minimizing the sugar losses (Kapoor and Sharma, 1990).

2.10   CONCLUDING REMARKS

Sugarcane a native of Asia and the most important sweetener-producing crop, is grown worldwide under tropical and subtropical climatic conditions. Besides sugar, it is used as raw material for the production of jaggery (*gur*), *khandsari*, bio-ethonol and various other distillery products. Indian farmers grow sugarcane as a cash crop, which grows well in medium deep to deep soils with adequate irrigation and drainage. Cultivars developed in India as inter-specific hybrids by crossing *S. officinarum* with *S. barberi* and *S. spontaneum* respond positively to agronomic management and high input levels. Asexual propagation with cane setts is adopted for commercial cultivation that necessitates careful seed selection from young (8–10 month), healthy and pure seed cane crop. Sugarcane planting in tropical states can be done round the year, however under subtropical states it is done during the month of October–November (autumn) or February–March (spring). The crop is planted in rows and spacing varies from 60 to 150 cm depending on location time, season and method of planting. The crop requires high dose of nutrients and water, as it is one of the highest biomass producing field crops. The most critical growth stage to maintain adequate nutrient and moisture level in the soil is tillering stage, which spans from 45 to 135 days after planting. Adequate tillering culminates into high number of millable canes that contributes most (40%) to the final crop yield. The fertilizer recommendation for sugarcane, therefore, varies from 150 to 400 kg N/ha for different sugarcane producing states in India. Similarly required number of irrigations too varies from 8 to 36 for different states. Sugarcane grown in tropical states (South India) requires higher nutrients and moisture than that of subtropical states (North India). The cane yield is also higher under tropical conditions mainly due to conducive climate for growth and comparatively longer growing season. Being a long duration crop sugarcane is most affected by weeds. Cultural, mechanical and chemical methods have been found useful for the control of weeds. Widely spaced planting of sugarcane provides sufficient rooms for growing of cereals, pulses, oilseeds, vegetables and spices as intercrops for sustained productivity and high profitability. Intercropping in sugarcane also reduces the weeds and improves resource utilization efficiency. Agronomic practices viz., hoeing, earthing-up and tying contribute significantly to crop yield and quality.

Protection of crop from diseases like red rot, smut and wilt is largely effected through adoption of resistant sugarcane varieties. As a matter of fact a sugarcane cultivar is released only if it possesses resistance against red rot and wilt. Seed selection and seed treatment with MHAT are important components of disease management. A number of borers and sucking insect-pests cause substantial yield loss, which could be effectively minimized by adopting integrated pest management involving cultural, mechanical, chemical and biological control methods. Use of natural enemies to control sugarcane borers and pyrilla has been found successful and is extensively adopted.

Ratooning is an integral component of sugarcane farming to raise overall productivity and profitability of the system. It is for this reason that most of the farmers go for at least one ratoon. In subtropical states time of ratoon initiation is crucial to ensure good ratoon stand and yield. As recommended, extreme winter and summer months are to be avoided for harvesting the crop to give rise to a new ratoon. Practices like stubble shaving, off barring, early earthing up and trash mulching are beneficial for good ratoon yield.

Sugarcane matures 10–12 months after planting and economic product, sugar, accumulates in stalks. Weather with wide difference in day and night temperatures favors sugar accumulation. Normally sugarcane crushing starts in October/November and goes up to April/May. In order to ensure good sugar recovery (>10%), harvested cane should be supplied to sugar mill as early as possible, preferably within 72 hours.

2.11 FUTURE RESEARCH THRUST

1. In-depth investigation on enhancing the fertilizer use efficiency in plant-ratoon system through appropriate application frequency of organic manures, crop residues, green manures and bio-manures.

2. Intensive investigation into the reasons for reduction in yield potential of successive ratoons even at higher level of manuring.

3. Development of Integrated Nutrient Management (INM) approach to supply secondary and micronutrients particularly S, Fe, Cu and Zn to correct their hidden hunger before visible symptoms appear in the crop.

4. Being a high water requirement crop, innovative irrigation management practices including micro-irrigation system may be developed for high irrigation water use efficiency (IWUE).

5. Development of technology package involving *in situ* chopping of stubble/trash, microbial inoculants capable of degrading cellulosic substances under low temperature conditions to facilitate sowing of wheat after ratoon harvest and modification in fertilizer schedule to avoid nutritional imbalance in rice-wheat-plant cane-ratoon-wheat cropping system.

6. Researches on developing 'Sugarcane based Integrated Farming System Modules' through inclusion of different components of Agriculture viz. dairy, duckery, poultry, apiary, pisciculture, vermicomposting, mushroom cultivation and horticultural crops (banana, papaya etc.) along with intercropping systems may hold promise to increase the net income of the farmers.

KEYWORDS

- **agro-techniques**
- **bio-fertilizer**
- **intercropping**
- **organics**
- **ratoon**
- **sugarcane**

REFERENCES

All Indian Coordinated Research Project on Sugarcane AICRPS, (2011). ICAR-IISR, Lucknow *Annual Report 2010–11,* pp. 11–15.

Anonymous, (2010). *Agricultural Statistics at a Glance* (2010*)*. Directorate of Economics and statistics, Department of Agriculture Cooperation and Farmers welfare, Ministry of Agriculture and Farmers welfare , Govt. of India, New Delhi, Col. No. 15.5 a & b. pp. 16–21.

Bhendia, M. L., Mann, H. S., & Gautam, O. P., (1964). *Proceedings of All India Conference of Sugarcane Research and Development Workers*, *5*, 275.

Burr, G. O., Hartt, C. E., Brodie, H. W., Tomimo, T., Kortschak, H. P., Takahashi, D., Ashton, F. M., & Coleman, R. E., (1957). The sugarcane plant. *Annual Review of Plant Physiology*, *8*, 275–307.

Gaur, A. C., (1992). In Fertilizer, Organic manures, Recyclable wastes and biofertilizers (Tondon, H. L. S., Ed.), EFCO, New Delhi, India, pp. 36–51.

Gunkel, G., Kosmol. J, Sobral, M., Rohn, H., Montenegro, S., & Aureliano, J., (2007). Sugarcane Industry as a source of water pollution. Case study on the situation in Ipojuca river, Pernambuco, Brazil. *Water Air Soil Pollut*. *180*, 261–269.

Hartemink, A. E., (2008). Sugarcane for bioethanol, Soil and environmental issues. *Advances in Agronomy*, *99*, 125–182.

IISR, (2011). Indian Institute of Sugarcane Research, Annual Report, 2010–11.

Irvine, J. E., & Benda, G. T. A., (1980). Sugarcane spacing. II Effect of inter and intra-row spacing on the plant. *Proceedings XVII Congress*, ISSCT, pp. 751–755.

Jafri, S. M. H., (1973). Studies on certain soil factors associated with solubility of phosphorus. PhD. Thesis, Agra University: Agra, pp. 175.

Kapoor, J., & Sharma, K. P., (1990). Effect of biocide on minimizing post-harvest deterioration of sugarcane. *Bharatiya Sugar*, *15*, 39.

Lal, M., & Singh, A. K., (2002). Importance of plant nutrient management in sugarcane. *Fertilizer News*, *47*(11), 91–98.

Lal, Menhi, & Singh, A. K., (2004). Technology Package for Sugarcane- based Intercropping Systems. ICAR- IISR, Lucknow, Technology Bulletin, 2004, 3–23.

Motiwale, M. P., & Singh, A. B., (1973). Annual Report, Indian Institute of Sugarcane Research, Lucknow, pp. 21.

NFCSF, (2016). National Federation of Cooperative Sugar Factories Limited. *Cooperative Sugar*, *42*(7), 57–98.

Panje, R. R., Mathur, P. S., & Seth, S. R., (1968). IISR technique to triple sugarcane yield in North India. *Indian Farming*, *18*, 4–7.

Saini, S. P., Sindhu, A. S. and Singh, P., (2012). Crop yield efficiency and economics of autumn and spring sown single bud sugarcane intercropped with pulse crops. *International Journal of Forestry and Crop Improvement*, 3(2), 60–65.

Sen, A., Prasad, J., & Prasad C. R., (1984). Studies on the soil application of micronutrients on sugarcane in relation to plant micronutrients balance, yield and commercial cane sugar, *Proceedings of Annual. Convention of STAI*, *48*, 69–75.

Shahi, H. N., (1998). Ratoon management. A prime factor for augmenting sugarcane and sugar production. *National Seminar on Sugarcane Ratoon Management*, Lucknow, Souvenir with Abstracts, pp. 7–16.

Sharma, M. P., & Singh, K., (1988). IISR Implements and machinery for sugarcane crops. *IISR Technical Bulletin No. 20*, 17–19.

Singh, A. K., & Mohan, (2000). Excerpts of Lectures, Summer School on Advances in Sugarcane Production Technology, 66–68.

Singh, A. K., Menhi Lal, & Arun Kumar Singh, (2006). Production potential and economic viability of sugarcane (*Saccharum spp*. Hybrids) – based crop diversification options in the North zone of subtropical India. *Indian J. Sug. Tech.*, *2*(1&2), 23–26.

Singh, A. K., Menhi Lal, & Prasad, S. R., (2006). Effect of row spacing and nitrogen on ratoonability of early maturing high sugar genotypes of sugarcane (*Saccharum* spp.) hybrids. *Indian J. Agricultural Sciences, 76*(2), 108–110.

Singh, A. K., Menhi Lal, Prasad, S. R., & Srivastava, T. K., (2007). Productivity and profitability of winter initiated sugarcane (*Saccharum* spp. Hybrid complex) ratoon through intercropping of forage legumes and nitrogen nutrition. *Indian Journal of Agronomy, 52*(3), 208–211.

Singh, A. K., Srivastava, T. K., & Menhi Lal, (2001). Effect of nitrogen management and potassium nutrition on yield and quality of seed cane under subtropical condition. *Indian J. Sugarcane Technol., 16*(1), 9–13.

Singh, B., Srivastava, S. C., & Ghosh, A. K., (1972). Comparative dynamics of sugarcane stalk population due to variations in row spacing, planting material and fertilizer levels. *Proc. STAI, 38,* 1–9.

Singh, G. B., Motiwale, M. P., Prasad, S. R., & Singh, A. B., (1984). Critical stages when sugarcane requires irrigation in subtropical India. *Indian Journal of Agricultural Sciences, 54*(5), 387–389.

Singh, K. D. N., (1993). Utilization of pressmud cake in sugarcane and reclamation of salt affected calcareous soils of Bihar. *Indian Sugar, 43*(6), 369–374.

Singh, K. P., Srivastava, T. K., Archna Suman, & Singh, P. N., (2010). Sugarcane productivity and soil health in a bio-nutrition based multi-ratooning system under subtropics. *Indian Journal of Agricultural Sciences, 80*(8), 87–89.

Singh, U. S., (1978). Nitrogen and Sugarcane. *Indian Sugar, 27,* 753–758.

Srivastava, A. K., Srivastava, Arun K., & Solomon, S., (2011). Sustaining sugarcane productivity under depleting water resources. *Current Science, 101, No. 5,* 25 Sept., 2011.

Srivastava, S. C., & Johari, D. P., (1979). The irrigated sugarcane in India., *Technical Bulletin* (1979), ICAR- *Indian Institute of Sugarcane Research*, Lucknow, p. 36.

Srivastava, T. K., & Dey, P., (2000). SWOT analysis of sugarcane cultivation under water logged conditions of eastern Uttar Pradesh. *Indian Sugar, XLIX, 11,* 921–926.

Srivastava, T. K., (2002). Critical period of crop weed competition in sugarcane ratoon. *Indian Journal of Weed Science, 34*(3&4), 318–319.

Srivastava, T. K., (2003). Bio-efficacy of sulfentrazone against nut-sedge (*Cyprusrotundus*) and other weeds in sugarcane. *Indian Journal of Weed Science, 35*(1&2), 83–86.

Srivastava, T. K., Chauhan, R. S., & Menhi Lal, (2005). Weed dynamics and their management in sugarcane under different preceding crops and tillage system. *Indian Journal of Agricultural Sciences, 75*(9), 254–260.

Srivastava, T. K., Sah, A. K., Gupta, R., & Singh, K. P., (2011). Water use efficient technologies for improving productivity and sustainability of sugarcane. *Farmers Participatory Action Research,* Final Report, pp. 17.

Srivatava, K. K., Narsimhan, R., & Shukla, R. K., (1981). A new technique for sugarcane planting. *Indian Farming, 31*(3), 15–17.

Suman, Archna, (2001). Paper presented in All India Seminar on Biotechnology as a Tool for Increasing Sugarcane and Sugar Productivity, IISR, Lucknow, October 30, 2001.

Tiwari, R. J., Bangar, K. S., Nema, G. K., & Sharma, R. K.(1998). *Journal of Indian Society of Soil Sciece, 46*(2), 243–245.

Van Dillewijn, C., (1952). *Botany of Sugarcane*. Chapter 8, Tillering, The Chronica Botanica Co. Book Department Waltham, Mass, U.S.A., 95–96.

Verma, R. S., Kumar, R. and Yadav, R. L., (1984). Status of research on intercropping with sugarcane in India, *Indian Sugar Crops Journal, 10*(2), 1–6.

Yadav, D. V., (2000). Nutrient management in sugarcane and sugarcane based cropping systems during 2000–2010. *Fert. News, 45*(4), 43–48.

Yadav, R. L., & Prasad, S. R., (1988). Moisture use characteristics of sugarcane genotypes under different available soil moisture regimes in alluvial entisol. *J. Agric. Sci. Camb., 110,* 5–11.

Yadav, R. L., (1992). *Ratooning of Sugarcane.* Chapter 4, Technology for ratoon crop production, Periodical Express Book Agency. New Delhi (India), 47–186.

Yaduvanshi, N. P. S., & Yadav, D. V., (1990). *Journal of Agricultural Sciences, 114,* 259–264.

Zende, G. K., & Kibe, M. M., (1981). *Proceedings of Biennial Conference of Sugarcane Research and Development Worker,* Dharwar, *34,* 43.

CHAPTER 3

SUSTAINABILITY AND EFFICIENCY OF SUGARCANE CULTIVATION IN INDIA

G. VLONTZOS[1] and P. M. PARDALOS[2]

[1]*Department of Agriculture Crop Production and Rural Development, School of Agricultural Sciences, University of Thessaly, Fytoko, 38446 Volos, Greece*

[2]*Department of Industrial and Systems Engineering, University of Florida, Center of Applied Optimization, Distinguished Professor, Paul and Heidi Brown Preeminent Professor, 401 Weil Hall, Gainesville, FL 32611-6595, USA, E-mail: pardalos@ise.ufl.edu*

CONTENTS

3.1   INTRODUCTION

3.1.1   GLOBAL SUGARCANE PRODUCTION

Sugarcane (*Saccharum officinarum*) is one of the several species of tall perennial true grasses in the genus Saccharum (Figure 3.1). The plant can reach up to 2–4 meters in height and is consisted of jointed, fibrous stalks, which are rich in sugar sucrose. Sucrose is used as raw material in human food industries in

FIGURE 3.1   Sugarcane.

order to produce sugar or after elaboration for ethanol production. Other major marketable products derived from sugarcane are molasses, rum, cachaça (the most popular alcoholic beverage in Brazil) and bagasse. Moreover, sugarcane reeds can be used to make pens, mats, screens and thatch.

Sugarcane originated in Southeast Asia and Melanesia and it is now cultivated in tropical and subtropical countries worldwide used for the production of sugar, ethanol and other industrial products. According to the Food and Agriculture Organization (FAO), nowadays sugarcane is the world's largest crop by production quantity. Especially in 2012, it was cultivated about 26×106 hectares (6.4×107 acres) in more than 90 countries all over the world with a worldwide harvest of 1.83×109 tonnes (Figure 3.2).

Brazil is, by far, the world's largest producer of sugarcane, accounting for the one third of world production (Brazil's annual production: 739.267 thousand metric tons). The south-central region of Brazil produces more than 90% of this national production output. Sugar is the main product sourced from the sugarcane cultivated in this country. However, production of ethanol from molasses is now growing more popular because of its use as a biofuel, and so a large quantity of Brazil's sugarcane harvest is used for ethanol production.

India is the second largest producer with an annual production of 341,200 TMTs of sugarcane. The states of Maharashtra, Uttar Pradesh, Punjab, and Bihar are accountable for the production of the greatest quantities in the country. In India, sugarcane is cultivated not only for the production of crystal sugar but also for processing jaggery (known as "Gur" in India and "panela" in other parts of the world) and a plethora of alcoholic beverages.

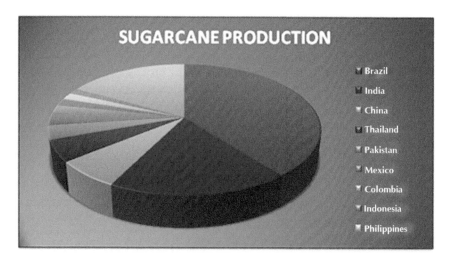

FIGURE 3.2 Top sugarcane producing countries (2012). *Source:* Food and Agricultural Organization of United Nations: Economic and Social Department—the Statistical Division.

China is a third largest sugarcane producer with the production of 125,536 TMTs; about 80% of which is cultivated in the South and Southwest region of the country. Taking into consideration of the rising domestic demand, China imports large quantities of sugar from other countries (especially Brazil, Thailand, Australia, Myanmar, Vietnam, and Cuba). Furthermore, China uses sugarcanes also for ethanol production; in order to be used as a fuel, as they are trying to fulfill high needs of increasing population.

Thailand, having exploited the ideal conditions of its fields and the improved cane varieties, increased its sugarcane annual production (100,096 TMTs) in order to manufacture sugar and molasses. Thailand exports sugar and molasses products to a large number of countries (especially China, Republic of Korea, Malaysia, and Japan).

Pakistan is the fifth largest sugarcane producer in the world (annual production of 63,750 TMTs). As sugarcane is a major crop in the country, it helps its export economy in a vital way. Mainly, Pakistan exports sugar to neighboring countries (Afghanistan, Tajikistan) and other central Asian countries as well.

In Mexico, more than 2 million people has employed in the cultivation of sugarcane because of its large annual production of 61,182 thousand metric tons. The major importer of Mexican sugar is the neighboring United States of America.

Colombia has the world record of sugar yields by hectare. The surface seeded in sugar cane in Colombia ascends to 170,000 hectares with an annually production of 34,876 TMTs of sugarcane and about 2.5 million tons of sugar. Sugarcane production and processing are located in the Cauca river valley. The growing climate and the density of mills and distilleries support Columbian economy for sugar cane production in Cauca over other regions of the country. The most significant importers of Colombian sugarcane products are Chile, Peru, Haiti, and the United States of America.

Indonesia's sugar cane production (33,700 TMTs per year) plays an important role in the country's economy. Despite its large sugarcane production, Indonesia imports sugar from other countries due to its old factories and the lack of refineries, which could enable the direct consumption of its own produced sugar.

Philippines is the ninth largest sugarcane producer in the world (annual production of 31,874 TMTs) growing it mainly on the islands of Negros, Luzon, Panay and Mindanao. Sugar canes in Philippines constitute a major cash crop, as they are precious not only for making sugar but also for producing renewable energy from its by-products.

The United States of America, with an annual production of 27,906 TMTs, remains an important producer of sugarcane and the fifth largest consumer of sugar as well. The majority of the country's production comes from Florida, Hawaii, Louisiana, and Texas. According to Foreign Agricultural Service (FAS) data, 84% of the refined sugar exports were shipped to Mexico, an important export destination since 2004. Other significant markets for U.S. sugar are Canada, Germany, and the Netherlands.

3.1.2 GLOBAL SUGAR TRADE

Sugarcane's agriculture arose as a result of the world demand for sugar, as cane accounts for 80% of sugar produced (most of the rest is made from sugar beets).

According to FAO, sugar trade is constantly increasing driven by a strong import demand by countries that may face a production shortfall.

In 2016, the quantity of sugar exported was about 56,000 metric tons (Figure 3.3) while the imported one was about 52,000 metric tons (Figure 3.4) as seen in the Tables 3.1 and 3.2, respectively.

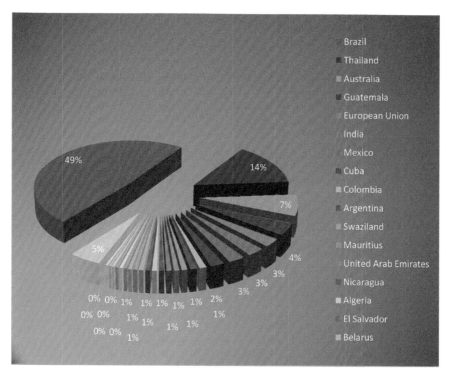

FIGURE 3.3 Sugar exporters for 2016 (Source: United States Department of Agriculture/ Foreign Agricultural Service November 2016).

3.2 SUSTAINABILITY ISSUES FOR SUGARCANE PRODUCTION

Sustainability is a continuous effort to achieve growth and development in a framework where natural resources are maintained over time (Biswas, 1994; Conway, 1985; Edwards, 1989). Based on this, it is too simplistic and leads to a misleading direction that sustainable agriculture is adequate to traditional one. Sustainable production methods require the adoption of innovative cultivation practices which increase the efficiency use of agricultural inputs, reducing by this way the quantities being used overtime.

The three pillars of agricultural sustainability can be described as (i) the maintenance of environmental soundness, (ii) stable crop and animal productivity, and (iii) social acceptability (Mnisi and Dlamini, 2012). The first principle accepts and requires adequate use of natural resources, reassuring their existence and availability overtime to future generations, as well as their improvement in cases where past uses resulted to degradation of them.

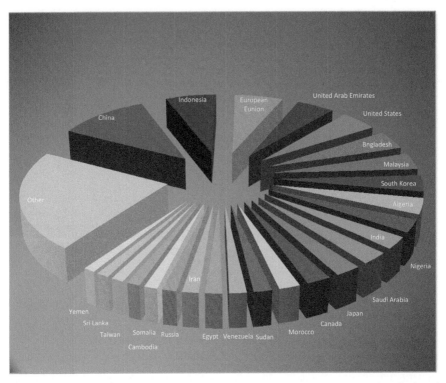

FIGURE 3.4   Sugar importers for 2016 (Source: United States Department of Agriculture/ Foreign Agricultural Service November 2016).

Preservation of natural resources must be combined with viable solutions on the productivity and income generation issue, in order the human involvement into the agricultural production process to be viable (Blair and Crocker, 2000; Peykani et al., 2010; Van Antwerpen and Meyer, 1996), motivating by this way labor to participate in this process. Finally, the social acceptability issue refers to both rural and urban societies, with different perspectives. Urban societies require evident and continuous preservation of natural environment, not being degraded due to agricultural production, while rural societies having a more direct contact with natural environment require equality referring to the standards of living and improved quality of life, similar with the one in urban regions (Yunlong and Smith, 1994). Different approaches appear when the prioritization issue appears. The major debate is about the importance of environmental preservation and the economic soundness of agricultural production, with urban-oriented stakeholders to insist on the

TABLE 3.1 Sugar Exports (1,000 Metric Tons, Raw Value)

Exports	2012/13	2013/14	2014/15	2015/16	May 2016/17	Nov 2016/2017
Brazil	27.650	26.200	23.950	24.350	26.100	27.120
Thailand	6.693	7.200	8.252	7.800	9.000	8.000
Australia	3.100	3.242	3.561	3.700	3.900	4.000
Guatemala	1.911	2.100	2.340	2.255	2.310	2.310
European Union	1.662	1.552	1.582	1.500	1.500	1.500
India	1.261	2.806	2.580	3.000	1.000	1.500
Mexico	2.091	2.661	1.545	1.280	1.606	1.405
Cuba	775	937	895	950	1.000	1.200
Colombia	542	900	835	600	700	600
Argentina	258	234	113	230	550	550
Swaziland	353	412	641	663	500	500
Mauritius	384	416	426	424	455	435
United Arab Emirates	569	739	350	435	435	435
Nicaragua	375	375	398	398	425	425
Algeria	522	478	480	380	475	400
El Salvador	480	404	481	380	385	385
Belarus	517	465	371	376	380	380
South Korea	383	361	305	339	340	350
Malaysia	317	315	258	287	225	295
South Africa	356	868	772	305	320	270
Morocco	0	8	39	234	30	250
Mozambique	261	233	237	240	240	240
Sudan	92	187	202	205	210	210
Egypt	400	350	263	200	200	200
Nigeria	200	200	200	200	200	200
Other	4.482	4.238	3.704	2.941	3.146	2.744
Total	55.634	57.881	54.780	53.672	55.632	55.904

Source: United States Department of Agriculture/Foreign Agricultural Service (November, 2016).

importance of agro-ecological maintenance, while the rural-oriented ones to insist on the necessity of stable productivity and financial autonomy of farmers (Bates and Sokhela, 2003; Brietbarth and Harris, 2008; McCarthy, 2007; SASA, 2011). This debate is more intense in developing countries,

TABLE 3.2 Sugar Imports (1,000 Metric Tons, Raw Value)

Imports	2012/13	2013/14	2014/15	2015/16	May 2016/17	Nov 2016/17
China	3.802	4.275	5.058	6.000	7.900	6.000
Indonesia	3.570	3.570	2.950	3.605	3.400	3.350
European Union	3.790	3.262	2.918	3.000	3.500	3.250
United Arab Emirates	2.583	2.108	2.366	2.460	2.480	2.480
United States	2.925	3.395	3.223	3.031	3.156	2.441
Bangladesh	1.547	2.085	1.982	2.284	2.350	2.210
Malaysia	1.966	1.897	2.063	2.009	1.915	1.990
South Korea	1.806	1.909	1.882	1.871	1.940	1.910
Algeria	2.014	1.854	1.844	1.850	1.850	1.850
Nigeria	1.450	1.470	1.465	1.470	1.430	1.430
India	1.722	1.078	1.000	1.352	1.000	1.400
Saudi Arabia	1.394	1.312	1.424	1.448	1.335	1.400
Japan	1.244	1.360	1.359	1.275	1.265	1.285
Canada	1.156	1.031	1.163	1.200	1.210	1.190
Morocco	985	762	857	1.010	890	1.050
Sudan	997	638	1.319	900	1.320	900
Venezuela	840	840	870	850	850	850
Egypt	1.050	1.208	1.330	880	830	830
Iran	1.553	1.629	264	791	1.450	800
Russia	735	1.020	1.100	750	750	700
Somalia	267	287	409	648	510	660
Cambodia	376	671	458	627	475	650
Taiwan	603	577	543	636	600	640
Sri Lanka	598	542	620	621	635	630
Yemen	563	749	680	602	680	600
Other	12.330	11.979	11.098	12.201	11.895	11.580
Total	51.866	51.508	50.245	53.371	55.616	52.076

Source: United States Department of Agriculture/Foreign Agricultural Service (November 2016).

where the urban-rural divide is more obvious, compared with the developed ones, due to considerable differences on income levels and quality of life. In order a common sense for sustainable agriculture to be achieved, there is a need for mutual understanding of the priorities of urban and rural societies

(Altieri, 1995; Conway, 1990; Edwards et al., 1993; Lynam and Herdt, 1989; Smith and McDonald, 1998; Tisdell, 1996). The intensity of this debate is increased when subsidy schemes are implemented to support in a direct or indirect way. Usually these subsidies are being paid by taxpayers' money. Interest groups usually form a set of requirements or preconditions strongly influenced by environmentally friendly concerns. The catalytic parameter which can reduce the intensity of this debate is the fact that both groups are sharing the same environment and use to a large extend the same natural resources, with the most important one to be the water (De Jager et al., 2001; Tellarini and Caporali, 2000). The realization of this can be achieved by organizing presentations and trainings, especially for farmers, presenting to them the holistic approach of production process, by introducing means and methods leading to viable productivity levels by protecting at the same time the environment. It has been proved too that introducing ways of protecting farmers' health due to their exposure to chemical inputs is a quite effective strategy to convince them to diversify their usual habits of handling them, adopting protocols which protect both of them and the environment.

Sustainability and productivity are concepts quite close, because the increase of the latter provides evidence for the improvement of the former. Productivity increase can be achieved by adopting new technology. For the sugarcane case that usually means increased mechanical planting and harvesting, usage of new varieties, implementation of new irrigation methods, and soil management focusing on preserving and increasing the organic matter and avoiding the erosion. Such changes to cultivation practices are able to increase production, without increasing in a linear way the usage of inputs (Webster, 1997). Sustainability in agriculture is an important precondition for stability of rural communities, because the risk undertaken during the production process is continuously decreased, providing opportunities to future generations to continue such production activities. Intensification of agricultural production usually is being accomplished via exploitation of natural resources, like soil and water, having as an outcome the degradation of both of them, without having in the short-term the ability to restore the damage being done. The consequences in such cases are not only economic ones, but they exceed to environmental and social aspects in an irreversible way.

Till now, the most important obstacles sugarcane production faces on a global basis are economic, social and environmental. The continuous stagnation of market prices, combined with the recent changes of agricultural policies worldwide, aiming to reduce farmer subsidies, have decreased farmers' motivation to cultivate sugarcane (Gowda and Jayaramaiah, 1998). Especially younger ones are seeking to cultivate alternative crops, which can provide higher revenues and better living and working conditions. The increased competitiveness globally has substantially raised the need for the industrialization of sugarcane farming, by adopting the principles of economies of scale, in rural areas where the precondition of the continuation of rural communities is the development of economies of scope. Intensified production process is not easy to be achieved with the co-existence of small farming, as well as with farmers not having the ability and the means to invest in new technology, like machinery and usage of new crop varieties.

The international treaties recently signed for global protection of the environment, and implementation of policies towards this direction, created a new framework for sugarcane production, striving this process to sugarcane-derived products, with the more demanding one to be the clean and renewable ethanol (World Bank, 2007). The continuous research on new varieties has delivered significant outcomes, providing plants with increased productivity and sugar recovery. Given the fact that farmers' income is highly dependent on production costs, special attention is being given on adoption of cultivation practices, which reduces these costs. More specifically, these crop management practices focus on new planting methods with the use of new planting machines, the adoption of modern irrigation technologies which increase the efficiency of water use, and implementation of environmentally friendly protocols for crop protection (Loganandhan et al., 2013). However, the demand for varieties both succeeding higher productivity of fiber, being used for energy production purposes, and sugar in many cases is quite controversial. Therefore, it is recommended that this outcome can be delivered only through a balance between the two desired sugarcane products. The strategic target for India is 20% blending of ethanol in the automobile fuel by 2017, increasing by this way the sustainable profile of this cultivation (ICAR, 2016). A holistic approach of sustainability of sugarcane cultivation lies upon several aspects, like land management, labor conditions the debate between food and energy, and practices applied for both cultivation and processing. In Brazil, there is demand for land use changes from pasture, to sugarcane cultivation mainly for ethanol production. The response to this challenge was the launching of agro-ecological zones for sugarcane, allowing by this directive the expansion of sugarcane production in areas where the agronomic, climate and environmental conditions were suitable for this. This expansion did not include the Amazon rainforest area. The sustainable benefit of this planning derives from the fact that the new land being utilized for sugarcane production was degraded pastures used for cattle ranching. Sugarcane requires minimum tillage because it is a perennial crop, which is replanted every five or six years. This structural characteristic captures larger amounts of CO_2, compared with previous uses of the same lands (Table 3.3).

3.3 RECENT TRENDS ON SUSTAINABILITY FACTORS FOR SUGARCANE PRODUCTION

Research activities are focused on genetic improvements for achieving taller sugarcane plants with increased sugar content. According to ICAR-Indian Institute of Sugarcane Research, Lucknow, specific varieties have been tested in sub-tropical India, providing promises for both increased cane productivity and sugar recovery. These positive characteristics of these varieties can lead to even better yields if modern machinery is used for production and processing. Due to the fact that there are significant climate differences across the country, these new varieties were tested in different regions, where traditionally sugarcane is cultivated, in order to quantify the impact of weather conditions on production, providing by this way the appropriate variety to specific regions of the country.

Increased productivity can be achieved also by applying high-density sugarcane farming. This cultivation protocol requires the application of hormones, mainly GA, during all the critical growth stages of the plants. These interventions increased yield significantly, providing additional income to farmers.

Quite promising for increased yields and further empowerment of sustainability is the application of the "sustainable sugarcane initiative" (SSI).

TABLE 3.3 Carbon Balance When Sugarcane Replaces Other Agricultural Activities

Biomass	Total carbon stocks (Mg/ha)	Carbon replace due to sugarcane replacement (Mg/ha)
Cotton	40.1	+21.8
Degraded pasture	42.0	+19.8
Maize	44.1	+17.7
Soybean	54.9	+6.6
Managed pasture	58.5	+3.3

Source: Neved do Amaral et al. (2008).

This is a cultivation protocol for sugarcane production, being developed after three years experiments carried out at the International Crop Research Institute for the Semi-Arid Tropics. These experiments took place from 2008–2011. The results focus on six principle components, which are raising nursery using budded chips, transplanting young seedlings (25–35 days old), maintaining wide spacing (5 × 2 feet) in the main field, providing sufficient moisture and avoiding flooding of fields, encouraging organic methods of nutrient, plant protection and other intercultural practices, and practicing intercropping for effective utilization of land (Loganandhan et al., 2013).

3.4 ACTION PLAN FOR SUSTAINABLE SUGARCANE PRODUCTION IN INDIA

3.4.1 STRENGTHS

The most important competitive advantage of the sector is the steady increase of sugar consumption in India. This trend accompanied with the superior quality of the product shapes a quite positive framework for further development of sugarcane production.

Production and trade statistics provide hints that neighboring markets, characterized by increased development and significant increase of purchasing power of their inhabitants, import considerable quantities of sugar. The dynamics of it, in accordance with the ability of the product to be the raw material for biodiesel production, is a quite promising framework for expansion of the cultivation to meet the growing demand.

The recent research findings provide new varieties to farmers, able to increase productivity and increase the efficiency of inputs utilization, satisfying by this way crucial preconditions of sustainability.

The ability provided to farmers to use new machinery during the production process increases productivity and improve the working conditions of them, a vital issue for the agricultural community of the country.

3.4.2 WEAKNESSES

One of the most important obstacles is high production costs due to lack of infrastructure. A large number of farms cultivate sugarcane in a traditional way, and therefore their productivity is declining. Irrigation infrastructure is largely absent, further lowering productivity.

The other crucial factor causing poor competitiveness is the large number of small agricultural holdings producing sugarcane. This status quo increases the cost of production, and dramatically reduces the bargaining power of producers when they negotiate the selling price of their product.

The majority of exports are in bulk. Despite the fact that there is space for creating added value through processing and packaging, this is not the case for Indian sugarcane producers.

3.4.3 OPPORTUNITIES

The continuous demand for biodiesel for both environmental and energy security reasons is a promising parameter for this cultivation. The recent genetic improvements increase the fiber production, being used for biodiesel production. This is a win-win situation for both farmers and consumers, because in a sustainable way a production process provides to the market two products characterized by high demand.

The continuous increase of purchasing power and the rapid adoption of urban lifestyle of large consumer groups increase the demand for both sugar and biodiesel. Recent studies diversify sugar being produced from sugarcanes from the one being produced from sugar beets, on the basis of nutritional characteristics, this product differentiation strategy is in favor of sugarcane sugar, on a global basis.

3.4.4 THREATS

The growing interest in sugarcane internationally has motivated farmers in regions with climates similar to India to plant sugarcane, even though there has not been any tradition or experience with this cultivation. Given the fact that there is no production and trading protection framework, the product is highly exposed to international competition, increasing by this way the risk undertaken by producers.

The fact that younger farmers prefer to cultivate alternative products, which provide better revenues threatens the future of this cultivation in India. There is a need for implementation of practices increasing productivity, making more sustainable the revenue of the product.

3.4.5 POLICY RECOMMENDATIONS

There is no doubt that sugarcane production is vital for the primary sector of India. The dual exploitation of the plant, for nutritional and energy purposes, motivate stakeholders to support and promote sugarcane cultivation. The sustainability issues for it mainly refer to natural resources management and their impact on both the environment and urban areas. Intensive research activities prepare and promote viable solutions to farmers to overcome economic and environmental obstacles, keeping productivity levels below the potential of it. New varieties and mechanical infrastructure on their own are not able to increase the efficiency status of the cultivation. Effective promotion strategy needs to be implemented, so farmers to be trained on how to use less inputs for more yield. In the long-term such a strategy would improve the competitive advantage of Indian agriculture, in a highly competitive trading environment. Based on these principles, sustainability in the sugarcane sector in India can be achieved with both rural and urban communities maintain natural resources for future generations.

KEYWORDS

- action plan
- efficiency
- productivity
- sustainability

REFERENCES

Agribusiness Handbook: Sugar Beet White Sugar. Food and Agriculture Organization, United Nations, 2009 Crop Production. Food and Agriculture Organization of the United Nations. Retrieved 2015–01–27.

Altieri, M., (1995). Agroecology: The Science of Sustainable Agriculture, *Westview Press*, Boulder, Colorado.

Bates, R., & Sokhela, P., (2003). The development of small-scale sugar cane growers: A Success Story?" In L. Nieuwoudt and J. Groenewald (eds.), *The Challenge of Change: Agriculture, Land and the South African Economy.* University of Natal Press: Pietermaritzburg, South Africa.pp105.

Biswas, M. R., (1994). Agriculture and environment: a review, 1972–1992. *Ambio, 23*(3), 192–197.

Blair, N., & Crocker G. J., (2000). Crop rotation effects on soil carbon and physical fertility of two Australian soils. *Aust. J. Soil Res., 38,* 71–84.

Briebarth T., & Harris, P., (2008). The role of corporate social responsibility in the football business: Toward the development of a conceptual model, European Sport Marketing Quarterly, http://www.managementmarketing.ro.

Conway, G. R., (1985). Agroecosystem analysis, *Agric. Admin., 20,* 31–55.

Conway, G. R., (1990). Agriculture and the environment: concept and issues. In: Huq, S., Rahman, A., Conway, G. R. (Eds.), *Environmental Aspects of Agricultural Development in Bangladesh.* University Press Ltd, Dhaka pp. 11–23.

De Jager, A., Onduru, D., van Wijk, Vlaming, J., & Gachini, G. N., (2001). Assessing sustainability of low-external-input farm management systems with the nutrient monitoring approach: a case study in Kenya. *Agric. Systems, 69,* 99–118.

Edwards, C. A., (1989). The importance of integration in sustainable agricultural systems. In: Paoletti, M. G., Stinner, B. R., Lorenzoni, G. G. (Eds.). Agricultural Ecology and Environment. *Agric. Ecosyst. Environ., 27,* 25–35.

Edwards, C. A., Grove, T., Harwood, R. R., & Colfer, J. P., (2003). The role of agroecology and integrated farming systems in sustainable agriculture, *Agric. Ecosyst. Environ., 46,* 99–121.

Gowda, M. J. C., & Jayaramaiah, K. M., (1998). Comparative evaluation of rice production systems for their sustainability. *Agric. Ecosyst. Environ., 69,* 1–9.

http://www.agri-outlook.org/.

http://www.fao.org/faostat/en/#home.

http://www.worldatlas.com/articles/top-sugarcane-producing-countries.html.

https://www.fas.usda.gov/data/sugar-world-markets-and-trade.

ICAR (2014–2015). Indian Institute of Sugarcane Research, Annual Report, Lucknow.

Loganandhan, N., Biksham Gujja, Vinod Goud, V., & Natarajan, U. S., (2013). Sustainable Sugarcane Initiative (SSI): A Methodology of 'More with Less,' *Sugar Tech., 15*(1), 98–102.

Lynam, J. K., & Herdt, R. W., (1989). Sense and sustainability: sustainability as an objective in international agricultural research. *Agricultural Economics, 3,* 381–398.

McCarthy, D. D. P., (2007). Moraine for Life Symposium: Stakeholders' Report. Oak *Ridges Moraine Foundation,* http://sig.uwaterloo.ca.

Mnisi, M. S., & Dlamini, C. S., (2012). The concept of sustainable sugarcane production: Global, African and South African perceptions, *African Journal of Agricultural Research, 7*(31), 4337–4343.

Peykani, G. R., Kelashemi, M. K., Shahbaza, H., & Akram, A. H., (2010). A determination of suitable sugarcane utilization system using Total Factor Productivity (TFP). Case study: Imam Khomeini cultivation and processing center in Khuzestan Province. *J. Agric. Sci. Tech., 12,* 511–521.

SASA, (2011). South African sugar industry and sustainable development, *http://www.sugarindustrydev.co.za.*

Smith, C. S., & McDonald, G. T., (1988). Assessing the sustainability of agriculture at the planning stage. *J. Environ. Management, 52,* 15–37.

Tellarini, V., & Caporali, F., (2000). An input/output methodology to evaluate farms as sustainable agroecosystems: an application of indicators to farms in central Italy. *Agric. Ecosyst. Environ., 77,* 111–123.

Tisdell, C., (1996). Economic indicators to assess the sustainability of conservation farming projects: an evaluation. *Agric. Ecosyst. Environ.*, *57*, 117–131.

Van Antwerpen, R., & Meyer, E., (1996). A review of soil degradation and management research under intensive sugarcane cropping. *Pro. S. Afr. Sug. Technol. Ass.*, *70*, 22–28.

Webster, J. P. G., (1997). Assessing the economic consequences of sustainability in agriculture. *Agric. Ecosyst. Environ.*, *64*, 95–102.

World Bank, (2007). Third Annual World Bank Conference on Environmentally Sustainable Development. World Bank, Washington, D. C., *www.wds.worldbank.org*.

Yunlong, C., & Smith, B., (1994). Sustainability in agriculture: a general review, *Agric. Ecosyst. Environ.*, *49*, 299–307.

CHAPTER 4

FROM CONVENTIONAL TO MOLECULAR APPROACHES: BUILDING BRIDGES FOR SUGARCANE GENETIC IMPROVEMENT

SANGEETA SRIVASTAVA[1] and PAVAN KUMAR[1,2]

[1]*Division of Crop Improvement, ICAR-Indian Institute of Sugarcane Research, Lucknow–226002, Uttar Pradesh, India,*
E-mail: Sangeeta.Srivastava@icar.gov.in, sangeeta_iisr@yahoo.co.in

[2]*Department of Biotechnology, Bundelkhand University, Jhansi – 284128, Uttar Pradesh, India*

CONTENTS

4.1 INTRODUCTION

Sugarcane an efficient photosynthetic crop in tropical and subtropical regions of world is cultivated for producing sugar and bio-fuel. Currently sugarcane is cultivated as a major commercial crop in South America, North/Central America, Asia, Africa, Australia, and Pacific Islands in more than 50.12 lakh hectares area of the world, yielding up to 3521.4 lakh tons of sugarcane and 245.5 lakh tons of sugar. India is the second largest sugar producer after Brazil (FAOS, 2015). Several by-products are obtained during sugar generation process such as, bagasse, molasses, alcohol, dextran, crude wax and glucose, etc., which are used in various bagasse-based and press mud-based industries. Sugarcane is one of the most promising crops as a source of green energy along with the production of sugar, alcohol and other by-products. Sugarcane bagasse is used for energy generation in sugar mills (Han and Wu, 2004; Pandey et al., 2000; Sun et al., 2004). The production of sugarcane is severely affected due to many environmental constraints and lack of suitable varieties. Commonly grown sugarcane varieties are derived from inter-specific hybridization followed by back crossing to attain nobilization, which takes long time in release of an elite variety. The sugarcane genome is of most complex type with high level of polyploidy (Glaszmann et al., 1997). Because of this, tagging, mapping and identification of genes is highly challenging in sugarcane. However, advent of molecular tools has greatly revolutionized the process of sugarcane crop improvement. Molecular markers such as amplified fragment length polymorphism (AFLP), restriction fragment length polymorphism (RFLP), and simple sequence repeats (SSR) have been very useful in gene mapping studies, linkage analysis, diversity analysis, and quantitative trait loci (QTL) studies. Advance molecular markers such as target region amplified polymorphism (TRAP), sequence related amplified polymorphism (SRAP), diversity array technology (DArT), and resistance gene analogues (RGAs) are now being frequently used in diversity and genetic analysis, and molecular breeding programmes. Sugarcane ESTs based studies and genome sequencing are also very useful in gene mapping, gene identification studies and transcriptome analysis (Aitken et al., 2016). Development of high throughput molecular technology (omics) would enable the genetic improvement of the sugarcane crop in a better and cost effective way.

4.2   ORIGIN AND CENTRE OF DIVERSITY OF SUGARCANE

Sugarcane belonging to the genus *Saccharum,* comprises of six species: *S. officinarum, S. spontaneum, S. sinense, S. barberi, S. robustum, S. edule,* of which *S. spontaneum* and *S. robustum* are considered as wild species and rest as cultivated species. Modern cultivated varieties of sugarcane are derived from the inter-specific cross between the species *S. officinarum,* and *S. spontaneum* with contribution from *S. sinense, S. robustum,* and *S. barberi* (Brandes, 1958; Daniels and Roach, 1987). It is theorized that sugarcane (*S. officinarum*) originated in New Guinea/Indonesia (Daniels and Roach, 1987) where it has been cultivated as garden plant or agriculture crop for chewing since about 8000 B.C. (Fauconnier, 1993). This view of origin in New Guinea is further supported by AFLP marker analysis (Aitken et al., 2006). It was also proposed that *S. officinarum* evolved from *S. robustum* (Daniels and Roach, 1987), an accepted view because the flavonoid pattern is similar in clones of *S. officinarum* and *S. robustum* (Williams et al., 1974). The regions of New Guinea/Indonesia are described as a rich source of *Saccharum* germplasm (Berding and Roach, 1987). The centre of origin and diversity of *S. spontaneum* is thought to be in northern India where cytoypes with small chromosome numbers are found (Mukherjee, 1957). Daniel and Roach (1987) opined that introgression of *Saccharum spontaneum* with *Miscanthus, Erianthus,* and *Sclerostachya*, led to the evolution of *S. barberi* and *S. sinense* in India and China, respectively.

4.3   TAXONOMY

Sugarcane belongs to the tribe *Andropogoneae*, subtribe *Saccharinae* and genus *Saccharum.* Mukherjee (1954, 1957) coined the term *Saccharum* complex to describe the subset of genera within the sub tribe *Saccharinae* that constituted the closely related interbreeding groups and contributed in the origin of sugarcane. Genera within the *Saccharum* complex were *Erianthus* (sec. Ripidium), *Sclerostachya* and *Narenga*. Daniels and Roach (1987) added *Miscanthus* in *Saccharum complex*. According to the recently accepted concept, *Saccharum* complex includes five genera, *viz., Saccharum,*

Erianthus (sec. *Ripidium*), *Miscanthus*, *Sclerostachya* and *Narenga*. The genus *Saccharum,* traditionally comprises of six species: *S. officinarum, S. spontaneum, S. sinense, S. edule, S. barberi,* and *S. robustum* (D' Hont et al., 1998). Based on the inter-fertility of these species, Irvine (1999) proposed to reduce the genus in two species: *S. officinarum* together with *S. robustum, S. sinense, S. edule,* and *S. barberi* as one species, and *S. spontaneum* as a separate species. Hodkinson et al. (2002) studied DNA sequence interrelationship of genera in the *Saccharum* complex and viewed that *Saccharum* and *Miscanthus* are more closely related to each other as compared to other genera. *Saccharum* complex theory is not well accepted now because recent molecular data do not support this theory (D'Hont et al., 2008).

4.4 BASIC CHROMOSOME NUMBERS

The basic chromosome number x (the haploid set of chromosomes in the cell) ranges from 5, 6, 8, 10 to 12 (Sreenivasan et al., 1987). The basic chromosome number of *S. spontaneum* and *S. officinarum* is 8 (although a number of variable cytotypes occur in *S. spontaneum*) and of *S. officinarum* and *S. robustum* is 10 (Panje and Babu, 1960; Piperidis et al., 2010). In rest three species, *S. sinense, S. barberi,* and *S. edule,* due to the fact that these are early inter specific hybrid cultivars, there has not been a consensus detected, but a study by Ming et al. (1998), proposed 10 as the basis chromosome number in these three. Off late, Fluorescence *in situ* hybridization (FISH) of ribosomal gene clusters has been used to discern the basic chromosome numbers in these species as $x = 10, 8,$ and 10, respectively. The $2n$ chromosome number is different in each *Saccharum* species and ranges from $2n = 80$ in *S. officinarum,* $2n = 40 – 128$ in *S. spontaneum,* $2n = 111 – 120$ in *S. barberi,* $2n = 81 – 124$ in *S. sinense,* $2n = 60 – 80$ in *S. edule,* and $2n = 60 – 80$ in *S. robustum.*

Modern sugarcane hybrids have been essentially derived from crosses between female *S. officinarum* $(2n = 80)$ and male *S. spontaneum* $(2n = 40 – 128)$ having two copies of the *S. officinarum* genome and one copy of the *S. spontaneum* genome in $2n + n$ fashion respectively (Bremer, 1961). That's why, they are highly aneuploidy in nature with inter-specific set of chromosomes. With the application of *in situ* hybridization techniques (genomic and fluorescent), it has been observed that chromosome of hybrid sugarcane constitutes of approximately 80% of *S. officinarum* and 10–20% *S.*

spontaneum and less than 5 to 17% recombinant chromosomes of these species (Cuadrado et al., 2004; D'Hont et al., 1996; Piperidis et al., 2001, 2010).

4.5 EXISTING GENETIC RESOURCES

Florida (USA) and Kerala (India) are the two world's largest collection center of sugarcane germplasm consisting of species level cones, genetic stocks and commercial cultivars which are used for making crosses in specific breeding programs. These center contain approximately 2000 accessions collected from 45 countries (Tai and Milller, 2001; Todd et al., 2014). The germplasm collection at Kannur Research Centre of SBI, Coimbatore consists of more than 1800 international and more than 1500 Indian collections. *Miscanthus* collections available as public and private both are collections associated with breeding programs. Sugarcane has been crossed with *Miscanthus* to transfer cold and drought resistance genes from *Miscanthus* into sugarcane. Although several hundreds of *S. spontaneum* clones are available in this world collection, only limited numbers of S. spontaneum clones have been used by various research stations for commercial varietal development even after a century of sugarcane breeding work. It is obvious that there is a large reserve of untapped genetic potential (Roach, 1989). The need to broaden the genetic base in order to introgress specific characters incorporating resistance to major biotic and abiotic stresses from wild and associated genera to cultivated sugarcane has raised the interest in pre-breeding. This has led to the development of several promising genetic stocks to be used as parents in future breeding programmes.

4.6 GENETIC IMPROVEMENT OF SUGARCANE

Basic breeding strategies for genetic improvement of sugarcane are to: collect, characterize, and maintain accessions of *Saccharum* species and related genera; to produce inter-specific/generic hybrids from a diverse range of parents; and to backcross them to efficiently concentrate useful genes while eliminating undesirable genes. Besides, to detect resistance and tolerance to biotic and abiotic stresses; to understand the mode of inheritance of tolerance to insect pests and environmental stresses; develop efficient selection criteria for important traits; synchronization of flowering and to use molecular approaches for identification and incorporation of desirable genes into elite cultivars and clones are some major breeding objectives. Modern sugarcane cultivars derive from a few crosses between *S. officinarum* and *S. spontaneum* and have been shown to be genetically very similar. *S. officinarum* is a plant with high sugar content in its stems but low productivity, and *S. spontaneum* has high tillering and biomass yield but low sugar accumulation. Breeding programmes have been able to increase yield and sucrose content by crossing cultivars but the improvement has reached a stagnation point. To continue the improvement of yield, it may be necessary to turn back to ancestral genotypes and broaden the genetic basis of crosses. Nowadays, *S. spontaneum* and *S. robustum* are also being used as parents, with the goal of designing a crop more amenable for cellulosic bio-fuel production, with increased stress tolerance and increased yield but less emphasis on stalk sugar concentration, the so-called 'energy cane.'

4.7 CONVENTIONAL BREEDING EFFORTS FOR SUGARCANE IMPROVEMENT

The conventional breeding of sugarcane involves selection and cloning of elite genotypes from segregating populations derived from inter-specific crossing of *Saccharum* species and diverse breeding strategies have been used for that in different sugarcane growing countries around the world according to their own convenience. Broadly, conventional breeding can be divided in to three steps: parental selection, hybridization or crossing and selection of elite genotype and their cloning. Crossing or hybridization results in genetic recombination, which is the basis of generating elite genotype with desired traits. The revolutionary work for practical sugarcane breeding was the discovery of the seed sets in sugarcane in 1888 by Dutch scientist Soltwedel (Java) and British scientists Harrison and Bovell (Barbados) and after this discovery, intergeneric and inter-specific hybridization were carried out in Java and India, which gave rise to the improved genotypes of sugarcane. In India, Sugarcane Breeding Institute, Coimbatore was started in 1912 for breeding sugarcane varieties; Indian Institute of Sugarcane Research, Lucknow was started in 1952 for crop production technologies, management of insect pests and diseases and coordinating various research activities in India. The AICRP (Sugarcane) was started in 1971 to develop appropriate variety, sugarcane production and protection technologies for different agro-ecological regions of the country. Vasant Dada Sugar Institute, Pune was established in 1975 to work out solutions for problem faced by co-operative sugar mills, particularly for Maharashtra state. The other important sugarcane research institutions involved in varietal development are: U.P. Council of Sugarcane Research, Shahjahanpur, U.P. and Regional Agricultural Research Station, Anakapalle, A.P. Besides there are sugarcane research stations in every sugarcane growing state with vast research contributions to their credit.

4.7.1 NOBILIZATION

Inter-specific hybridization with *S. spontaneum* conferred desirable attributes, like vigor, increased, tillering, resistance to diseases and pests, tolerance to adverse environments and ratooning ability potential, but it also introduced undesirable attributes such as low sugar, high starch, high fiber, reduced stalk girth, increased flowering, etc., Such unavoidable genetic consequences led to a series of direct or modified backcrosses to *S. officinarum* as the recurrent female parent in a process known as 'nobilization' to achieve agronomically adapted genotypes with high sugar yield increased resistance to diseases along with increase in yield and improved ratooning depending upon flowering synchrony, compatibility, etc. (Roach, 1972). During nobilization *S. officinarum*, transferred $2n$ chromosomes to the F_1 and BC_1, whereas *S. spontaneum* transferred n chromosomes (Bremer, 1961). Therefore, nobilized clones were characterized by a high chromosome number ($2n = 100 - 130$), with roughly 80% of the chromosomes derived from *S. officinarum*, and the remaining ones from *S. spontaneum* either directly or via *S. barberi* (Roach, 1969). Using *S. officinarum* and *S. spontaneum* chromosomes as probe, it is possible to characterize the chromosomes of commercial cultivars into two parental components through fluorescent *in situ* hybridization (D'Hont et al., 1996; Srivastava and Gupta, 2004). Moreover, inter-specific as well as intra-chromosomal recombination can also be elucidated in modern cultivars. Chromosome transmission in commercial hybrids x *S. spontaneum* clones in F_1, BC_1 to BC_3 generations or *S. officinarum* x *S. robustum* crosses is strictly of $n + n$ type (Burner and Legendre, 1993).

4.7.2 INTER-SPECIFIC HYBRIDIZATION

Inter-specific hybridization has provided a major breakthrough in sugarcane improvement. Before the advent of interspecific crosses, the improvement of sugarcane relied on the selection of naturally occurring variants of noble cane, i.e., *S. officinarum* obtained by collecting expeditions to its center of origin in New Guinea. Although true viable seeds capable of giving rise to seedlings and mature plants in sugarcane were known in 1858, it took almost 30 years to start raising sugarcane seedlings on experimental basis. Active breeding programmes were initiated in Java, Barbados, British Guyana, Mauritius, Queensland and Reunion during the 1890s. In Hawaii, India and other parts of the world, breeding programmes started in the early 19th century. Kobus in 1897 crossed a *S. barberi* clone "Chunnee" with *S. officinarum* and then by backcrossing the progeny to *S. officinarum* for eliminating undesirable traits from *barberi* clone, obtained 'Sereh' disease resistant varieties. This marked the onset of interspecific hybrids and end of the noble cane era. The early success of interspecific hybridization led to the inter-crossing of other species to produce tri-species hybrids that proved very successful in subtropical areas in India (Daniels and Roach, 1987). The Dutch breeders used a highly vigorous, disease-free, naturally occurring F1 hybrid, Kassoer, to produce POJ 2878 (BC_2) of the 'POJ' series in 1921. This clone occupied 90% of the cane area in Java within 8 years and became the java wonder cane. The importance of the wild species *S. spontaneum* was realized after its successful hybridization with the cultivated species, *S. officinarum*. Dr. C.A. Barber, who is the in charge of Sugarcane Breeding Station, Coimbatore carried out crosses between *S. officinarum* and *S. spontaneum* (Coimbatore form, $2n = 64$), which led to the production of first commercial hybrids Co 205 and Co 285, that yielded fifty percent more than the indigenous cultivars belonging to *S. barberi* and had resistance against environmental stress, and replaced the indigenous cultivated varieties of northern India.

4.7.3   INTER-GENERIC HYBRIDIZATION

Intergeneric hybridization was used in sugarcane to introgression commercially useful characteristics and to increase hybrid vigor in order to broaden the genetic base (Tai, 1989). The first intergeneric hybrid of sugarcane was made by C.A. Barber in 1913 at Coimbatore when he crossed *S. officinarum* var. Vellai (2n=80) with *Narenga prophyrocoma* ($2n = 30$) and found two types of hybrids ($2n = 95$ and 55). In 1938, E.K. Janaki Ammal crossed the same variety of *S. officinarum* with *Zea mays* and obtained first *Saccharum-Zea* hybrid with $2n = 52$ chromosomes. Since then, several intergeneric hybrids have been made by crossing *S. officinarum* with *Erianthus, Sclerostachya, Miscanthus, Sorghum, Imperata* and *Zea mays* (Sreenivasan et al., 1987; Srivastava, 2000). Hybridization of *Erianthus* with sugarcane has resulted in introgression of genes for cold tolerance and red rot resistance and the hybrids have shown great potential since they surpass most of the existing types in plant vigor (Nair and Mary, 2006). Many unsuccessful attempts have been made to introduce these characters into modern sugarcane cultivars. It is difficult to identify true *Saccharum-Erianthus* hybrids through morphological tools, hence isozyme and molecular markers have been applied to specifically detect the *E. arundinaceus* genome in intergeneric hybrids between *S. officinarum* and *E. arundinaceus* (Pathak et al., 2005). Low and high sugar yielding genotypes from this progeny were differentiated using RAPD markers (Srivastava et al., 2011). Genomic *in situ* hybridization (GISH) and fluorescent *in situ* hybridization (FISH) were used to identify the parental genomes in intergeneric hybrid of *S. officinarum* and *E. arundinaceus* (D'Hont et al., 1995).

4.7.4 MODERN COMMERCIAL HYBRIDS

The first sugarcane variety for subtropical India Co 205 was released in 1918 and it became an immediate success, replacing the indigenous cultivars, especially, "Katha" in large areas in sub-tropical India. Since then, many sugarcane genotypes were developed through inter-specific hybridization of *S. officinarum* and *S. spontaneum* followed by backcrossing with the *S. officinarum* parent which resulted in the development of improved sugarcane varieties with high yield and tolerance to stress. The basic breeding concept of a combination of *S. spontaneum* and *S. officinarum* for increasing sucrose content while retaining disease resistance in commercial hybrid cultivars reduces *S. spontaneum* genetic component in commercial hybrid cultivars. Several celebrated sugarcane varieties such as Co 290, Co 312, Co 419 (the *Wonder Cane of India*), Co 527, Co 740, Co 997, CoC 671, Co 1148, CoJ 64, CoS 767, Co 86032, CoLk 8102, CoLk 94184 have been released. Some of the varieties such as POJ 2878 from Indonesia and Co 213 and Co 290 from India find their place in parentage of most of the sugarcane varieties developed world over. Though modern cultivars of sugarcane are derived from the inter-specific crosses from basic germplasm but there has been merely a few generations for chromosome recombination chances, hence modern sugarcane population has narrow genetic base (Lima et al., 2002; Raboin et al., 2008; Roach, 1989).

4.8 MUTATION BREEDING

Induced mutagenesis may be an effective way to improve sugarcane crop by enhancing genetic variation. Price and Warner (1959) described the prospects for sugarcane improvement by induced mutations. Mutational breeding has been used to develop desired trait in sugarcane by inducing mutation through tissue culture practices. Physical and chemical mutagens have been used for mutagenesis to obtain beneficial modification in varieties (Patade and Suprasanna, 2008; Suprasanna et al., 2010). Several breeders have described the successful application of induced mutations for disease resistance in sugarcane, e.g., red rot resistance (Srivastava et al., 1986) and sugarcane mosaic virus resistance (Breanax, 1975; Dermodjo, 1977). Recently, Patade and Suprasanna (2008) and Nikam et al. (2015) described that radiation induced *in vitro* mutagenesis and selection can be used to develop salt resistant lines of sugarcane. Tolerance to salinity (NaCl) was enhanced when *in vitro* mutagenesis was combined with Gamma rays (Nikam et al., 2015).

4.9 APPLICATION OF CELL AND TISSUE CULTURE

Tissue and cell culture of sugarcane has existed since about 1965. Initial activities focused on the production of soma clonal variants and somatic hybrids, which expanded further to micropropagation to eliminate Fiji virus and *in vitro* cultures as a tool for germplasm conservation. Tissue culture technology is required for all gene transfer systems that are based on the production of suitable target cells followed by the regeneration of transgenic plants. Much of the recent tissue culture research has been directed towards genetic transformation activities. Some of the *in vitro* culture approaches being used in sugarcane improvement are succinctly mentioned below.

4.9.1 MICRO-PROPAGATION

As sugarcane is vegetatively propagated by nodal cuttings, and consequently, micro-propagation offers a practical and rapid approach for large production of clonal material. Mainly three main routes of plant regeneration through tissue culture have been identified as, (i) axillary shoot, (ii) adventitious shoot, and (iii) somatic embryogenesis (Maretzki, 1987). The first route is considered as the safest for production of clones. Plants can be produced either by direct regeneration from apical or axillary meristems (Taylor and Dukic, 1993) or from immature leaf tissues (Geijskes et al., 2003). As compared to other plant species, sugarcane plants propagated *in vitro* from meristem tissue are genetically and phenotypically more stable than those produced from callus (Hendre et al., 1983). Hendre et al. (1983) standardized an apical meristem culture method used for the rapid multiplication of mosaic virus-free plants of variety Co 740. Sugarcane plants produced by micro-propagation are free from disease, highly growing and better to seed cane in yield and recovery of sugar in field agronomic practices (Sood et al., 2006).

4.9.2 SOMATIC EMBRYOGENESIS

Somatic embryogenesis, an *in vitro* method of regeneration has been detected from a large number of commercial sugarcane clones and can be achieved directly or indirectly from the leaf tissues (Guiderdoni and Demarly, 1988; Guiderdoni et al., 1995; Manickavasagam and Ganapathi, 1998). Embryogenic callus can be maintained for several months without losing its regeneration potential (Fitch and Moore, 1993).

4.9.3 SOMACLONAL VARIATION

Sugarcane was amongst the first plants in which soma clonal variation was reported (Heinz and Mee, 1969; Larkin and Scowcroft, 1981). Somaclonal variation occurring in high frequency in callus cultures is projected as a promising approach to overcome constraints of conventional breeding and has been used for improvement of sugarcane crop. A number of soma clonal variants have been identified for desired traits, e.g., *Fiji* disease resistance (Krishnamurthi and Tlaskal, 1974), eye spot disease resistance (Ramos et al., 1996), drought tolerance (Wagih et al., 2004) and red rot resistance (Singh et al., 2008). ICAR-IISR, Lucknow developed many soma clones from sugarcane varieties such as CoLk 8102, CoJ 64, CoLk 8001, BO 91, CoS 767 and Co 7717, etc., for traits like higher NMC, better cane weight, red rot resistance, better flowering behavior, higher sucrose content, earlier maturity, tolerance to top borer and wilt, and greater vigor and productivity. Many of these changes proved epigenetic and were not stable in the successive clonal generations, particularly, those related to agronomic characters.

4.9.4 TRANSGENIC APPROACH

Breakthrough in genetic transformation of sugarcane came with the use of micro projectiles (biolistic method) by Bower and Birch (1992). Its success was dependent on the formation of callus and regeneration of plants, which can vary with genotype and culture condition (Kaeppler et al., 2000). Highly unstable gene expression has been experienced in sugarcane transgenic plants produced biolistic method, which may sometimes lead to gene silencing induced by high copy number transgene integration in genome. The incidence of allele silencing is more frequent in sugarcane owing to its highly polyploid nature as compared to other diploid plants (Birch et al., 2010). Another method, i.e., *Agrobacterium tumefaciens* mediated transformation is more efficient because of stable transgene expression and low copy number integration in genome (Dai et al., 2001) but, the biolistic method of gene transfer is more useful for gene pyramiding and when multiple expression cassettes are required for co expression. Other important issues are selection of transformed tissue and promoters for gene expression. Antibiotic resistance genes (Bower et al., 1996) or herbicide resistance genes (Falco et al., 2000) or phosphomannose isomerase (PMI) and reporting genes such as glucuronidase (GUS) (Liu et al., 2003) or green fluorescent protein (GFP) (Elliott et al., 1998) are used for the selection of transformed sugarcane. The constitutive promoters such as 35SCaMV (Arvinth et al., 2010) or UBI (polyubiquitin) promoter from maize (Christensen et al., 1996) or rice (Liu et al., 2003) are used for gene expression. Bar gene was the first gene of agronomic interest to be introduced in sugarcane for herbicide resistance, followed by several genes including the coat protein gene for sugarcane mosaic virus and Fiji disease resistance, Bt for resistance to borers, Proteinase inhibitor for resistance to cane grubs, alb D for resistance to leaf scald disease, trehalose for drought tolerance and capsid gene to confer resistance to pest and pathogen including bacteria and viruses (Arencibia et al., 1998; Falco et al., 2003; Smith et al., 1992, 1996, 1999; Weng et al., 2010; Zhang et al., 2006).

4.10 MOLECULAR INTERVENTIONS FOR SUGARCANE IMPROVEMENT

Modern sugarcane varieties are highly heterozygous, with several different alleles at each locus. The large and complicated genomic organization and high level of polyploidy present special challenges for sugarcane genetic analysis, and generally slows rates of gain in crop improvement program through conventional breeding approach. Molecular breeding approaches have been introduced in the last two decades to improve the breeding efficiency and has been recently used to improve various traits in sugarcane such as increased biomass production, tolerance to biotic, abiotic stress, sugar content and acceptable fiber content. The techniques and molecular approaches used for sugarcane improvement in last two and a half decade include: use of markers for diversity analysis, development of ESTs and identification of genes and trait linked markers for marker assisted breeding, etc.

4.10.1 MOLECULAR MARKERS APPLICATIONS

Molecular marker systems have been used for assessment of genetic diversity of germplasm collections and other elite genotypes of sugarcane (Ming et al., 2006; Srivastava and Gupta, 2008; Swapna and Srivastava, 2012). Probe based markers including nuclear ribosomal DNA using RFLP, low-copy RFLP probes and chloroplast DNA probes were used initially to distinguish *S. spontaneum* from other *Saccharum* species, to trace the origin of *S. officinarum* in New Guinea and of *S. barberi* and *S. sinense* as hybrids of *S. officinarum* and *S. spontaneum* and, to differentiate *S. spontaneum* clones of India (Burnquist et al., 1992; Glaszmann et al., 1990; Lu et al., 1999; Nair et al., 1999). With the advent of PCR based markers, RAPD were used widely in sugarcane for diversity analysis, varietal identification, etc. Later, SSRs or microsatellites from enriched genomic libraries and EST SSRs emerged as the popular marker system for DNA finger printing, differentiation of different species level and cultivated clones, genetic diversity analysis, etc. (reviewed in Ming et al., 2006; Swapna and Srivastava,

2012). Refinement of SSR approach such as SSCP-SSRs (Srivastava et al., 2005) made assaying of polymorphism more efficient or use of sugarcane enriched genomic microsatellites (SEGMS), and unigene derived microsatellites (UGMS) was more capable of assaying the inherent polymorphism and provided further understanding of abundance of transposons in the sugarcane genome. AFLP was also successful in detecting large number of polymorphic loci (Lima et al., 2002). With the advent of high throughput sequencing techniques, single nucleotide polymorphism (SNP) markers have evolved as a system of choice in sugarcane to study the copy number of genes and variation in gene sequences of sucrose synthesis and accumulation, metabolic pathways, red rot resistance, etc. Other markers like ISSR, TRAP, SRAP, SCOT, DArT, CISP, cDNA AFLP (Alwala et al., 2006; Andru et al., 2008; Carmona et al., 2004; Huang et al., 2015; Khan et al., 2011; Srivastava and Gupta, 2008, Wu et al., 2013), etc., have also been successfully used in this crop to analyze the diversity. Molecular markers have also been used as potential tool in identification of candidate genes in sugarcane (Andru et al., 2011). Several resistance genes and gene analogues have been identified, which can be used to develop disease resistant varieties (Hameed et al., 2015; Srivastava et al., 2016). Sequence Characterized Amplified Region (SCAR) markers were used to screen of drought tolerant sugarcane genotype (Srivastava et al., 2012).

4.10.2   SUGARCANE ESTS AND GENE IDENTIFICATION

Despite the absence of a sequenced genome and the complexities associated with the presence of about 8 to 12 copies of each gene, functional genomics of sugarcane has made considerable progress. Studies on sugarcane gene expression based mainly on expressed sequence tags (ESTs) information were first implemented in South Africa (Carson and Botha, 2000, 2002) with the small collection of ESTs from sugarcane meristematic tissue. The largest collection of ESTs was generated from SUCEST (Sugarcane EST) a consortium of Brazilian researchers, which sequenced around 238,000 ESTs from 26 cDNA libraries (Vettore et al., 2001, 2003), covering various developmental stages and different organs and tissues of sugarcane (Arruda, 2001). A sugarcane computational environment (SUCEST-FUN Database) has been developed for storage, retrieval and integration of gene catalogues, genome sequencing, transcriptome, expression profiling, physiology measures and transgenic plant data and 282,683 ESTs are currently cata-logued in this. Sugarcane ESTs have also been generated in USA and com-pared with EST libraries of *Sorghum* and *Arabidopsis* to look for common genes (Ma et al., 2004). Nearly 26000 tissue specific and 1069 ESTs from tissue infected with red rot have been developed in India by ICAR-IISR, Lucknow in collaboration with Delhi University South Campus (Gupta et al., 2010). ESTs are important tool in mapping of genes (Pinto et al., 2010). EST derived marker such as EST-RFLP have been used to target the sugar content (Da Silva and Bressiani, 2005). ESTs have also been successfully used in gene expression and RGAs studies in sugarcane. Several EST data bases have been successfully analyzed by many workers to develop molec-ular markers for yield, sugar content and other related quantitative as well as qualitative characters and the information has further been utilized in molecular mapping studies in this crop (Banerjee et al., 2015; Ming et al., 2006; Singh et al., 2013;).

4.10.3 SUGARCANE CHLOROPLAST GENOME SEQUENCING

The whole nucleotide sequence of the chloroplast genome of sugarcane (*Saccharum officinarum*) has been determined. It is a circular double stranded DNA molecule, and the size is 141,182 bp. Chloroplast genome is composed of a large single copy of 83,048 bp and a small single copy of 12,544 bp along with a pair of inverted repeat regions of 22,795 bp each (Asano et al., 1994). A comparative study among monocots indicated that the chloroplast genome of sugarcane was very similar to maize but not to rice or wheat (Asano et al., 1994).

4.11 RECENT MOLECULAR APPROACHES: A PARADIGM SHIFT

The notable progress in molecular tools in past two decades has resulted in to the availability of plethora of genetic markers especially robust SSRs and SNP markers, which are being extensively used by sugarcane researchers. Recent progress in this era has witnessed development of mapping popula-tions and deployments of the genetic markers/QTLs in sugarcane improve-ment programmes. Besides, association mapping or linkage disequilibrium analysis has unraveled non-random association of alleles between different loci within the sugarcane genome.

4.11.1 QTLS AND MARKER ASSISTED SELECTION

Marker assisted selection and QTLs reduce the time of selection, and consequently improve the breeding efficiency. Trait linked markers, once identified, can help in the selection of desirable genotypes from a large population, thereby reducing the size of population to be handled in field at a very early stage itself. Quantitative trait loci or QTL are defined as a region of chromosome that contains alleles or genes controlling a trait, which is commonly used in breeding programme. Many species/genus specific and trait specific markers have been identified and are being used to identify inter-specific/generic hybrids in this crop and analysis of red rot disease resistant/moderately resistant and susceptible elite sugarcane (Alix et al., 1998; Besse and McIntyre, 1998; Nair et al., 2006; Virupakshi and Naik, 2008). Several DNA markers have been found to be closely linked to QTLs of important traits in sugarcane such as RAPD marker linked to fibre content, eye spot disease (Msomi and Botha, 1994; Mudge et al., 1996) and rust disease (Barnes and Botha, 1998), RFLP, and AFLP markers linked to Brown Rust disease (Asnaghi et al., 2004; Daugrois et al., 1996), AFLP marker linked to Brix, stalk length, stalk diameter and number of stalks (Hoarau et al., 2002), AFLP marker linked to smut disease (Raboin et al., 2003), RFLP marker linked to sugar content and short day flowering (Guimaraes et al., 1997; Ming et al., 2001), and AFLP marker linked to red rot disease (Selvi et al., 2006). A number of QTLs have been detected in sugarcane which are associated with tiller number and suckering (Jorden et al., 2004), sugar content (Aitken et al., 2006; Hoarau et al., 2002; Ming et al., 2001, 2002) and yield related traits such as stalk weight, stalk diameter and number of stalk (Aitken et al., 2008). QTL mapping for sugar content was carried out in in a segregating population of a selfed hybrid R570 and two interspecific populations (Hoarau et al., 2002; Ming et al., 2002), and it was mapped with *Sorghum* linkage group. Studies have been performed on a sugarcane genotype to identify QTLs which regulate the *Pachymetra* root rot and brown rust resistance (McIntyre et al., 2005), resistance to *Sugarcane yellow leaf virus* using DaRT and AFLP markers (Debibakas et al., 2014), agro-morphological character, sugar yield, disease resistance and bagasse content using the DaRT, AFLP, and SSR markers (Gouy et al., 2015); and, yield and sucrose traits using the SSR marker (Banerjee et al., 2015). QTL mapping studies can help resolve the genetics of quantitative traits in a better way. The progress in the area of eQTLS (expression QTLs) has further advanced the understanding of sugarcane molecular structure to a new horizon.

4.11.2 COMPARATIVE GENOMICS

Comparative mapping is a way of using the concept of utilizing the information available from a much simpler crop to understand the genetics of a more complex crop by utilizing conserved DNA sequences of both, e.g., sugarcane and sorghum chromosomes are highly synthetic with perfect collinearity at several locations between the genomes of *Sorghum* and *Saccharum,* and highly conserved order of the DNA sequences (Dufour et al., 1997; Glaszmann et al., 1997; Guimaraes et al., 1997). Comparative studies using molecular markers demonstrated synteny between maize and *Saccharum* for several chromosome segments and large chromosomal rearrangements like translocations, centric fusion, etc., could also be detected in some cases and these were assumed to have played a significant evolutionary role. *Sorghum* is the closest diploid relative of sugarcane for comparative studies in the area of functional genomics, molecular diversity, cross-transferability of molecular markers and comparative genome mapping, etc. Comparison of sugarcane BAC clones with *Sorghum* chromosomes aligned many sequences in the sugarcane BAC clones to *Sorghum* chromosomes.

4.11.3 WHOLE GENOME SEQUENCING OF SUGARCANE

The sugarcane genome sequencing initiative (SUGESI) has selected the sugarcane variety R570 as the reference genome for sequencing because partial genome of R570 had already been sequenced and there were a number of other tools available for R570 such as a high-density genetic map and a bacterial artificial chromosome (BAC) library. Two strategies were chosen to generate the genome sequence; BAC-by-BAC sequencing and whole genome shotgun approach for generation of sugarcane genome sequence. The groups of four countries (Australia, Brazil, South Africa and France) in SUGESI have contributed to the sequencing and assembly of the BAC clones. The 2767 BAC clones sequenced till date are available in SUGESI (Aitken et al., 2016). The genomics studies offer the knowledge needed to assign a physiological function to a gene. However, the journey from genotype to phenotype still needs a more integrated approach from molecular data to sugarcane physiology and production, thus setting the basis for modeling the regulatory pathways that connect genes, metabolites and physiological processes.

4.11.4 TRANSCRIPTOMIC STUDIES IN SUGARCANE

Biotic and abiotic stresses are major problem for sugarcane growers and sugar industry because they cause major loss in terms of yield. To overcome these stresses in sugarcane transcriptomics studies have proved very helpful. Major high throughput techniques involved in transcriptomics studies are microarray, cDNA-AFLP and serial analysis of gene expression (SAGE) etc. Transcriptional response of methyl jasmonate in sugarcane was studied by using cDNA microarray (Bower et al., 2005; Souza et al., 2001) and by nylon arrays containing 1536 ESTs from cDNA libraries (Rosa et al., 2005). Three hundred differentially expressed transcripts were reported on red rot infection (Prathima et al., 2013). Using differential display-PCR, Rahul et al. (2014) found 241 transcripts that were differentially expressed upon inoculation of *C. falcatum*. Using cDNA-AFLP, 62 differentially regulated genes were identified in sugarcane (Orlando et al., 2005). Que et al. (2011), reported 136 differentially expressed transcripts, in a resistant variety using cDNA-AFLP and of these 40 TDFs were consistent, 34 TDFs were newly induced and 6 transcripts were significantly up regulated after inoculation. Wu et al. (2013) reported 2015 differentially expressed transcripts of which 1125 were up regulated and 890 were down regulated in pathogen-inoculated samples. Using cDNA-AFLP technique, Medeiros et al. (2014), carried out a transcriptional profile study in SCMV resistant sugarcane variety challenged with sugarcane mosaic virus and identified 392 TDFs. Gupta et al. (2010) identified 25 clusters of genes associated with water deficit stress in sugarcane using real time reverse transcription-PCR profiling of selected EST clusters. Using microarray technology based on the expression profile of 15,593 sugarcane genes, Li et al. (2016) found that 1501 gene genes were differentially expressed of which 821 genes were up regulated and 680 genes were down regulated in sugarcane variety (GT21) under different water level parameter. Microarray and RNAseq analysis followed by qPCR revealed 10 genes as differentially expressed with three genes related to drought tolerance. Nogueira et al. (2003) identified 34 cold responsive ESTs a of which 20 were novel cold responsive genes (COR) including cellulose synthase,

ABI3 interacting protein 2, a negative transcription regulator, phosphate transporter and others identified as unknown genes in *Saccharum* species (cv SP80-3280) using ESTs profiling studies. The expression of SsNAC23 (member of NAC transcription factor) was induced in sugarcane plants when exposed to low temperatures (4^0C) and thioredoxin acted as interacting protein (Ditt et al., 2011). Several miRNA were expressed during abiotic stresses conditions in sugarcane (Gentile et al., 2015). To identify differentially expressed miRNA in sugarcane under cold stress (4°C), Thiebaut et al. (2012) used 12 miRNA (miR156, miR159, miR160, miR166, miR167, miR169, miR172, miR319, miR393, miR394, miR408, miR528) and their expression profile was assayed by stem loop RT-PCR. The expression of miR319 was more during cold stress in both roots and shoots. Srivastava and Sunkar (2013) also explored the role of miRNA in response to drought in sugarcane.

4.12 MOLECULAR CYTOGENETICS

Over the past two decades, techniques of *in situ* hybridization have allowed the examination of the numbers and sizes of the sites of rRNA genes in sugarcane. This has helped in identification of true hybrids in inter-generic and inter-specific crossing programmes especially between *S. officinarum* and *E. arundinaceus*, differentiate parental chromosomes in interspecific hybrids using rDNA multigene families, tracking down the introgression of alien chromosomes segments, exact genome constitution of modern cultivars and exchange between chromosomes derived from different species (Cuadrado et al., 2004; D'Hont et al., 1996; Srivastava, 2003; Srivastava and Gupta, 2004). The physical location of 25S and 5S rDNA genes by FISH (fluorescent *in situ* hybridization) provided useful cytological markers to differentiate chromosomes of different species of sugarcane. The 5S were always interstitial in position while the 25S rDNA occurred at terminal (*S. officinarum*) as well as interstitial positions (*S. spontaneum*) (Srivastava and Gupta, 2004). Comparative *in situ* hybridization of two sugarcane clones My5514 ($2n = 102 - 106$) and C236-51 ($2n = 113 - 117$) revealed 16% of total chromosomes being inherited from *S. spontaneum* (Cuadrado et al., 2004). Srivastava and Gupta (2004) observed 14–18% *S. spontaneum* specific chromosomes in commercial cultivars of subtropical India. Approximately 10% recombinant chromosomes (D'Hont et al., 1996) have been observed through GISH. Size heterogeneity between S. officinarum homologous chromosomes carrying the 18S ± 5.8S ± 25S and 5S ribosomal genes confirmed that remodeling occurred by chromosomal interchange events, at least in these homologous chromosomes (Caudardo et al., 2004).

4.13   CONCLUDING REMARKS

Starting from initial hybridization and selection leading to the development of several elite varieties, sugarcane crop improvement ushered into the era of molecular markers. Molecular breeding tools have evolved at a fast pace and simple PCR based markers given way to the present day highly advanced technologies of Next Generation Sequencing. The investigation at the chromosomal level has reached new dimensions with individual nucleotides being studied and manipulated. Complex metabolic pathways are being studied at length enabling much needed manipulations at crucial stages. Cutting-edge molecular marker systems have generated high-resolution genetic maps and helped resolve genetic linkage and markers trait relationship for sugarcane. The bioinformatics tools and software that aid in analysis and interpretation of huge data generated are also very essential to undertake molecular breeding, functional genomics, transcriptomics or related studies. Advance software are being developed and used to tackle these problems. The complex genetic architecture and genomic peculiarities of this crop specifically necessitate suitable modifications in the existing approaches and programmes and that is why newer information is being added each passing day for sugarcane crop improvement.

KEYWORDS

- eSTs
- genetic improvement
- genomics
- molecular markers
- saccharum
- trait linked markers
- transcriptomics

REFERENCES

Aitken, K.S., Li, J-C., Jackson, P., Piperidis, G. & McIntyre, C.L. (2006). AFLP analysis of genetic diversity within Saccharum officinarum and comparison with sugarcane cultivars. Australian *Journal of Agricultural Research, 57*(11), 1167–1184.

Aitken, K.S., Jackson, P.A. & McIntyre, C.L. (2006). Quantitative trait loci identified for sugar related traits in sugarcane (*Saccharum* spp.) cultivar x *Saccharum officinarum* population. *Theor. Appl. Genet., 112*(7), 1306–1317.

Aitken, K., Karno, K., Bonnett, G.D., McIntyre, L.C. & Jackson, P.A. (2008). Genetic control of yield related stalk traits in sugarcane. *Theor. Appl. Gene., 117*(7), 1191–1203.

Aitken, K., Berkman, P., & Rae, A. (2016). The first sugarcane genome assembly: How can we use it? *Proc. Aust. Soc. Sugar Cane Technol., 38,* 193–199.

Alix, K., Baurens, F. C., Paulet, F., Glaszmann, J. C., & D' Hont, A. (1998). Isolation and characterization of a satellite DNA family in the *Saccharum* complex. *Genome, 41*(6), 854–864.

Alwala, S., Andru, S., Arro, J. A., Parco, A.S., Kimbeng, C.A. & Baisak, N. (2006). Target region amplification polymorphism (TRAP) for assessing genetic diversity in sugarcane germplasm collections. *Crop Sci., 46*(1), 448–455.

Arencibia, A. D., Carmona, E. R., Tellez, P., Chan, M., Yu, S., Trujillo, L. E., & Oramas, P. (1998). An efficient protocol for sugarcane (*Saccharum* spp. L.) transformation mediated by *Agrobacterium tumefaciens. Transgenic Res., 7*(3), 213–222.

Arvinth, S., Arun, S., Selvakesavan, R. K., Srikanth, J., Mukunthan, N., Kumar, P. A., Premachandran, M. N., & Subramonian, N. (2010). Genetic transformation and pyramiding of aprotinin-expressing sugarcane with cry1Ab for shoot borer (*Chilo infuscatellus*) resistance. *Plant Cell Rep., 29*(4), 383–395.

Asnaghi, C., Roques, D., Ruffel, S. et al. (2004). Targeted mapping of a sugarcane rust resistance gene (Bru1) using bulked segregant analysis and AFLP markers. *Theor. Appl. Genet., 108*(4), 759–764.

Banerjee, N., Siraree, A., Yadav, S., Kumar, S., Singh, J., Kumar, S., Pandey, D.K. & Singh, R.K. (2015). Marker trait association study for sucrose and yield contributing traits in sugarcane (*Saccharum* spp. hybrid). *Euphytica, 205*(1), 185–201.

Barnes, J. M., & Botha, F. C. (1998). Progress towards identifying a marker for rust resistance in sugarcane variety NCo376. *Proc. S. Afi. Sug. Technol. Ass., 72,* 149–151.

Berding, N., & Roach, B. T. (1987). Germplasm collection, maintenance and use. In: *Sugarcane improvement through breeding.* Heinz, D. J. (ed.). Elsevier Press, Amsterdam, pp. 143–210.

Besse, P., & McIntyre, L. C. (1998). Isolation and characterization of repeated DNA sequences from *Erianthus* spp. (*Saccharinae: Andropogoneae*). *Genome, 41*(3) 408–416.

Birch, R. G., Bower, R. S., & Elliott, A. R. (2010). Highly efficient, 5-sequence specific transgene silencing in a complex polyploid. *Trop Plant Biol. 03,* 88–97.

Bower, N. I., Casu, R. E., Maclean, D. J., Reverter, A., Chapman, S. C., & Manners, J. M. (2005). Transcriptional response of sugarcane roots to methyl jasmonate. *Plant Sci., 168*(3), 761–772.

Bower, R., & Birch, R. G. (1992). Transgenic sugarcane plants via micro projectile bombardment. *The Plant J., 02*(3), 409–416.

Bower, R., Elliott, A. R., Potier, B. A. M., & Birch, R. G. (1996). High-efficiency, micro projectile-mediated cotransformation of sugarcane, using visible or selectable markers. *Mol. Breed., 02,* 239–249.

Brandes, E. W. (1958). Origin, classification and characteristics. In: *Sugarcane (Saccharum officinarum* L.). Artschwager, E., & Brandes, E. W. (eds.), Agric. Handbook (U.S. Dept. Agric.) *122*, pp. 1–35.

Breanx, R. D. (1975). Radio sensitivity and selection for mosaic resistant variety in sugarcane. *Proc. Int. Soc. Sugar Cane Tech.*, *4*, 97–100.

Bremer, G. (1961). Problems in breeding and cytology of sugarcane. *Euphytica*, *10*(1), 59–78.

Burner, D. M., & Legendre, B. L. (1993). Chromosome transmission and meiotic stability of sugarcane (*Saccharum* spp.) hybrid derivation. *Crop Science*, *33*, 600–606.

Burnquist, W. L., Sorrells, M. E., & Tanksley, S. (1992). Characterization of genetic variability in *Saccharum* germplasm by means of restriction fragment length polymorphism (RFLP) analysis. *Proc. of the Intern. Soc. of Sugar Cane Technol.*, *21*, 355–365.

Carmona, E., Vargas, D., Borroto, C. J., Lopez, J., Fernandez, A. I., Arencibia, A., & Borras-Hidalgo, (2004). O cDNA-AFLP analysis of differential gene expression during the interaction between sugarcane and *Puccinia melanocephala*. *Plant Breed*, *123*(5), 499–501.

Christensen, A. H., & Quail, P. H. (1996). Ubiquitin promoter based vectors for high-level expression of selectable and/or screenable marker genes in monocotyledonous plants. *Transgenic Res.*, *05*, 213–218.

Cuadrado, A., Acevedo, R., Moreno Diaz DeLa Espina, S., Jouve, N., & Dela Torre, C. (2004). Genome remodelling in three modern *S. officinarum* x *S. spontaneum* sugarcane cultivars. *J. Exp. Bot.*, *55*(398), 847–854.

D'Hont, A., Grivet, L., Feldmann, P., Glaszmann, J.C., Rao, S. & Berding, N. (1996). Characterization of the double genome structure of modern sugarcane cultivars *(Saccharum* spp.) by molecular cytogenetics. *Mol. Gen. Genet.*, *250*(4), 405–413.

D'Hont, A., Ison, D., Alix, K., Roux, C., & Glaszmann, J. C. (1998). Determination of basic chromosome numbers in the genus *Saccharum* by physical mapping of ribosomal RNA genes. *Genome*, *41*(2), 221–225.

D'Hont, A., Rao, P. S., Feldmann, P., Grivet, L., Islam-Faridi, N., Taylor, P., & Glaszmann, J. C. (1995). Identification and characterization of sugarcane intergeneric hybrids, *Saccharum officinarum* x *Erianthus arundinaceus*, with molecular markers and DNA in situ hybridization. *Theor. Appl. Genet.*, *91*(2), 320–326.

D'Hont, A., Souza, G. M., Menossi, M., Vincentz, M., Van-Sluys, M. A., Glaszmann, J. C., & Ulian, E. (2008). Sugarcane: a major source of sweetness, alcohol, and bio-energy. In: Moore, P. H., Ming R. (eds.) Plant genetics and genomics: crops and models. *Springer* New York, *01*, 483–513.

Da, Silva, J. A., & Bressiani, J. A. (2005). Sucrose synthase molecular marker associated with sugar content in elite sugarcane progeny. *Genet. Mol. Biol.*, *28*(2), 294–298.

Dai, S. H., Zheng, P., Marmey, P., Zhang, S. P., Tian, W. Z., Chen, S., Beachy, R.N. & Fauquet, C. (2001). Comparative analysis of transgenic rice plants obtained by Agrobacterium-mediated transformation and particle bombardment. *Mol. Breed*, *07*(1), 25–33.

Daniels, J., & Roach, B. T. (1987). Taxonomy and evolution. In: *Sugarcane Improvement through Breading*. Heinz, D. J. (ed.). Elsevier, Amsterdam, Netherlands, *11*, 07–84.

Daugrois, J. H., Grivet, L., Roques, D. et al. (1996). A putative major gene for rust resistance linked with a RFLP markers in sugarcane cultivar 'R570.' *Theor. Appl. Genet.*, *92*(8), 1059–1064.

Debibakas, S., Rocher, S., Garsmeur, O., Toubi, L., Roques, D., D'Hont, A., Hoarau, J-Y. & Daugrois, J. H. (2014). Prospecting sugarcane resistance to sugarcane yellow leaf virus by genome-wide association. *Theor. Appl. Genet.*, *127*(8), 1719–1732.

Dermodjo, S. (1977). Induction of mosaic disease resistance in sugarcane by gamma ray irradiation. *Int. Soc. Sugar Cane Tech,. Sug. Breed.* Newsletter, *39*, 4–7.

Ditt, R. F. et al. (2011). Analysis of the stress-inducible transcription factor SsNAC23 in sugarcane plants, *Sci. Agric. (Piracicaba, Braz.)*, *68*(4), 454–461.

Dufour, P., Deu, M., Grivet, L., D'Hont, A., Paulet, F., Bouet, A., Lanaud, C., Glaszmann, J. C., & Hamon, P. (1997). Construction of a composite sorghum genome map and comparison with sugarcane, a related complex polyploid. *Theor. Appl. Genet.*, *94*(3), 409–418.

Elliott, A. R., Campbell, J. A., Brettell, R. I. S., & Grof, C. P. L. (1998). Agrobacterium mediated transformation of sugarcane using GFP as a screenable marker. *Aust J Plant Physiol.*, *25*, 739–743.

Falco, M. C., & Silva-Filho, M. C. (2003). Expression of soybean proteinase inhibitors in transgenic sugarcane plants: effects on natural defense against *Diatraea saccharalis*. *Plant Physiol. Biochem.*, *41*(8), 761–766.

Falco, M. C., Neto, A. T., & Ulian, E. C. (2000). Transformation and expression of a gene for herbicide resistance in Brazilian sugarcane. *Plant Cell Rep.*, *19*, 1188–1194.

Fitch, M. M. M., & Moore, P. H. (1993). Long term culture of embryogenic sugarcane callus. *Plant Cell Tiss. Organ Cult.*, *32*(3), 335–343.

Geijskes, R. J., Wang, L., Lakshmanan, P., McKeon, M. G., Berding, N., Swain, R. S., Elliott, A. R., Grof, C. P. L., Jackson, J. A., & Smith, G. R. (2003). Smartsett™ seedlings: tissue cultured seed plants for the Australian sugar industry. *Sugarcane Int.*, May/June, 13–17.

Gentile, A., Dias, L. I., Mattos, R. S., Ferreira, T. H., & Menossi, M. (2015). Micro RNAs and drought responses in sugarcane. Front. *Plant Sci.*, *06*, 58.

Glaszmann, J. C., Dufour, P., Grivet. L., D'Hont, A., Deu, M., Paulet, F., & Hamon, P. (1997). Comparative genome analysis between several tropical grasses. *Euphytica*, *96*(1), 13–21.

Glaszmann, J. C., Lu, Y. H., & Lanaud, C. (1990). Variation of nuclear ribosomal DNA in sugarcane. *J. Genet. Breed*, *44*(3), 191–197.

Gouy, M., Rousselle, Y., Thong Chane, A., Anglade, A., Royaert, S., Nibouche, S. et al. (2015). Genome wide association mapping of agro-morphological and disease resistance traits in sugarcane. *Euphytica*, *202*(2), 269–284.

Guiderdoni, E., & Demarly, Y. (1988). Histology of somatic embryogenesis in cultured leaf segments of sugarcane plantlets. *Plant Cell Tiss. Organ Cult.*, *14*(2), 71–88.

Guiderdoni, E., Merot, B., Eksomtramage, T., Paulet, F., Feldmann, P., & Glaszmann, J. C. (1995). Somatic embryogenesis in sugarcane (*Saccharum* species). In: Y. P. S. Bajaj (ed.): *Biotechnology in Agriculture and Forestry*. Somatic embryogenesis and synthetic seed II. Berlin, Springer, *31*, 92–113.

Guimaraes, C. T., Sills, G. R., & Sobral, B. W. S. (1997). Comparative mapping of Andropogoneae: *Saccharum* L. (sugarcane) and its relation to sorghum and maize. *Proc. Natl. Acad. Sci.* USA, *94*(26), 14261–14266.

Gupta, V., Raghuvanshi, S., Gupta, A., Saini, N., Gaur, A., Khan, M. S., Gupta, R. S., Singh, J., Duttamajumder, S. K., Srivastava, S., Suman, A., Khurana, J. P., Kapur, R., & Tyagi, A. K. (2010). The water-deficit stress- and red-rot-related genes in sugarcane. *Funct. Integr. Genomics*, *10*(2), 207–214.

Hameed, U., Pan, Y. B., & Iqbal, J. (2015). Genetic Analysis of Resistance Gene Analogues from a Sugarcane Cultivar Resistant to Red Rot Disease. *J. Phytopathol*, *163*(9), 755–763.

Han, G., & Wu, Q. (2004). Comparative properties of sugarcane rind and wood strands for structural composite manufacturing. *For Prod. J.*, *54*(12), 283–288.

Heinz, D. J., & Mee, G. W. P. (1969). Plant differentiation from callus tissue of *Saccharum* species. *Crop Sci.*, *09*, 346–348.

Hendre, R. R., Iyer, R.S., Kotwal, M., Khuspe, S.S. & Mascarenhas, A.F. (1983). Rapid multiplication of sugarcane through tissue culture. *Sugarcane*, *01*, 05–07.

Hoarau, J. Y., Grivet, L., Offmann, B., Raboin, L. M., Diorflar, J. P., Payet, J., Hellamann, M., D'Hont, A., & Glaszmann, J. C. (2002). Genetic dissection of a modern sugarcane cultivar (*Saccharum* spp.). ll. Detection of QTLs for yield components. *Theor. Appl. Genet.*, *105*(6), 1027–1037.

Hodkinson, T. R., Chase, M. W., & Renvoize, S. A. (2002). Characterization of a genetic resource collection for *Miscanthus* (Saccharinae, Andropogoneae, Poaceae) using AFLP and ISSR PCR. *Ann. Bot.*, *89*(5), 627–636.

Huang, N., Zhang, Y.Y., Xiao, X.H., Huang, L., Wu, Q.B., Que, Y.X. & Xu, L.P. (2015). Identification of smut-responsive genes in sugarcane using cDNA-SRAP. *Genet. Mol. Res.*, *14*(2), 6808–6818.

Jordan, D. R.,   Casu, R.E., Besse, P., Carroll, B.C., Berding, N. & McIntyre C.L. (2004). Markers associated with stalk number and suckering in sugarcane colocate with tillering and rhizomatousness QTLs in sorghum. *Genome*, 47(5), 988–993.

Kaeppler, S. M., Kaeppler, H. F., & Rhee, Y. (2000). Epigenetic aspects of somaclonal variation in Plants. *Plant Mol. Biol.* 2000, *43*(2), 179–188.

Khan, M. S. Yadav S. Srivastava S. Swapna M. Chandra A., & Singh R K. (2011). Development and Utilization of CISP Marker in sugarcane. *Australian Journal of Botany*, *59*, 38–45.

Krishnamurthi, M., & Tlaskal, J. (1974). *Fiji* disease resistant *Saccharum officinarum* var. pindar subclones from tissue cultures. *Proc. Int. Soc. Sugar Cane. Technol.*, *15*,130–137.

Larkin, P. J., & Scowcroft, W. R. (1981). Somaclonal variation-a novel source of variability from cell cultures for plant improvement. *Theo. Appl. Gene.*, *60*(4), 197–214.

Li, Changning, Qian, N., Solanki, M. K., Liang, Q., Xie, J., Liu, X., Li, Y., Wang, W., 1, Yang, L. T.& Li, Y.R. (2016). Differential expression profiles and pathways of genes in sugarcane leaf at elongation stage in response to drought stress. *Scientific Rep.*, *06*, 25698.

Lima, M. L. A., Garcia, A.A.F., Oliverira, K.M., Motsuoka, S., Arizono, H., De Souza Jr.C.L. & de Souza, A.P. (2002). Analysis of genetic similarity detected by AFLP and coefficient of parent genome types of sugarcane (*Saccharum* spp.). *Theor. Appl. Genet.*, *104*, 30–38.

Liu, D., Oard, S. V., & Oard. J. H. (2003). High transgene expression levels in sugarcane (*Saccharum officinarum* L.) driven by the rice ubiquitin promoter RUBQ2. *Plant Sci.*, *165*,743–750.

Ma, H. M., Schulze, S., Lee, S., Yang, M., Mirkov, E., Irvine, J., Moore, P., & Paterson A. (2004). An EST survey of the sugarcane transcriptome. *Theor. Appl. Genet.*, *108*(5), 851–63.

Manickavasagam, M., & Ganapathi, A. (1998). Direct somatic embryogenesis and plant regeneration from leaf explants of sugarcane. *Indian J. Exp. Biol.*, *36*, 832–835.

Maretzki, A. (1987). Tissue culture: Its prospects and problems. In: *Sugarcane Improvement Through Breeding*. D. J. Heinz (ed.). Elsevier Science Publishers, Amsterdam. 343–384.

McIntyre, C. L., Whan, V. A., Croft, B., Magarey, R., & Smith, G. R. (2005). Identification and validation of molecular markers associated with pachymetra root rot and brown rust resistance in sugarcane using Map and association based approaches. *Mol. Breeding.*, *16*(2), 151–161.

Medeiros, C. N. F., Gonçalves, M. C., Harakava, R., Creste, S., Nóbile, P. M., Pinto, L. R., Perecin, D., & Landell, M. G. A. (2014). Sugarcane transcript profiling assessed by cDNA-AFLP analysis during the interaction with *Sugarcane mosaic virus*. *Advances in Microbiology*, *4*(9), 511–520.

Ming, R., DelMonte, T. A., Hernandez, E., Moore, P. H., Irvine, J. E., & Paterson, A. H. (2002). Comparative analysis of QTLs affecting plant height and flowering among closely-related diploid and polyploid genomes. *Genome*, *45*(5), 794–803.

Ming, R., Liu, S. C., Moore, P. H., Irvine, J. E., & Paterson, A. H. (2001). QTL analysis in a complex autopolyploid: genetic control of sugar content in sugarcane. *Genome Res.*, *11*(12), 2075–2084.

Msomi, M., & Botha, F. C. (1994). Identification of putative molecular markers linked to the fiber using bulk segregants analysis. SASTA'94, *68*, 41–45.

Mudge, J., Anderson, W. R., Kehrer, R. L., & Fairbanks. D. J. (1996). A RAPD genetic map of *Saccharum officinarum*. *Crop Sci.*, *36*, 1362–1366.

Mukherjee, S. K. (1957). Origin and distribution of *Saccharum*. *Botanical Gazette*, *119*, 55–61.

Nair, N. V., Nair, S., Sreenivasan, T. V. et al. (1999). Analysis of genetic diversity and phylogeny in *Saccharum* and related genera using RAPD markers. *Genet. Resour. Crop Evol.*, *46*(1), 73–79.

Nair, N. V., Selvi, A., Sreenivasan, T. V., & Pushpalatha, K. N. (2006). Characterization of intergeneric hybrids of *Saccharum* using RAPD markers. *Genet. Resour. Crop Evol.* *53*(1), 163–169.

Nikam, A. A., Devarumath, R. M., Ahuja, A., Babu, H., Shitole, M. G., & Suprasanna, P. (2015). Radiation-induced in vitro mutagenesis system for salt tolerance and other agronomic characters in sugarcane (*Saccharum officinarum* L.). *The Crop Jour.*, *03*(1), 46–56.

Nogueira, F. T. S., De Rosa, J. V. E., Menossi, M., Ulian, E. C., & Arruda, P. (2003). RNA expression profiles and data mining of sugarcane response to low temperature. *Plant Physiol.*, *132*(4), 1811–1824.

Orlando, B. H. et al. (2005). Identification of sugarcane genes induced in disease-resistant soma clones upon inoculation with *Ustilago scitaminea* or *Bipolaris sacchari*. *J. Plant physiol.*, *43*(12), 1115–1121.

Pandey, A., Soccol, C. R., Nigam, P., & Soccol, V. T. (2000). Biotechnological potential of agro-industrial residues-I: sugarcane bagasse. *Bio. Resour. Technol.*, *74*(1), 69–80.

Panje, R., & Babu, C. N. (1960). Studies in *Saccharum spontaneum*; distribution and geographical association of chromosome numbers. *Cytologia*, *25*, 152–172.

Patade, V. Y., & Suprasanna, P. (2008). An *in vitro* radiation induced mutagenesis-selection system for salinity tolerance in sugarcane. *Sugar Tech.*, *11*(3), 246–251.

Pathak, A. D., Srivastava, S., Gupta, P. S., Saxena, V. K., & Misra, S. K. (2005). *Ganne ke antarjateeya sankarit paudhon ka sharkara hetu aakalan*. Hindi Seminar Shodh Patra Sankalan. (Yadav R. L. et al. eds.), Sept. 29–30, IISR, Lucknow, pp. 31–32.

Pinto, L. R., Garcia, A. A. F., Pastina, M. M. et al. (2010). Analysis of genomic and functional RFLP derived markers associated with sucrose content, fiber and yield QTLs in a sugarcane (*Saccharum* spp.) commercial cross. *Euphytica*, *172*(3), 313–327.

Piperidis, G., & D'Hont, A. (2001). Chromosomal composition analysis of various *Saccharum* interspecific hybrids by genomic *in situ* hybridization. *Proceedings of International Society of Sugarcane Technologists*, *24*, 565–566.

Piperidis, G., Piperidis, N., & D'Hont, A. (2010). Molecular cytogenetic investigation of chromosome composition and transmission in sugarcane. *Mol. Genet. Genomics*, *284*(1), 65–73.

Prathima, P. T., Raveendran, M., Kumar, K. K., Rahul, P. R., Ganesh Kumar, V., Viswanathan, R., Ramesh Sundar, A., Malathi, P., Sudhakar, D. & Balasubramaniam, P. (2013). Differential regulation of defense related gene expression in response to red rot pathogen *Colletotrichum falcatum* infection in sugarcane. *Appl. Biochem. Biotechnol.*, *171*(2), 488–503.

Que, Y. X., Lin, J. W., Song, Su, L. P. X. X., & Chen, R. K. (2011). Differential gene expression in sugarcane in response to challenge by fungal pathogen *Ustilago scitaminea* revealed by cDNA-AFLP. *J. Biomed. Biotechnol.*, *160*, 934.

Raboin, L. M., Hoarau, J. Y., Costet, L., Telismart, H., Glaszmann, J. C., & D'Hont, A. (2003). Progress in genetic mapping of sugarcane smut resistance. *Proc. S. Afr. Sug. Technol. Ass.*, *77*, 134–141.

Raboin, L. M., Pauquet, J., Butterfield, M., D'Hont, A., & Glaszmann, J. C. (2008). Analysis of genome-wide linkage disequilibrium in the highly polyploid sugarcane. *Theor. Appl. Genet.*, *116*, 701–714.

Rahul, P. R., Kumar, V. G., Viswanathan, R., Ramesh Sundar, A., Malathi, P., Prasanth, C. N., & Pratima, P. T. (2014). Defense Transcriptome Analysis of Sugarcane and Colletotrichum falcatum Interaction Using Host Suspension Cells and Pathogen Elicitor. *Sugar Tech.*, *18*(1).

Ramos Leal, M. A., Maribona, R. H., Ruiz, A., Korneva, S., Canales, E., Dinkova, T. D., Izquierdo, F., Coto, O., & Rizo, D. (1996). Somaclonal variation as a source of resistance to eyespot disease of sugarcane. *Plant Breeding*, *115*, 37–42.

Roach, B. T. (1969). Cytological studies in *Saccharum*. Chromosome transmission in interspecific and intergeneric crosses. *Proc. ISSCT*, *13*, 901–920.

Roach, B. T. (1989). Origin and improvement of the genetic base of sugarcane. *Proc. Aust. Soc. Sugar Cane Technol.*, *11*, 34–47.

De Rosa Jr, V.E., Nogueira, F.T.S., Menossi, M., Ulian, E. and Arruda, P. (2005). Identification of methyl jasmonate responsive genes in sugarcane using cDNA arrays. *Braz. J. of Plant Physiol.*, *17*(1), 173–180.

Selvi, A., Nair, N. V., Mohapatra, T., Kartikprabhu, T., & Sunderavelpandian, K. (2006). Identification of molecular markers for red rot resistance in a complex polyploid sugarcane. In: *Proceedings of Second National Plant Breeding Congress*. March 1–3, TNAU, Coimbatore, 269–270.

Singh, R. K., Jena, S. N., Khan, M. S., Yadav, S., Banarjee, N., Raghuvanshi, S., Bhardwaj, V., Dattamajuder, S. K., Kapur, R., Solomon, S., Swapna, M., Srivastava, Sangeeta, & Tyagi, A. K. (2013). Development, cross-species/genera transferability of novel EST-SSR markers and their utility in revealing population structure and genetic diversity in sugarcane, Gene (doi: pii: S0378–1119 (13)00420–4.10.1016/j.gene.2013.03.125).

Sood, N., Gupta, P. K., Srivastava, R. K., & Gosal, S. S. (2006). Comparative studies on field performance of micro propagated and conventionally propagated sugarcane plants. *Plant Tissue Cult. Biotech.*, *16*, 25–29.

Souza, G. M., Simoes, A. Q., Oliveira, K., Humberto M.G., Fiorini, L., Felipe, dos Santos, G., Nishiyama, Jr, Milton, Y. & da Silva, A. (2001). The sugarcane signal transduction (SUCAST) catalogue: prospecting signal transduction in sugarcane, *Genet. Mol. Bio.*, *24*(1–04), 25–34.

Sreenivasan, T. V., Ahloowalia, B. S., & Heinz, D. J. (1987). Cytogenetics. In: *Sugarcane Improvement through Breeding*. Heinz, D. J. (ed.). Elsevier, Amsterdam, Netherlands. Pp. 211–253.

Srivastava S. (2000). Cytogenetics of Sugarcane. In: *50 Years of Sugarcane Research in India*. Shahi H. N., Shrivastava, A. K., & Sinha, O. K. (eds.) IISR, Lucknow. Pp. 55–71.

Srivastava S., Gupta, P. S., & Srivastava, B. L. (2005). Molecular genetic diversity of sugarcane genotypes of subtropical India: SSCP-PCR analysis of simple sequence repeats. *Proc. ISSCT, 25*, 602–606.

Srivastava, S., Pathak, A. D., & Gupta, P. S. (2011). Molecular characterization based on RAPD markers of high and low sugar intergeneric hybrids of *Saccharum* and *Erianthus*. *Tropical Agr. Trinid, 88*(4), 186–192.

Srivastava, B. L., Bhat, S. R., Pandey, S., Tripathi, B. S., & Saxena, V. K. (1986). Plantation breeding for red rot resistance in sugarcane. *Sugarcane, 05*, 13–15.

Srivastava, M. K., Li, C. N. & Li, Y. R. (2012). Development of sequence characterized amplified region (SCAR) marker for identifying drought tolerant sugarcane genotypes. *Amer. J. Crop. Sci., 06*(4), 763–767.

Srivastava, S., & Gupta, P. S. (2004). Fluorescent *in-situ* detection of r-DNA sites on somatic chromosomes of different species in an inter-specific hybrid of sugarcane. *Indian J. Sugarcane Technol.*, 19(1&2), 55–57.

Srivastava, S., & Gupta, P. S. (2008). Inter simple sequence repeat profile as a genetic marker system in sugarcane. *Sugar Tech, 10*, 48–52.

Srivastava, S., & Sunkar, R. (2013). Emerging role of microRNA in drought stress tolerance in the biofuel, bioenergy crop sugarcane. *J. Biotechnol. Biomater, 3*(3), 56.

Srivastava, S., Pathak, A. D., Yadav, S., Kumar, P., & Kumar, R. (2016). Identification of resistance gene candidates in sugarcane by PCR with degenerate oligonucleotide primers. Souvenir International Congress on Post-harvest technologies of Agricultural produce for Sustainable Food and Nutritional Security, Integral Institute of Agricultural Science and Technology (IIAST), Lucknow, November 10–12.

Suman, A., Kimbeng, C. A., Edme, S. J. & Veremis, J. (2008). Sequence related amplified polymorphism (SRAP) markers for assessing genetic relationship and diversity in sugarcane germplasm collection. *Plant Genet. Res.*, 6(3), 222–231.

Suman, A., S., Pan, Y., Thongthawee, S., Burner, D. M., & Kimbeng, C. A. (2011). Genetic analysis of the sugarcane (*Saccharum* spp.) cultivar 'LCP 85–384.' I. Linkage mapping using AFLP, SSR, and TRAP markers. *Theor. Appl. Genet.*, 123(1), 77–93.

Sun, J. X., Sun, X. F., Zhao, H., & Sun, R. C. (2004). Isolation and characterization of cellulose from sugarcane bagasse. *Polym. Degrad. Stab.*, 84(2), 331–339.

Suprasanna, P., Jain, S. M., Ochatt, S. J., Kulkarni, V. M., & Predieri, S. (2010). Applications of *in vitro* techniques in mutation breeding of vegetatively propagated crops. *In: Plant mutation*. (ed.) Q. Shu, Chap., 28, 369–383.

Swapna, M., & Srivastava, S. (2012). Molecular marker applications for improving sugar content in sugarcane. *Springer Briefs in Plant Science* – eBook. Springer New York Heidelberg Dordrecht London, ISBN 978-1-4614-2256-3, p.49.

Tai, P., & Miller, J. (2001). A core collection for *Saccharum spontaneum* L. from the world collection of sugarcane. *Crop Sci.*, 41, 879–885.

Taylor, P. W. J., & Dukic, S. (1993). Development of an *in vitro* culture technique for conservation of *Saccharum* spp. hybrid germplasm. *Plant Cell Tiss*. Organ Cult., 34, 217–222.

Todd, J., Wang, J., Glaz, B., Sood, S., & Ayala-Silva, T. (2014). Phenotypic characterization of the Miami World Collection of sugarcane (*Saccharum* spp.) and related grasses for selecting a representative core. *Genet. Resour. Crop. Evol.*, *61*, 1581–1596.

Vettore A. L. da Silva F. R. Kemper E. L. et al. (2003). Analysis and functional annotation of an expressed sequence tag collection for tropical crop sugarcane. *Genome Res.*, *13*(12), 2725–2735.

Vettore, A. L., Da Silva, F. R., Kemper, E. L., & Arruda, P. (2001). The libraries that made Sucest. *Genet. Mol. Biol.*, *24*, 01–07.

Virupakshi, S., & Naik, G. S. (2008). ISSR analysis of chloroplast and mitochondrial genome can indicate the diversity in sugarcane genotypes for red rot resistance. *Sugar Tech.*, *10*, 65–70.

Wagih, M. E., Ala, A., & Musa, Y. (2004). Regeneration and evaluation of sugarcane soma clonal variants for drought tolerance. *Sugar Tech.*, *06*(1–02), 35–40.

Weng, L. X., Deng, H. H., Xu, J. L., Li, Q., Zhang, Y. Q., Jiang, Z. D., Li, Q. W., Chen, J. W., & Zhang, L. H. (2010). Transgenic sugarcane plants expressing high levels of modified cry1Ac provide effective control against stem Borers in field trials. *Transgenic Res.*, *20*(4), 01–14.

Williams, C. A., Harborne, J. B., & Smith, P. (1974). The taxonomic significance of leaf flavonoids in *Saccharum* and related genera. *Phytochemistry*, *13*, 1141–1149.

Wu, J. M., Li, Y. R., Yang, Li-Tao, Fang, F. X., Song, H.-zhong, Tang, H.-Qin, Wang, M., Weng, M-Ling (2013). cDNA-SCoT: A novel rapid method for analysis of gene differential expression in sugarcane and other plants. *Amer. J. Crop. Sci.*, *07*(5), 659–664.

Zhang, S. Z., Yang, B. P., Feng, C. L., Chen, R. K., Luo, J. P., Cai, W. W., & Liu, F. H. (2006). Expression of the *Grifola frondosa* trehalose synthase gene and improvement of drought-tolerance in sugarcane (*Saccharum officinarum* L.). *J. Integr. Plant Biol.*, *48*(4), 453–459.

CHAPTER 5

UTILIZATION OF SUGARCANE GENETIC RESOURCES FOR ENHANCED RESILIENCE IN DIVERSE CONDITIONS

GULZAR S. SANGHERA,[1] ARVIND KUMAR,[2] A. ANNA DURAI,[3] and K. S. THIND[1]

[1]*PAU, Regional Research Station, Kapurthala, Punjab, 144601, India, E-mail: sangheragulzar@gmail.com*

[2]*UPCSR, Sugarcane Research Institute, Shahanjanpur – 242001, UP, India*

[3]*ICAR- Sugarcane Breeding Institute, Coimbatore – 641007, Tamilnadu, India*

CONTENTS

ABSTRACT

Sugarcane [*Saccharum* spp. complex] improvement programmes worldwide are closely linked with the exploration, collection, and utilization of sugarcane genetic resources. Its genetic resources comprise six species and the related genera comprising of *Erianthus, Miscanthus, Narenga,* and *Sclerostachya* form the most valuable and essential basic raw materials to meet the current and future needs of its improvement programmes. Apart from the basic germplasm several historical and commercial hybrids developed through inter-specific and inter-generic hybridization involving cultivated and wild species of *Saccharum* over the years at different cane breeding stations has broadened the genetic base of this crop and further enhanced the basis of its varietal improvement. In consequent to the sustained efforts by several national and international agencies, a large collection of sugarcane germplasm is available today representing the native variability available in the *Saccharum* complex. During the last three decades, a growing awareness has been witnessed to collect and conserve these fast depleting, irreplaceable resources for the good of the present and future generations. At the same time, it has been accepted that the success of the entire genetic resources activities is dependent upon the descriptive information of the conserved material, which enables plant breeders to make decisions regarding the material to be used in breeding programmes. Therefore, an attempt has been made to compile the information available regarding the characterization and utilization of sugarcane genetic resources for development of improved varieties for various traits like yield, quality, biotic, and abiotic stresses in this chapter that may help sugarcane breeders to exploit these resources efficiently in future sugarcane improvement.

5.1 INTRODUCTION

Man's interest in agriculture started about 10,000 years ago and, during this long period, transition from 'gathering' to 'growing' of plants occurred. In this process, a wide array of crop variability got generated by natural means and through both conscious and unconscious selection (Frankel, 1984; Gepts, 2006). Gradually, a new wealth of variability also got generated/ adapted and diversified by crop introductions in the exotic environment or through migration of human population. Associated with this process was the keenness of human mind to explore the rich global diversity of plant wealth in general so as to judiciously tap the potential of useful flora (Gepts, 2006; Guo et al., 2000). Crop genetic resources are the basic materials that are essential for development of improved crop varieties designed to combine high yield potential with superior quality, resistance to diseases and pests, and also better adaptation to abiotic stress environments (Gilbert et al., 1999; Hammer, 2003). Their continued availability to plant breeders is necessary not only for sustaining advances in crop productivity but also for stabilizing production in the country. These resources of known or potential use to man constitute a broad spectrum of diverse gene pools representing assemblage of landraces, primitive cultivars, varieties of traditional agriculture as well as wild and weedy relatives of crop plants (Chaudhuri, 2005; Moose, 2008). In the last two decades or so, much attention has been drawn to indigenous locally adapted cultivars in particular because of the useful genetic variation they contain as an invaluable resource for present and future plant breeding, and the rapid rate at which they are disappearing through replacement by high yielding varieties (Hammer, 2003). In addition, the natural habitats of wild relatives of crop plants are continuously getting eroded threatening survival of these populations. Indian national programme on genetic conservation aims at exploring and collecting, classifying, evaluating, conserving and documenting this natural heritage for its current and future use. All these operations constitute a chain of activities that are now better understood and carried out by the national and international centers mandated with such responsibilities (Chaudhuri, 2005). The last 30 years have seen the great upsurge of this activity, with more awareness generated by the FAO, IBPGR, the IARCs and also by the IUCN, UNESCO and the WWF in their concern for conservation of biodiversity with particular reference to *in-situ* aspects (FAO, 1996; Maxted et al., 2009). Equally important in this context has been the phenomenal growth in biotechnology during the past two decades, which has also created new awareness about the value of plant genetic resources since sexual process of fertilization and recombination was no longer a pre-requisite to shuffling of desirable traits (Hodgkin et al., 2001; Moose, 2008; Rao, 2004). A broad outline of sugarcane genetic resources available at SBI, Coimbatore and other research institutes in India, their characterization and utilization for development of improved varieties for various traits like yield, quality, biotic, and abiotic stresses are presented in the first chapter that may help sugarcane breeders to exploit these resources efficiently in their future sugarcane improvement programmes.

5.2   THE GENUS SACCHARUM AND ITS USEFULNESS

The *Saccharum* genus was believed to consist of six major species, including two wild species *S. spontaneum and S. robustum* and four cultivated species, *S. officinarum, S. barberi, S. sinense* and *S. edule* (D'Hont et al., 1998; Irvine, 1999). However, there were controversial reports by Irvine 1999 mentioning the existence of only two *Saccharum* species: viz. *S. officinarum* and *S. Spontaneum*. The *Saccharum* genus together with related genera, such as Erianthus, Miscanthus, *Narenga*, and *Sclerostachya* were referred to as the "*Saccharum* Complex" (Mukherjee, 1999). Sugarcane [*Saccharum* spp.] is a perennial grass, belonging to the Poaceae family and Andropogoneae tribe, which is grown widely in tropical and subtropical regions. It is the highest yielding crop worldwide (Henry et al., 2010) and accounts for approximately 75% of the world sugar production (Bull et al., 1963; Dillon et al., 2007). The origin of modern sugarcane cultivars is from inter-specific hybridizations of domesticated species *S. officinarum* [$2n = 80$, $x = 10$], which is characterized by high sugar and low fiber content (Daniels et al., 1987) and the wild species *S. spontaneum* [$2n = 40–128$, $x = 8$], which is resistant to biotic and abiotic stresses (Panje et al., 1960; Silva et al., 1993). Modern sugarcane genotypes are highly polyploid and aneuploid with multiple alleles at each locus. The genome composition of sugarcane cultivars has been estimated as 85% from *S. officinarum* and 15% from *S. spontaneum* (D'hont et al., 1995). The genome complexity in *Saccahrum* spp. has made sugarcane and energy cane breeding cumbersome. The genotypes utilized over decades in earlier breeding programs are a limited number of *S. spontaneum* and *S. officinarum* clones, which has resulted in a narrow genetic base of sugarcane cultivars (Lima et al., 2002).

The classification of the *Saccharum* spp. based on morphology, chromosome numbers and geographic distribution has been a matter of debate for a long time. Two duplicated '*Saccharum* complex' germplasm collections known collectively as the "World Collection of Sugarcane and Related Grasses" [WCSRG] were generally utilized of them, one WCSRG is maintained in Coimbatore, India and the other in Miami, FL, USA. The WCSRG may contain significant genetic diversity and many valuable alleles for numerous morphological traits, biomass yield components, adaptations to biotic and abiotic stresses (Table 5.1), and many other quality traits (Berding et al., 1987). Earlier studies on genetic

TABLE 5.1 Genetic Diversity for Some Morphological, Biomass Yield Components and Adaptations to Biotic and Abiotic Stresses

Trait(s)	Clones/Accessions/Land race
Sucrose content	M 336, PR 1000, CP33-224, Co 281, PR1140, 00-0402, 00-1805, 01-0031, 01-0047, 02-0288, Chin, Chunnee, Raksi, Burra Chunnee, Baraukha
Cold resistance	CP 1165
Salt tolerance	Co 453
Drought tolerance	PR 980, Co 312, Katha, Kalkya, Khadya, Bansi, Sunnabile
Lodging resistance	Q27
Erectness	CP38-34, CP66-346, CP52-68
Smut resistance	Co 419, Co 453, Co 603 (pistil parent)
Red rot resistance	Co 475, Co 980, Co 1227, Bo91
Leaf scald, Fiji disease and Mosaic resistance	Co 290, Co 475, US 1694
Ratooning ability	CP44101, CP701133, CP 721210, CP 73351, CP 74383, F 160, G6888, G 75393, Co 1148, N 12, NCo 310, NCo 376, PR 980, Q 113, SP701143, Pindar
Tillering ability	CP 65357, CP 73351, H 700/44, N12, NCo 376, Phil 56226, SP 701143, Tuc 6818, Tuc 6819
Wide adaptation	PoJ 2878; NCo 310

diversity analysis in selected clones in this collection have provided limited information (Brown et al., 2007; Tai et al., 2002). In addition, limited numbers of clones in the WCSRG have been used for sugarcane and energy cane improvement. This large genetically diverse collection with vast potential remains unutilized. With its large number and genetically complex accessions, it is a formidable task to fully characterize and use the WCSRG in breeding programs. A core collection that is a condensed assembly of the entire collection with maximized genetic diversity and minimized redundancy is essential for its utilization (Brown, 1989). Hence, it is important to characterize the genetic variation among the domestic cultivars and the available genetic resources in order to exploit them and accelerate sugarcane and energy cane improvement. A germplasm collection with high genetic diversity would enable breeders to broaden the genetic base of parental lines and thereby facilitate genetic gains of sugarcane and energy cane cultivars (Cooper et al., 2001; Ming et al., 2006).

5.3   COLLECTION AND CONSERVATION OF SUGARCANE GERMPLASM: AN OVERVIEW

In India, sugarcane germplasm collections were initiated by C.A. Barber in 1912. He surveyed the northern part of India from Punjab to Bihar, and collected both native cultivated and introduced clones. By 1915, the total collection assembled at Coimbatore reached 292 clones, of which 180 were introduced tropical clones with thick stalks, and 112 were indigenous subtropical clones with thin stalks. Barber (1916a) studied the north Indian canes and classified them into five groups, which were subsequently placed under two species: *S. barberi* [for Indian canes] and *S. sinense* [for canes of Chinese origin]. Systematic approach to collect *Saccharum* genetic resources from India begun after the establishment of SBI, Coimbatore (1912), Barber collected 112 North Indian sugarcane clones from subtropical regions of the country during 1912 for comparative evaluation and selective introduction. The SES [*Spontaneum* Expedition scheme] was launched for collection of *Saccharum* Germplasm from India and other Asian countries and over 800 clones were collected. Collection efforts were further strengthened during 1999 under NATP on Plant Biodiversity and a total of 535 clones were collected across the country up to 2004. A large number of accession representing the wide range of variability for Saccharum and related genera present in the country. *S. spontaneum* was found to be the most variable form in terms of morphology and cytotypes. Short and medium tall forms are common in most parts of the country, while tall, thick, juicy type of *S. spontaneum* was found in North East (Barber, 1916b). Consequent to the sustained efforts by several national and international agencies, a large collection of sugarcane germplasm is available represent the native variability in the *Sacccharum* complex. These collections have been conserved in the two world collections in USA and India. The USDA-World collections of sugarcane Germplasm maintain 1100 accessions at Miami, Florida. The world collection in India is maintained at SBI, Regional center, Kannur specifically for the disease and pest free maintenance of the sugarcane Germplasm. A part of this collection, particularly the wild species such as *S. spontaneum*, *Erianthus* spp. and other related genera are being maintained at SBI, Coimbatore (Table 5.2). Nearly 4000 accessions of *Saccharum*, related genera, manmade historical, and commercial hybrids are currently being maintained in India.

TABLE 5.2   Status of the Indian Collection of Sugarcane Germplasm

S. No.	Germplasm	At Kannur		At Coimbatore
		International collections	Indian collections	
1.	S. officinarum	764	-	8
2.	S. berberi	43	-	-
3.	S. sinense	29	-	-
4.	S. robustum	145	-	-
5.	S. spontaneum	67	398	977
6.	Foreign Hybrids	585	-	-
7.	Related Genera	150	82	322
8.	Indian Hybrids	-	1158	-
9	Indo American clones	-	130	-
	Total	**1783**	**1549**	**1299**

5.3.1   SUGARCANE WILD SPECIES/GENERA FOR ECONOMIC TRAITS

A major breakthrough in sugarcane improvement was achieved through the use of the wild species viz., *S. spontaneum* in breeding. The modern sugarcane varieties are complex interspecific hybrids involving two or more species of *Saccharum* and the high productivity and adoptability of the hybrids have been attributed to the *S. spontaneum* of the genome. The wild Germplasm available had been characterized over the years for the various attributes and potential sources for important traits (Anonymous, 1987; Roach and Daniels, 1987) (Table 5.3). *Saccharum spontaneum* and *E. arundinaceus* have been identified as the most potent wild sources for the varietal improvement of sugarcane. *S. spontaneum* is considered as a source for high productivity, adaptability and tolerance to pest and diseases. It is also endowed with the natural ability to withstand severe adverse conditions including cold, salinity, drought and waterlogging. *S. robustum*, the immediate progenitor of *S. officinarum* as a source of fiber, yield and waterlogging resistance. *Erianthus* spp. is characterized by high biomass production, multiple pest resistant and tolerance to drought waterlogging and salinity. Besides *E. arundinaceus* is also an important source for fiber, now being used as a substitute for wood pulp in the paper industry.

TABLE 5.3 *Saccharum* Species and Related Genera Which May Impart Tolerance/ Resistance to Abiotic Stresses and Nutrient Use Efficiency in Sugarcane

Characteristics	Genera/species	References
Tolerance/resistance to drought *Erianthus* spp.	*S. spontaneum, Narenga* spp.	Krishnamurthi, 1989; Roach and Daniels, 1987; Sreenivasan, T. V. and Sreenivasan, 2000; Sreenivasan et al., 2001
Tolerance/resistance to waterlogging	*S. robustum* and *S. spontaneum*	Krishnamurthi, 1989
Tolerance/resistance to cold performance at high altitudes	*Miscanthus* spp., *Miscanthus nepalensis, Erianthus fulvus* *S. spontaneum, S. barberi*	Anonymous, 1987; Brandes et al., 1939 Earle, 1928
Tolerance/resistance to salinity	*Erianthus* spp., *S. barberi, S. sinense, S. robustum*	Ramana Rao et al., 1985; Sreenivasan et al., 2001
High nutrient use efficiency	*S. spontaneum* [IK 76-20, SES 24, IS 760164], *S. robustum* [51 NG 27], *S. sinense* [*Khadya*], *S. officinarum* [UB-16]	Shrivastava et al., 2001
Low nutrient requirement	*S. spontaneum, Erianthus* spp.	Krishnamurthi et al., 1989
Robust growth under low input conditions	*Erianthus* spp.	Sreenivasan et al., 2001

5.3.2 SUGARCANE INDIGENOUS LANDRACES FOR ECONOMIC TRAITS

Indigenous canes growing in India had desirable features like tolerance to drought, waterlogging, wider adaptability, ratooning, early ripening, and high yield (Table 5.4). Among these, Khagri grew under 6 feet water for over three months. Salt-tolerant clones have also been identified in *S. barberi:* Katha [Coimbatore], Kewali-14-G, Khatuia-124, Kuswar, Lalri, Nargori and Pathari (Ramana Rao et al., 1985), in *S. sinense*: Khakai, Panshahi, Reha, Uba –Seedling (Ramana Rao et al., 1985), and in *S. robustum:* IJ-76-422, IJ-76-470, 28 Ng-251, 57-Ng-201, 57-Ng-231, Ng-77-34, Ng-77-55, Ng-77-136, Ng-77-34, Ng-77-55, Ng-77-160, Ng-77-167, Ng-77-170, Ng-77-221 and Ng-77-237 (Ramana Rao et al., 1985). Use of *S. spontaneum* imparts

resistance, and stress tolerance (Nair, 2011). However, several countries experience sugar yield plateaus. The static sugar yields may be overcome by intensive breeding for new cultivars. Recently developed genomic resources and acquired molecular tools in sugarcane have the potential to further improve sugar yields, but these must be used as tools in support of traditional crop improvement (Sanghera et al., 2016). Evaluation of interspecific hybrid clones involving different *Saccharum* species and intergeneric hybrids involving *Erianthus* clones indicated differential maturity/sugar accumulation pattern. This gives an indication of different juice quality genes in *Saccharum barberi/ sinense, S. spontaneum, S. officinarum/robustum and Erianthus*. Pyramiding sucrose genes from these sources seems to have potential to improve the quality traits of sugarcane varieties (Jackson, 2005; Krishnamurthi, 1989). Subtropical India is characterized by low sugarcane productivity due to weather extremes and other stressful growing conditions, and thus, varieties with economic attributes and adaptation to stressful conditions are need of the hour. Sugarcane varieties with abiotic tolerance and red rot resistance are to be developed through genetic resources, pre-breeding, varietal development and biotechnological interventions (Shrivastava and Srivastava, 2012).

Concerted breeding efforts made for sugarcane improvement since the discovery of fertility in the seeds, were limited to crossing only among the *Saccharum officinarum* clones during the first phase. The resultant hybrids though rich in sucrose content, lacked vigor, ratooning ability and resistance to pests and diseases. After realizing the potential of indigenous canes to adapt to diverse environments, resistance to insect pests and diseases, tolerance to abiotic stress and ratooning ability, the second phase of sugarcane breeding involved interspecific hybridization between *S. officinarum* clones and other *Saccharum* species (Nair, 2011; Sreenivasan, 2004). Introduction of POJ varieties in the breeding programme resulted in many good sugarcane varieties such as Co 213, Co 244, Co 312 and Co 313, which were successful from 1920 to 1940 and helped in establishing the sugar industry of North India in the 1940s (Srivastava and Srivastava, 2000). However, only four clones, viz. Chunnee, Katha, Saretha and Kansar figured in the parentage of most of the commercial varieties (Sreenivasan, 2004). Globally, sugarcane breeders develop better-adapted and high yielding varieties to meet the requirements of the sugar industry. Likewise, the sugarcane research institutes in India initiated breeding programs, which were expected to deliver locally bred varieties adapted to the diverse sugarcane growing environments. This initiative established points on the need to breed locally adapted varieties to avoid over-reliance on imported varieties (Surat, 2009). The genetic variability within the local sugarcane germplasm widely grown under different production systems has not been explored. Consequently, it is vital to broaden the germplasm base for breeding locally adapted varieties.

5.4.1 GENETIC STOCKS FOR HIGH SUGAR AND ECONOMIC TRAITS

Breeding population in sugarcane tends to show a steady improvement for cane yield and there have been very little increase in juice sucrose content for several decades. An improvement in sugar content of sugarcane increases sugar yields with only a marginal increase in costs of production. This makes gains in sugar content economically more beneficial than corresponding increases in cane yield, means that increased sugar content is an important objective of sugarcane breeding programs (Jackson, 2005). However, comparisons of cultivars released in different years indicate that sugarcane breeding programs have delivered increased sugar yields via improvements in cane yield, with much smaller contributions from sugar content (Legendre, 1995). Studies in Lousiana breeding population to improve sugar yields via indirect selection in the first clonal testing stage of sugarcane [*Saccharum* spp.] improvement programs indicated selection emphasis based on high juice brix and low pith increased sucrose content. Maintaining high levels of sucrose content in high cane-yielding selections should best increase sucrose yield (Gravois et al., 1991). Successful efforts to improve sucrose content through the adoption of different selection strategies, coupled with the choice of appropriate parents, have been reported in many countries (Legendre, 1995). During last decade, efforts have been made through development of pre-breeding stocks with improved sucrose content in sugarcane (Shanthi, and Alarmelu, 2010) who recorded average juice sucrose of 22.0% with 13.40% improvement over base population of different cross combinations and 8% over zonal check CoC 671. They identified some potential pre-breeding stocks with early high sugar content, which could be exploited in future. A list of high sugar varieties/landraces/genetic stocks having desirable economic attributes is presented in Table 5.5. With the advent of biotechnological and genetic engineering tools, genes responsible for high sugar has been mapped recently that could be exploited for the improvement of this important trait in sugarcane (Sanghera et al., 2016; Shrivastava, and Srivastava, 2012).

TABLE 5.5 List of High Sugarcane Varieties/Landraces/Genetic Stocks Having High Sugar and Other Desirable Attributes

Trait [s]	Elite Clones/Varieties/Accessions	Land race/Wild relative spp/genra
A. Agro-morphological traits of interests		
1. Cane diameter	B 52107, B 67150, BJ7015, CP 47155, CP 57603, CP 70330, CR 74250, D 158/41, D 38/57, H624671, H 736110, M377/56, Phil 6607, Phil 7495, Q 80, R 570, ROC 1, ISH clones, Co 0238, CoS 88230	Kaba, Mali, Vatu, Yasawa
2. Erectness	B 52298, B 59162, B 67150, B73385, BJ 6808, BJ 7015, CP 57603, CP 701133, GT 549, H 736110, M377/56, Phil 7495, PR 980, R 570, Tuc 681, UP 49, CoSe 01434, CoSe 08279	Kaba, Mali
3. Cane height	B 52298, B 59162, B 67150, B 73385, BJ 6808, BJ 7015, CP 57603, CP 701133, H 736110, M 377/56, Phil 6723, Phil 7495, PR 980, R 570, Tuc 681	Kaba, Mali
4. Tillering ability	CP 65357, CP 73351, H 700/44, N12, NCo 376, Phil 56226, SP 701143, Tuc 6818, Tuc 6819	
5. Ratooning ability	CP44101, CP701133, CP 721210, CP 73351, CP 74383, F 160, G6888, G 75393, Co 1148, N 12, NCo 310, NCo 376, PR 980, Q 113, SP701143	Pindar
B Quality traits		
1. High sucrose content	CoS 510, CoS 96268, CoS 95255, CoC 671, CoC 671, Co885, Co 7201, Co 7220, Co 8015, Co 8214, Co 8301, Co 8316, Co 8334, Co 7224, Co 775, Co 62198, Co 0118, CoJ 64, 00-0402, 00-1805, 01-0031, 01-0047, 02-0288	

5.4.2 SUGARCANE VARIETIES/GENETIC STOCKS FOR ABIOTIC STRESSES

An environmental factor that limits crop productivity or destroys biomass is referred to as stress or disturbance (Grime, 1979). Abiotic stress is the primary cause of crop loss worldwide, reducing average yields for most major crop plants by more than 50%. Low temperature, drought, and high salinity are common stress conditions that adversely affect plant growth and crop production (Xiong et al., 2002). Among the abiotic factors that have shaped and continue shaping plant evolution, water availability is the most important, while light is the best-studied environmental factor in plant research with respect to molecular details. Plant growth, productivity, and distribution are greatly affected by environmental stresses such as high and low temperature, drought, and high salinity. In response to abiotic stresses, plants undergo a variety of changes at the molecular level (gene expression) leading to physiological adaptation (Mantri, 2012; Zheng, 2010). Salinity and drought are the major abiotic stresses, which severely affect yield and quality in many regions of the world endangering the food security. The situation has become more serious with concerns of global climate change. Therefore, studies on abiotic stress tolerance have become one of the main areas of research worldwide. In this direction, efforts are being made to breed tolerant varieties using conventional breeding and contemporary biotechnological tools. Recent advances in this area include unraveling the physiological, biochemical, and molecular mechanism of abiotic stress tolerance and corresponding development of tolerant cultivars through transgenic technology or molecular breeding (Ashraf, 2010; Patade et al., 2011). Desired tolerant genotypes/varieties so developed need to be screened at laboratory as well as field levels for functional validation. Numerous physio-biochemical indicators for tolerance screening have been suggested. In addition, indirect selection using molecular markers linked to desired loci is being deployed for accelerating the production of stress-tolerant varieties (Ashraf et al., 2013; Ribaut et al., 2007). Research into the physiology and metabolism of so-called extremophiles has foster better understanding of the evolutionary processes that have created the diversity of life, as it exists on earth, and also has economic implications for agricultural biotechnology and the development of novel products. On the other hand, sugarcane production is expected to reduce by 30% in the future due to climate change, as revealed in a recent four-year study conducted by the World Bank (http://www.sriindia.net). The capacity to sequence genomes and the availability of novel molecular tools have now catapulted biological research into eras of genomics and post-genomics, creating an opportunity to apply genomic techniques to extremophile models (Amtmann et al., 2005; Sanghera et al., 2016), which is a dire need of time to feed the increasing global population through significant increase in agricultural production. Even though sugarcane can survive long dry periods, it demands a fair amount of water for optimal yield, leading to the use of irrigation in many areas. However, the lack of genetic and molecular information about drought tolerance mechanisms and inheritance in sugarcane has limited the development of improved cultivars (Abbas et al., 2013, 2014; Sanghera et al., 2016). There is a need to distinguish genes definitely associated with the response to water deficit, which hold an adaptive function to water deprivation and in stress environments. In this regard, interspecific crosses involving *S. officinarum, S. barberi, S. robustum* and *S. spontaneum* are carried out and the progenies are evaluated for identifying superior hybrids. These hybrids are intercrossed, backcrossed and crossed with the commercial varieties to develop genetic stocks and new varieties. Intergeneric crosses of *Saccharum* with other related genera like *Erianthus, Sclerostachya* and *Narenga* also had been attempted to improve productivity and adaptability of the new varieties (Nair, 2007). Crosses involving *Erianthus* had been found to be promising in terms of productivity and better adaptability. *Erianthus* spp. show better biomass potential, resistance to major diseases and pests and a high level of tolerance to abiotic stresses and these traits are being transferred to commercial varieties through intergeneric crosses (Abbas et al., 2013; Nair et al., 2006). The superior genetic stocks with broad and diverse genetic base developed through interspecific and intergeneric crosses are simultaneously inducted into the regular breeding programmes as parents (Nair et al., 2006). These prebreeding activities ensure that a broad and diverse genetic base having tolerance to water stress, salt and water logging (Table 5.6) is maintained in the breeding pool.

5.4.3 GENETIC STOCKS FOR DURABLE RESISTANCE TO BIOTIC STRESSES

Genetic resistance to pests and diseases is a *sine qua non,* [i.e., indispensable and essential condition] in plant breeding. Pests and pathogens often conquer new territories and are well known to dynamically evolve towards breaking resistances, always posing new challenges (Walker, 1987). Indeed, biotic stresses are of special concern in sugarcane breeding programs, because they may cause great economical impact in plantations with susceptible cultivars

TABLE 5.6 List of High Sugarcane Varieties/Landraces/Genetic Stocks Having Tolerance to Abiotic Stresses

Stress (s)	Elite clones/varieties	Related species/genra
1. Water Stress	B 41227, B 43337, B 59162, B 63118, B 64278, Co 213, Co 331, Co 421, Co 8208, Co 85007, Co775, D 41/46, F 140, F 177, M 134/32, M 13/56, NCo 310, PM 72, PR 980, PR 1048, Co 1148, ISH 007, ISH 135, ISH 148, ISH 261, ISH 273	Erianthus Spp, spontaneum, Dhaur-Alig
2. Salt	H 6982235, NCo 310, NCo 376, Co 331, Co 997, Co 1148, Co 6806, Co 86011, Co 7717, Co 7219, Co 8208	Mali
3. Water logging	B 52298, NCo 310, Co 740, Co 775, Co 951, Co 975, Co 6304, Co 8231, Co 8232, Co 8145, Co 8371, Co 9906, Bo 91, CoSe 96436, UP 9530, F 160, H 736110, Q 113, Tuc 6727, ISH 007, ISH 135, ISH 175, ISH 261	Waya, Dhaur-Alig, Pararia-Shaj
4. Cold/frost	Co 1148, CP 65357, CP 70321, L 6014, N 12	Pindar
5. Input efficiency	Co 331, Co 1148, NCo 310, NCo 376, H 681158	Mali

(Agnihotri, 1996; Jayashree et al., 2010). Examples of biotic stresses to which sugarcane breeders, geneticists, pathologists, entomologists have been paying attention, depending on the location of the breeding program are fungal diseases such as rusts [especially the brown rust [*Puccinia melanocephala*] and the orange rust [*Puccinia kuehnii*]], smut [*Ustilago scitaminea*], red rot [*Glomerella tucumanensis*], eye spot [*Helminthosporium sacchari*], pokkah boeng [*Fusarium moniloforme*], and "pineapple disease" [*Ceratocystis paradoxa*]. The main bacterial diseases are "ratoon stunting disease" [*Leifsonia xyli*] and "leaf scald" [*Xanthomonas albilineans*], and important viral diseases are *Sugarcane mosaic virus* (SCMV) and *Sugarcane yellow leaf virus* (SCYLV) (Prabhu et al., 2008; Rao et al., 2001). Identification of genetic resistance for these diseases is important to allow incorporation of resistance traits as goals in breeding programs to reduce production threats (current and potential), as well as reduce fungicide spraying (Palaniswami et al., 2011; Sundar et al., 2009).

In the same context, insects are also potential threats to sugarcane production, either directly or as disease vectors. Among the sugarcane pests, moth-borers are the most widely prevalent and at the same time the most difficult to control. About a dozen species are known to occur in, India. Of these top-borers and shoot-borers are, found in almost all the important cane-growing areas, of the country, whereas others such as stalk borers, Gurdaspur borers, etc. occur only in particular regions. The damage by the individual species ·of borers is usually more pronounced at particular stages of the crop growth, and in certain specific portions of the shoot or cane. On this basis, the borer species have been grouped into three broad categories, viz. (i) top-borers, (ii) shoot-borers, and (iii) cane-borers (David et al., 1986). Other sugarcane pests include root froghopper [*Mahanarva fimbriolata*, Hemiptera: Cercopidae], the sugarcane weevil [*Sphenophorus levis*, Coleoptera: Curculionidade], longhorn beetle [*Migdolus fryanus*, Coleoptera: Cerambycidae], sugarcane borer [*Diatreae saccharallis*, Lepdoptera: Pyralidae], and the stem borer [*Telchin licus*, Lepdoptera: Castniidae]. Whereas aphids are of little concern as pests *per se*, two species [*Melanaphis sacchari* and *Sipha flava,* Hemiptera: Aphididae] are SCYLV vectors. Sugarcane resistance against these insects is beneficial where the virus is a potential danger (Christy et al., 2009). Biological control of sugarcane pests by using natural enemies is a viable crop management technique in some cases, such as the fungus *Metarhizium anisopliae* that controls the root froghopper. However, the incorporation of genetic resistances against pests of economical or potential impact is indisputably the best option, when available (Arvinth et al., 2010; Chen et al., 1989). *Narenga* clones are resistant to almost all the diseases, pests and root parasites and tolerated drought, whereas *Miscanthus* clones are high-yielders, resistant to diseases and tolerant to cold. Their use in breeding could impart these characteristics to the progeny (Table 5.7). Such crosses have been attempted in India, Fiji, Hawaii and Taiwan (Krishnamurthi, 1989). Downy mildew [*Peronosclerospora sacchari*] resistance genes have been successfully transferred from *Miscanthus* to sugarcane (Chen et al., 1989).

Diverse plant genetic resources provide options to plant breeders to improve the quality, diversity and performance of crops for various qualitative and quantitative attributes, resistance to abiotic and biotic stresses, besides an efficient nutrient management through development of improved varieties with desired characteristics. In this context, use of certain *S. spontaneum* clones has led to perceptible improvement of sugarcane varieties with respect to desirable agronomic traits.

TABLE 5.7 List of Sugarcane Varieties/Landraces/Genetic Stocks Having Tolerance to Biotic Stresses

Stress (s)	Elite clones/varieties	Landraces/ Genetic Stocks
I. Diseases		
1. Red rot	Co 1336, Co 62198, Co 62399, Co 6304, CoS 659, CoS 698, Bo 70, ISH 111, ISH 193, ISH 268, ISH 268, ISH 400, ISH 425, ISH 438, BO 91	SES 3, SES 4, SES 183, SES 275, SES 594, Baragua
2. Smut	B 41227, B 52107, B 52298, B 66134, BJ 6808, Co 462, Co 6806, Co 1001, CP 5659, CP 63588, CP 70321, CP 70330, CP 701133, CP 721210, F 148, H 736110, M 43148, N 14, N 17, POJ 2878, Q 83, Q 96, Q 102, SP 701143	Ragnar
3. Rust	B 62163, B 64278, BJ 65282, Co 997, CP 5243, CP 5659, CP 701133, H 700144, NCo 310, POJ 2878, Q 110, Q 113	Manna
4. Leaf scald	B 62163, BJ 6808, BJ 7015, CP 721210, M 93/48, M 964/66, M 574/62, NCO 310, NCo 376, Q 110, Q 113, R 526	Pindar, Ragnar
5. Yellow spot	B 62163, B 63118, B 72177, B 73385, Co 740, CP 57603, F 160, Ja 605, M13/56, NCo 310, Q 90, Q 99	Kaba, Manna
6. Mosaic	CP 70321, CP 721210, CP 74383, CP 76331, F 160, N 52219, Phil 6723, POJ 2878, Co 775	
7. Ratoon Stunting Disease [RSD]	B 52298, BJ 6808, G 6888, H 606909, PR 1048, PR 1117,	Ragnar
II. Insect-pests		
1. Borer complex	B 49119, B 51131, B 59136, BO 17, Co 331, Co 975, Co 6806, CoJ 46, CB 44105, CP 70321, CR 74250, N 12, NCo 310, NCo 376, B 63118	

5.5 CONCLUSION AND FUTURE OUTLOOK

Traditionally, the main focus on sugarcane breeding had been on *sugar yield*. However, recently, a new sugarcane genotype concept is emerging, focusing on *biomass production* to enable better explore ethanol or energy production. Within this new concept, breeding programs must be reoriented to strengthen its efforts on the development of new cultivars that fit this new variety profile. For this, it is essential to quickly answer to question related to biometrics (stalk number, diameter, height) and processing (sucrose content, reducing sugars, fiber content). Surely, new germplasm resources must be explored by sugarcane breeding programs. The implementation of a parallel introgression program, aiming at broadening the genetic base of sugarcane cultivars for sugar content and/or biomass production, will definitively bring great contributions for increases on yield, ensuring a more sustainable cultivation of sugarcane. Gains on important traits, such as vigor (robustness), will contribute to biomass production and may be found within *S. spontaneum* accessions and related genera, such as *Miscanthus* and *Erianthus*. New resources and tools are constantly been made available for sugarcane such as better understanding of its genome, genetics, physiology, molecular biology, new markers associated with traits of agronomical relevance, and new analysis tools. This has motivated scientists to look into the diversity for desirable traits not only in *Saccharum* species, but also at the level of the Saccharum complex comprising *Erianthus, Sclerostachaya, Narenga*, etc. The new generation ISH clones and incorporation of *Erianthus* in sugarcane breeding programmes have shown promise. Of late, identification of candidate genes for tolerance towards various biotic and abiotic stresses has opened up more avenues to impart climate resilience in elite sugarcane genotypes. Breeding programs should take advantage of these tools and incorporate in their selection pipelines to generate superior new cultivars that respond to current and future needs of the industry and the hope of the general society.

KEYWORDS

- **genetic resources**
- **plant stresses**
- **pre-breeding**
- ***Saccharum spp.* plant stress**
- **transporters**

REFERENCES

Abbas S. R., Ahmad, S. D., Sabir, S. M., & Shah, A. H., (2014). Detection of drought tolerant sugarcane genotypes [*Saccharum officinarum*] using lipid peroxidation, antioxidant activity, glycine-betaine and proline contents. *Journal of Soil Science and Plant Nutrition, 14*(1), 233–243.

Abbas, S. R., Ahmad, S. D., Sabir, S. M., Wajid, A., Aiya, B., Abbas, M. R., & Sabir, H. S., (2013). Screening of drought tolerant genotypes of sugarcane through biochemical markers against polyethylene glycol. *International Journal of Scientific and Engineering Research, 4*, 980–988.

Agnihotri, V. P., (1996). Current sugarcane disease scenario and management strategies. *Indian Phytopathology, 49,* 10–126.

Al-Janabi, S. M., Honeycutt, R. J., McClelland, M., & Sobral, B., (1993). A genetic linkage map of *Saccharum spontaneum* L. 'SES 208.' *Genetics, 134,* 1249–1260.

Amtmann, A., Bohnert, H. J., & Bressan, R. A., (2005). Abiotic stress and plant genome evolution. Search for new models. *Plant Physiol., 138,* 127–130.

Anonymous, (1987). Maintenance of germplasm. *Annual Report of the Sugarcane Breeding Institute,* Coimbatore. Pp. 16–18

Arvinth, S., Arun, S., Selvakesavan, R. K., Srikanth, J., Mukunthan, N., Ananda Kumar, P., Premachandran, M. N., & Subramonian, N., (2010). Genetic transformation and pyramiding of aprotinin-expressing sugarcane with *cry1Ab* for shoot borer [*Chilo infuscatellus*] resistance. *Plant Cell Reports, 29*(4), 383–395.

Ashraf, M., & Foolad, M. R., (2013). Crop breeding for salt tolerance in the era of molecular markers and marker-assisted selection. *Plant Breeding. 132*(1), 10–20.

Ashraf, M., (2010). Inducing drought tolerance in plants: recent advances. *Biotechnol Adv., 28,*169–183.

Barber, C. A. (1916a). The classification of indigenous canes. *Agricultural Journal of India, 11,* 371–376.

Barber, C. A. (1916b). Studies in Indian sugarcanes. No. 2. Memoirs of the Department of Agriculture of India, *Botany Series, 8,* 103–199.

Berding, N., & Roach, B. T., (1987). Germplasm collection, maintenance, and use. In: *Sugarcane Improvement Through Breeding.* Heinz, D. J., ed. Amsterdam: Elsevier, pp. 143–210.

Brandes, E. W., Sartoris, G. B., & Grassl, C. O., (1939). Assembling and evaluating wild forms of sugarcane and related plants. In: *Proceedings of the International Society of Sugar Cane Technologists, 6,* 128–153.

Brown, A., (1989). Core collections: a practical approach to genetic resources management. *Genome, 31,* 818–824.

Brown, J. S., Schnell, R., Power, E., Douglas, S. L., & Kuhn, D. N., (2007). Analysis of clonal germplasm from five *Saccharum* species: *S. barberi, S. robustum, S. officinarum, S. sinense,* and *S. spontaneum*. A study of inter- and intra species relationships using microsatellite markers. *Genet Resour Crop, 54,* 627–648.

Bull, T., & Glasziou, K., (1963). The evolutionary significance of sugar accumulation in *Saccharum. Aust J Biol Sci., 16,* 737–742.

Chaudhuri, S. K., (2005). Genetic erosion of agrobiodiversity in India and intellectual property rights: interplay and some key issues. *Patentmatics, 5*(6), 1–10.

Chen, Y. H., & Lo, C. C., (1989). Disease resistance and sugar content in *Saccharum Miscanthus* hybrids. *Taiwan Sugar, 36*(3), 9–12.

Christy, L. A., Arvinth, S., Saravanankumar, M., Kanchana, M., Mukunthan, N., Srikanth, J., Thomas, G., & Subramonian, N., (2009). Engineering sugarcane cultivars with bovine pancreatic trypsin inhibitor [aprotinin] gene for production against top borer [*Scirpophaga excerptalis*walker]. *Plant Cell Reports, 28,* 175–184.

Cooper, H. D., Spillane, C., and Hodgkin, T., (2001). Broadening the genetic base of crops: An overview. In: *Broadening Genetic Base of Crop Production* (Cooper, HD, Spillane, C, and Hodgkin, T (Eds.) Wallingford: CABI Publishing, pp. 1–23.

D'Hont, A., Ison, D., Alix, K., Roux, C., & Glaszmann, J. C., (1998). Determination of basic chromosome numbers in the genus Saccharum by physical mapping of ribosomal RNA genes. *Genome, 41,* 221–225.

D'Hont, A., Rao, P., Feldmann, P., Grivet, L., Islam-Faridi, N., et al. (1995). Identification and characterisation of sugarcane intergeneric hybrids, *Saccharum officinarum* x *Erianthus arundinaceus*, with molecular markers and DNA in situ hybridization. *Theor Appl Genet., 91,* 320–326.

Daniels, J., & Roach, B. T., (1987). Taxonomy and evolution. In: *Sugarcane Improvement Through Breeding.* Heinz, D. J., ed. Amsterdam: Elsevier, pp. 7–84.

David, H., & Nandagopal, V., (1986). Pests of Sugarcane, Distributions, symptomology of attack and identification. In: *Sugarcane Entomology in India*, pp. 1–30.

Dillon, S. L., Shapter, F. M., Henry, R. J., Cordeiro, G., Izquierdo, L. et al., (2007). Domestication to crop improvement: genetic resources for Sorghum and *Saccharum* [Andropogoneae]. *Ann Bot.,* 100, 975–989.

Earle, F. S., (1928). *Sugarcane and Its Culture*, John Wiley & Sons, New York, pp. 355.

FAO (1996). Report on the state of world's plant genetic resources – International Technical Conference on Plant Genetic Resources, Leipzig, Germany, FAO, Rome.

Frankel, P. H., (1984). Genetic perspective of germplasm conservation. In: *Genetic Manipulations: Impact on Man and Society.* Arber, W., Llimensee, K., Peacock, W. J., & Starlinger, P. (eds.). Cambridge: Cambridge University Press, pp. 161–170.

Gepts, P., (2006). Plant genetic resources conservation and utilization: the accomplishments and future of a societal insurance policy. *Crop Science, 46,* 2278–2292.

Gilbert, J. E., Lewis, R. V., Wilkinson, M. J., & Caligari, P. D. S., (1999). Developing an appropriate strategy to assess genetic variability in plant germplasm collections. *Theoretical and Applied Genetics, 98,* 1125–1131.

Gravois, K. A., Milligan, S. B., & Martin, F. A., (1991). Indirect selection for increased sucrose yield in early sugarcane testing stages. *Field Crops Research, 26*(1), 67–73.

Grime, J. P., (1979). *Plant Strategies and Vegetation Processes.* Chichester, New York: John Wiley & Sons Ltd. pp. 222.

Guo, H., Padoch, C., Fu, Y., Dao, Z., & Coffey, K., (2000). Household level agro-biodiversity assessment. *PLEC News and Views, 16,* 28–33.

Hammer, K., (2003). A paradigm shift in the discipline of plant genetic resources. *Genetic Resources and Crop Evolution, 50,* 3–10.

Henry, R. J., & Kole, C., (2010). Genetics, Genomics and Breeding of Sugarcane. In series: Genetics, Genomics and Breeding of Crop Plants (Henry, RJ & Kole, C; Eds.); vol (10) Science Publishers, Enfield, NH. pp. 276

Hodgkin, T., Roviglioni, R., De Vicente, M. C., & Dudnik, N., (2001). Molecular methods in the conservation and use of plant genetic resources. In: Dore, C., Dosba, F., Baril, C. (eds.). Proceedings of the International Symposium on Molecular Markers for Characterization of Genotypes and Identifying Cultivars in Horticulture. *Acta Hort., 546,* 107–118.

Irvine, J. E., (1999). *Saccharum* species as horticultural classes. *Theor Appl Genet., 98,* 186–194.

Jackson, P. A., (2005). Breeding for improved sugar content in sugarcane. *Field Crops Res., 92*(3), 277–290.

Jayashree, J., Selvi, A., & Nair, N. V., (2010). Characterization of resistance gene analog polymorphism in sugarcane cultivars with varying levels of red rot resistance. *Electronic Journal of Plant Breeding, 1*(4), 1191–1199.

Krishnamurthi, M., (1989). Utilization of germplasm to improve sugarcane varieties through conventional and unconventional methods. In: *Sugarcane Varietal Improvement,* Naidu, K. M., Sreenivasan, T. V., & Premachandran, M. N. (eds.), pp. 163–176.

Legendre, B. L., (1995). Potential of increasing sucrose content of sugarcane varieties in Louisiana through breeding. *Proc. Int. Soc. Sugar Cane. Technol., 21,* 367–377.

Lima, M., Garcia, A., Oliveira, K., Matsuoka, S., Arizono, H. et al., (2002). Analysis of genetic similarity detected by AFLP and coefficient of parentage among genotypes of sugar cane [*Saccharum* spp]. *Theor Appl Genet., 104,* 30–38.

Mantri, N., Patade, V., Penna, S., Ford, R., & Pang, E. C. K., (2012). Abiotic stress responses in plants-present and future. In: Ahmad, P., Prasad, M. N. V. (eds.). *Abiotic Stress Responses in Plants: Metabolism to Productivity.* Springer, Science Business Media, NY, USA, pp. 1–19.

Maxted, N., & Kell, S. P., (2009). Establishment of a global network for the in situ conservation of crop wild relatives: status and needs. FAO Commission on Genetic Resources for Food and Agriculture, Rome. 266 pp.

Ming, R., Moore, P. H., Wu, K., D Hont, A., Glaszmann, J. C. et al., (2006). Sugarcane improvement through breeding and biotechnology. In: Janick, J., ed. Plant Breeding Reviews. Hoboken: John Wiley & Sons, pp. 15–118.

Moose, S. P., & Mumm, R. H., (2008). Molecular plant breeding as the foundation for 21st century crop improvement. *Plant Physiology, 147,* 969–977.

Mukherjee, S. K., (1957). Origin and distribution of *Saccharum. Bot Gaz.,* 55–61.

Nair, N. V., & Mary, S., (2006). RAPD analysis reveals the presence of mainland Indian and Indonesian forms of *Erianthus arundinaceus* [Retz.], Jeswiet in the Andaman–Nicobar Islands. *Curr. Sci., 90,* 1118–1122.

Nair, N. V., (2007). Sugarcane genetic resources: An Indian perspective. In: *Sugarcane: Crop Production and Improvement*, Singh, S. B. et al., (eds.). Houston: Studium Press, pp. 1–20.

Nair, N. V., (2011). Sugarcane varietal development programmes in India: an overview. *Sugar Tech., 13*(4), 275–280.

Nair, N. V., Selvi, A., Sreenivasan, T. V., Pushpalatha, K. N., & Sheji, M., (2006). Characterization of intergeneric hybrids of *Saccharum* using molecular markers. *Genetic Resources and Crop Evolution, 53*(1), 163–169.

Palaniswami, C., Bhaskaran, A., Gopalasundaram, P., Rakkiyappan, R., & Viswanathan, R., (2011). Precision farming in sugarcane: Evolving site specific management practices for increasing the sugarcane productivity, Annual Report for 2010–2011, *Sugarcane Breeding Institute*, Coimbatore, pp. 51–52.

Panje, R., & Babu, C., (1960). Studies in Saccharum spontaneum distribution and geographical association of chromosome numbers. *Cytologia, 25,* 152–172.

Patade, V. Y., Bhargava, S., & Suprasanna, P., (2011). Transcript expression profiling of stress responsive genes in response to short-term salt or PEG stress in sugarcane leaves. *Mol Biol Rep.,* doi: 10.1007/s11033-011-1100-z.

Prabhu, G. R., Kawar. P. G., & Teertha, Prasad, D., (2008). Differentiation filtration approach for isolation and enrichment of sugarcane grassy shoot phytoplasma. *Sugar Tech.*, *10*, 274–277.

Ramana Rao, T. C., Sreenivasan, T. V., & Palanichami, K., (1985). Catalogue on sugarcane genetic resources – II *Saccharum barberi*, Jeswiet, *Saccharum sinense*, Roxb. Amend. Jeswiet, *Saccharum robustum* Brandes et Jeswiet ex., *Saccharum edule* Hassk. Sugarcane Breeding Institute, Coimbatore, India.

Rao, G. P., Gaur, R. K., Singh, M., Viswanathan, R., Chandrasena, G., & Dharamwardhaanhe, N. M. N. N., (2001). Occurrence of *Sugarcane yellow leaf virus* in India and Srilanka. *Proceedings of International Society of Sugar Cane Technologists*, *24*, 469–470.

Rao, N. K., (2004). Plant genetic resources: Advancing conservation and use through biotechnology. *African Journal of Biotechnology*, *3*(2), 136–145.

Ribaut, J. M., & Ragot, M., (2007). Marker-assisted selection to improve drought adaptation in maize: the backcross approach, perspectives, limitations, and alternatives. *J Exp Bot.*, *58*(2), 351–360.

Roach, B. T., & Daniels, J., (1987). A review of the origin and improvement of sugarcane. In: *Proceedings of the Copersucar International Sugarcane Breeding Workshop*, Brazil, pp. 1–32.

Rott. P., Davis, M.L. (2000) Red Stripe (Top rot). In: *A Guide to Sugarcane Diseases*. Rott, P., Bailey, R. A., Comstock, J. C., Croft, B. J., & Saumtally, S., (Eds.). CIRAD-ISSCT, CIRAD Publication Services, France. pp. 58–62.

Sanghera, G. S., Kashyap L., & Kumar, R., (2016). Genetic improvement of sugarcane through non-conventional approaches. In: *Genetics and Molecular Biology in Crop Improvement* (ed. P. C. Trivedi), pp. 1–25.

Shanthi, R. M., & Alarmelu, S., (2010). Development of pre-breeding stocks with improved sucrose content over two selection cycles in sugarcane. *Electronic Journal of Plant Breeding*, *1*(4), 660–665.

Shrivastava, A. K., & Srivastava, S., (2012). Sugarcane: physiological and molecular approaches for improving abiotic stress tolerance and sustaining crop productivity. In: *Improving Crop Resistance to Abiotic Stress* (eds. Tuteja, N., et al.), Wiley-Blackwell, Germany, *2*, pp. 885–922.

Shrivastava, A. K., Shahi, H. N., Kulshreshtha, N., Shukla, S., & Darash, R., (2001). Nutrient uptake characteristics of *Saccharum* species. XIV *International Plant Nutrition Colloquium, Hannover*, Germany, abstr. No. *S1*. A. 218.

Silva, J. A., Sorrells, M. E., Burnquist, W. L., & Tanksley, S. D., (1993). RFLP linkage map and genome analysis of *Saccharum spontaneum*. *Genome*, *36*, 782–791.

Sreenivasan, T. V., & Sreenivasan, J., (2000). Inter-generic hybrids for sugarcane improvement. Sugarcane Breed. *Inst. Newsl.*, *18*(2), 1–2.

Sreenivasan, T. V., (2004). Improving indigenous sugarcane of India. *Sugar Tech.*, *6*(3), 107–111.

Sreenivasan, T. V., Amalraj, V. A., & Jebadhas, A. W., (2001). *Catalogue on Sugarcane Genetic Resources IV. Erianthus Species*, SBI, Coimbatore, India, pp. 98.

Srivastava, H. M., & Srivastava, S., (2000). Sugarcane breeding and varietal improvement during last fifty years [1947–97] in India. In: *50 Years of Sugarcane Research in India*. Shahi, H. N., Srivastava, A. K., & Sinha, O. K. (eds.), IISR, Lucknow. pp. 306

Srivastva, A. K., & Srivastva, S., (2016). Diversity of the germplasm of *Saccharum* species and related genera available for use in directed breeding programmes for sugarcane improvement. *Current Science*, 111(3), 471–482.

Sundar, R. A., Viswanathan, R., & Nagarathinam, S., (2009). Induction of systemic acquired resistance [SAR] using synthetic signal molecules against *Colletotrichum falcatum* in *Saccharum officinarum. Sugar Tech., 11*(3), 274–281.

Surat, S., (2009). Adoption gaps and constraints analysis of sugarcane cultivation in Buland-shahr District of Uttar Pradesh (PhD Dissertation). C.C.S. University, India. pp. 146.

Tai, P., & Miller, J., (2002). Germplasm diversity among four sugarcane species for sugar composition. *Crop Sci., 42,* 958–964.

Walker, D. I. T., (1987). Breeding for disease resistance. In: *Sugarcane Improvement Through Breeding*. Heinz, D. J. (ed.), Elsevier, Amsterdam, pp. 455–502.

Xiong, L., Schumaker, K. S., & Zhu, J. K., (2002). Cell signaling during cold, drought, and salt stress. *Plant Cell, 14,* S165–S183.

Zheng, J., Fu, J., Gou, M., Huai, J., Liu, Y., Jian, M., Huang, Q., Guo, X., Dong, Z., Wang, H., & Wang, G., (2010). Genome-wide transcriptome analysis of two maize inbred lines under drought stress. *Plant Mol Biol., 72,* 407–423.

CHAPTER 6

SOURCE-SINK DYNAMICS IN SUGARCANE: PHYSIO-GENOMICS PERSPECTIVES

AMARESH CHANDRA

Division of Plant Physiology and Biochemistry,
ICAR-Indian Institute of Sugarcane Research, Lucknow–226002, India,
E-mail: amaresh_chandra@rediffmail.com

CONTENTS

6.1   INTRODUCTION

Sugarcane (*Saccharum* spp. hybrids) a C_4 plant contributes >70% sugar in the world and rest mainly from sugar beet and corn. Based on physiological limit it has been estimated that sugarcane including its unselected clones can accumulate sucrose as high as 600 mg/g sucrose in a given volume of cane stalk (dry matter basis) (Inman-Bamber et al., 2008). However, in general, the sucrose content in mature, field-grown sugarcane stalks ranges from 500–560 mg/g (dry matter basis). These values are still much higher than those achieved commercially (350–400 mg/g dry matter), indicating the existence of a complex mechanism for both sucrose accumulation and its control in stalk (Inman-Bamber et al., 2008). Development stages of the crop along with growing conditions and variety/genotype determine the sucrose level and thus variation exist. Significant work has gone to this direction and role of some pivotal genes/enzymes like three different types of invertases, sucrose synthase, sucrose-phosphate synthase have been contemplated playing important role in metabolism of sucrose in sugarcane. Despite of all this available information, metabolism of sucrose in sugarcane is a complicated process warrant many genes, regulatory sequences and transcription factors playing significant role. In recent past, source-sink dynamics, which exist in sugarcane has been re-looked with molecular perspectives (Chandra et al., 2011; McCormick et al., 2006). Thus controlling sucrose in sugarcane proposes a synergistic look with respect to change in photosynthetic rate, expression/ regulation of genes, transcription factors linked with sucrose biosynthesis, transport and storage in context of source-sink communication.

Extensive research has been directed towards basic understanding of accumulation of photoassimilates in defined and undefined storage plant organs. Bio-physiological study pertaining to carbon source (leaf)-sink (specialized storage organs, culm parenchyma tissue in sugarcane) communication and its relationship has been better addressed in C_3 over C_4 crops. Nevertheless, sugarcane (*Saccharum* spp. hybrids), is the best reference crop where source-sink relationships study has been largely concentrated owing to its ability of accumulating a highly economical product, i.e., sucrose in stalk. Several authors have opined accumulation of sucrose in sugarcane is principally regulated at the level of sink (Chandra et al., 2015). Furthermore, futile cycle of degradation-synthesis of sucrose occurring in culm is also one factor which controls and regulate the sucrose concentration, hence all these cellular activities account for variable accumulation of sucrose in the cane stalk. A better understanding of sucrose synthesis and accumulation in sugarcane and its modulation through exogenous or endogenous means leading to higher sucrose productivity would be a boon for sugarcane farmers, millers and associated industries.

With extensive industrialization, agricultural land is diminishing day-by-day thereby agriculture has suffered irrevocably. Increasing areas even for cash crop like sugarcane is meager, as it needs better quality land and more water. Increasing sucrose content per unit area is the only option left to the planners, growers and executers. At global level, contrary to the demand of high sucrose bearing genotypes, sugarcane improvement for sucrose content in last fifty years have been largely through the increase of cane yield rather sucrose content per unit mass (Jackson, 2005). Even the transgenic approach where mostly single gene or single enzyme of stalk sucrose metabolism was targeted has generated mixed success (Lakshmanan et al., 2005).

Many researchers have argued that through conventional breeding a plateau has been achieved for level of sucrose accumulation in cane stalk. Contrary to that it has been also argued that genetic stocks so far used were possessing narrow genetic base. Many researchers still believe that significant potential exists for further gains in sucrose content without affecting other economical traits. In this direction the noteworthy work (Wu and Birch, 2007) where level of sugar content has been doubled through novel incorporation of a bacterial gene (sucrose isomerase, SI) producing sucrose isomer isomaltulose (IM), clearly showed alternate pathway of addressing such issue.

The present chapter deals the current understanding of source-sink communication in sugarcane keeping in view to how to maneuver to increase the sucrose content in cane stalk including role of plant growth regulators having ability to perturb the source-sink communication being active in crop like sugarcane where sugar is the main economical product accumulated in stalk parenchyma. Studies including transgenic work established the negative relationship that exists between leaf photosynthetic rates and stalk sucrose content, the sugar-sensing and signaling mechanisms that have emerged as potential regulators of this relationship.

6.2   PHOTOSYNTHESIS AND SOURCE-SINK DYNAMICS

In general, partitioning of photo-assimilate between the competing sink of vegetative growth, fiber and sucrose is under strict control of physiological genetic control, hence the sizes of any of these can be manipulated at breeding level so as to increase the sink for sucrose accumulation. It has been well narrated that mature sugarcane has the capacity to store up to 62% sucrose on a dry weight basis or 25% sucrose on a fresh weight basis under favorable conditions. The most commercial sugarcane varieties are hybrid of *S. officinarum* and *S. spontaneum* and these two species naturally distinct in many agronomical traits including that *S. officinarum*, a high sucrose accumulating species photosynthesize two-thirds of those low sucrose accumulating *S. spontaneum* spp. pave the way to better visualize how one is superior in one trait than those of other. Enhanced sugar accumulation in sugar booster line (transgenic bearing SI gene) was accompanied with increased rate of photosynthesis, electron transport reflecting photosynthetic efficiency, sugar transport and most importantly sinks strength. Doubling the total sugar content in mature internodes of an elite high sugar cultivar eliminates osmotic limits and osmotic sensing as primary constraints behind the previous concentration ceiling in sugarcane. Even decline in rate of photosynthesis at the maturity of the crop presumably direct that the accumulation of sucrose in stalk (culm) is precisely a result of source-sink interaction governed by many genes, regulatory sequences and thereby the signals being operational in the process (Figure 6.1). The hypothesis that the control of sucrose in cane stalk is more directed by sink strength, supported by much recent experimentation, has further augmented the need for looking into the expression of some pivotal genes, *vis-a-vis* physiological behavior of crop, for better understanding of their functionality in stalk (sink) and leaf (source) tissues (Chandra, 2011; Chandra et al., 2011; McCormick et al., 2006, 2009).

Some of the high sugar accumulating early maturing variety like CoJ 64 accumulates 18–20% (fresh weight basis) sucrose in its mature stalk. The physiological characteristics of sugarcane, however, still assuage for higher level of sucrose accumulation. Being C_4 crop the basic physiology supports the higher activity of ribulose biphosphate under enriched conditions of CO_2.

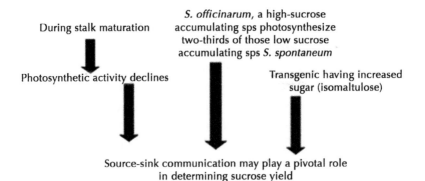

FIGURE 6.1   Photosynthesis and sucrose content in cane stalk showing inverse relationships.

An important parameter like apparent free space of stem parenchyma reflects the level of sucrose can be accumulated by a variety. Results have indicated a positive relationship between free space in the internode parenchyma tissue and Pol%, thus supports the role of free space in sucrose accumulation and possibly this can be used as a physiological indicator for sucrose yield. Therefore, it is possible that by increasing sink size (duration of stalk formation), internodes' length, better and enhance potentiality for assimilate portioning towards sucrose in the storage sink, sucrose yield can be increased.

Apart from increasing the source activity (increase leaf nitrogen content and specific leaf area) overall photosynthetic efficiency encompassing radiation use efficiency, water use efficiency and nutrient use efficiency should be targeted to increase the sucrose yield. Fast rate of growth which is achieved by enhancing the relative growth rate, net assimilation rate, leaf area ratio, leaf area index, leaf area duration and crop growth rate will eventually lead to higher physiological efficiency of the crop. These criterions should be used in breeding for higher sucrose accumulation. Physiological traits like decreased leaf plastochron, decreased leaf width, improved photosynthate partition early to leaves/late to storage as well as increased tillering, if selected as probable candidates in screening of varieties meaningful increase in yields and efficiency of yields have been achieved.

6.3   SOURCE-SINK RELATIONS AND IMPACT OF CLIMATE CHANGE

Since sugarcane contributes >70% of the world sugar, any impact of climate change on sugar yield will not only have impact on sugar but also on worldwide biofuel production. In general increased/elevated CO_2 has got low response in C_4 crop like sugarcane than those of C_3 crops. Apart from elevated CO_2 other variables associated with climate change like temperature and rainfall will also generate useful information to better understand the source-sink dynamics ultimately influencing sugar production. Despite of low response from C_4 crop, elevated CO_2 (ca 700 ppm) causes 30% increase in photosynthesis (De Souza et al., 2008). It is also evident from the work so far reported that mature and young tissues responded differently to increased CO_2 level. Enhanced photosynthetic rates declined sharply with leaf age in increased CO_2 scenario in sugarcane (Vu et al., 2006). It is possible that initial enhancement of sugarcane photosynthesis by CO_2 enrichment is finally diminished by sink demand (Watt et al., 2014). The elasticities as observed in source-sink relations indicate that initially demand is high in young sink tissues and therefore photosynthesis limits the production of carbon, but with enhanced CO_2 scenario change and lead to increased foliar sucrose. With maturity, demand is reduced and thus the photosynthesis rate reduced.

Results have demonstrated that elevated CO_2 influence the soil moisture through reduced transpiration and stomatal conductance while increases water use efficiency and therefore elevated CO_2 mimic the effects of better water and nutrient supply. In drought condition, leaf transpiration rate and stomatal conductance of sugarcane is reduced while water use efficiency increases indicated ameliorative effect of elevated CO_2. Therefore sugarcane crop grown under elevated CO_2 will finally help in delaying the adverse effect of drought and protect the crop from drought. Rubisco activity which usually decreases in drought raises the possibility that diminished sink demand resulting from drought might increases leaf sugars.

6.4   SOURCE-SINK COMMUNICATION AND FUNCTIONAL GENOMICS

In recent past, source-sink dynamics, which precisely exist in sugarcane has been re-looked with molecular perspectives (Chandra et al., 2011; McCormick et al., 2006). Thus controlling sucrose in sugarcane needs synergistic look wherein change in rate of photosynthesis, expression/regulation of genes, transcription factors linked with sucrose biosynthesis, transport and storage in context of source-sink communication must be visualized together. Sink strength has been opined as the major factor to be targeted by physiological or molecular interventions including water and temperature management so as to improve the sucrose yield. It is possible that when sink strength is maneuvered, source capacity as challenged. In past several authors have attempted to limit the source activity by partially shading or defoliating the major leaves leaving single leaf having capability to maintain normal supply of carbon based on sink demand. Such experimentations have provided ample evidence that sink regulates photosynthesis in sugarcane.

McCormick et al. (2006, 2009) have opined differential behavior of expression of genes especially those associated with carbohydrate metabolism having association with source-sink perturbation in sugarcane. It is also proposed that there is a existence of feedback mechanism operating between source and sink but merely studying regulation of some pivotal genes during stalk development and sucrose accumulation, deciphering genes having differential expression in high and low sugar bearing varieties as well as genotypes showing contrasting behaviors towards abiotic stress may not pin point precise nature of feedback mechanism that exist in sugarcane. In light of these facts, it is perceived that both physiological responses and expression of genes associated with carbohydrate metabolism must be studied together to reveal the regulatory mechanism and their role in source and sink tissues and communication per se.

Several authors have reviewed the role of various enzymes namely, sucrose synthase (SuSy), sucrose-phosphate synthase (SPS), the three invertases namely cell wall (CWI), neutral (NI) and soluble acid (SAI) in both leaf and stem tissues of sugarcane (Batta and Singh, 1986; Burg and Bieleski, 1962). Recently complexities of invertases and their expression in defined cells (leaf, apo-plastic cell wall and vacuoles) have been reviewed (Chandra et al., 2012). It has been argued that invertases work in tandem and balance the level of sucrose in cane stalk (Chandra et al., 2012). Burg and Bieleski (1962) have opined that in sugarcane, sucrose is accumulated in storage tissues, against a concentration gradient, using energy provided by respiration. This was said to be accompanied by a continuous cleavage and synthesis of sucrose, during accumulation of sucrose in storage tissue (Batta and Singh, 1986). SPS being important enzyme controls the sucrose synthesis in sugarcane leaves (Moore, 1995) and moreover it also controls the flux of carbon into sucrose (Huber and Huber, 1996) and thus manipulate the level of photosynthates to be passed on from source to sink tissues (Jang and Sheen, 1994). The primary role of sucrose synthase has been in the direction of sucrose cleavage in stalk/sinks tissues, ultimately helps in improving sink strength. SPS belongs to multi-gene family and have been mostly regulated at post-transcriptional level. Regulation of SPS through up/down expression of SPS genes have been also reported (Doehlert and Huber, 1985; Huber and Huber, 1996). It is allosteric enzyme being controlled by activator molecules like glucose-6-phosphate (Glu-6-P) and inhibitor inorganic phosphate and covalent modification via reversible phosphorylation. Unlike the four SPS isoforms/gene families reported in wheat (Castleden et al., 2004), five isoforms have been reported in *Saccharum* spp. (Grof et al., 2006). Expression of these isoforms are found different in various plant tissues especially leave and stalk and that to under mature and immature stalk tissues indicating specific role of each of these isoforms. The sequential action of sucrose synthase and sucrose phosphate synthase regulates the sucrose accumulation in sugarcane (Hatch et al., 1963).

Apart from sucrose synthase, invertase (β-fructofuranosidase) is also involved in breaking down of sucrose into free hexoses. There are three classes of invertase and they are soluble acid invertase (SAI), cell wall bound (CWI) and neutral invertase (NI). CWI is bound to cell wall, but is totally absent in the cytosol (Walker and Pollock, 1993). SAI is soluble form and mostly located in vacuole and involved in regulating the sucrose content in leaves (Gutierrez-Miceli et al., 2005; Scholes et al., 1996). Usually, sucrose synthesized in the cytosol, which is later either exported to different tissues or stored in the vacuole (Winter et al., 1994). In normal course, when cane are matured the level of SAI is usually low while immature tissues depict high level of SAI leading high reducing sugar helpful in growth and respiration. Low acid invertase activity also favors sucrose accumulation. There is ample evidence that show that CWI plays a key role in phloem uploading and assimilate uptake, especially in sink tissues. Under such case a steep concentration gradient of sucrose from source to sink is found (Eschrich, 1980; Koch, 2004). Neutral invertase activity is predominantly found high in immature and young tissues where sucrose is low (Bosch et al., 2004). Having major role in regulating the level of glucose and sucrose in various plant tissues, these three invertases are presumably involved in maintaining a strong source-sink relationship (Roitsch et al., 2003).

In normal course, partitioning of carbon in sugarcane sink tissues is in the form of (i) sugar pool (rapid synthesis and degradation), (ii) water insoluble compounds (primarily fiber and also proteins) whereas in non mature tissue, proteins and fiber are the competing sinks with sucrose for incoming carbon, and finally in (iii) respiratory pathways. Allocation towards (ii) and (iii) decreases with tissue maturation and concomitant rise towards sugar pool (i) in stem parenchyma cell. The maximum level of accumulation of partitioned carbon into sucrose is 0.7 M in sugarcane culm. Study in sugarcane have indicated that some physiological traits namely delayed leaf senescence, high photosynthetic activity along with higher CWI activity are linked with high total sugar phenotype of a sugarcane. Figure 6.2 indicate that the level sucrose in sink tissues is highly regulated through source-sink communication. In general the observed sucrose content on dry matter basis is 350–400 mg/g, however, lines are available having capacity to accumulate 500 to 560 mg/g sucrose. It has been also reported that some physio-biochemical

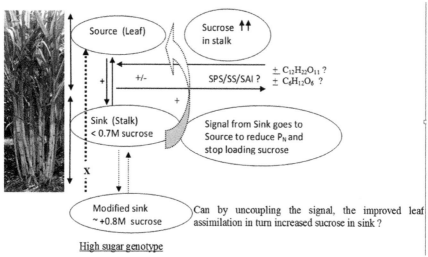

FIGURE 6.2   Source-sink communication plays pivotal role in controlling sucrose in cane stalk. (Adapted with permission from Chandra, A., (2011). Physio-biochemical and molecular approaches to enhance sucrose content in sugarcane: Indian initiatives. Sugar Tech., 13, 315–321. © 2011 Springer Nature.)

processes, loading and unloading of sucrose in leaf and culm, three phasic metabolic activity of sucrose in parenchymatous cells (apparent free space, metabolic space and vacuolar storage space), developmental constraints such as duration and timing of maturation may control the ceiling of apparent sucrose concentration in the culm (Chandra et al., 2011). In light of this as well as continuous cleavage and synthesis of sucrose within the storage pool it is possible that feedback regulation by bio-sensors like hexoses play important role in maintaining the physiological threshold of sucrose in stalk.

6.5   PERTURBATION OF SOURCE-SINK AND PLANT GROWTH REGULATORS

To improve sugar yield over cane yield, it is imperative to address the feedback inhibition system of source-sink dynamics. It is also conjectured whether perturbation of source-sink communication will direct to identify the crucial point, which controls the levels of sucrose to be accumulated in stalk. If existence of source-sink scenario in sugarcane is taken to supply and demand mode, how and when demand will control the supply. Also, excess source available with crop will have any bearing on the controlling factor of source. The existence of difference in cane yield between sub-tropical and tropical part of the country has any relevance with source-sink relationship or it is totally weather-controlled mechanism. It is also to be seen whether the sugar booster transgenic line will deliver the new insights into the mechanism by which plants regulate sugar accumulation. This is further going to be interesting to visualize the impact of climate change especially temperature (both low and high) and high level of CO_2 on the storage capacity of sucrose in juice and yield of the crop which might address the source-sink scenario in sugarcane. It is possible to utilize the new molecular tools and techniques to decipher the possible mechanism involved in controlling of source-sink dynamics. Situation is more challenging while addressing the situation in light of modern sugarcane crop, which is usually hybrid and has got complex genome. In this direction molecular markers developed so far including functional CISP (Chandra et al., 2013) will have some bearing.

The main driving factors that co-regulate the crop yield are the photosynthetic rate of the leaves and strong sink demand, which in turn controlled by source (McCormick et al., 2006). In sugarcane, source tissues were found to have a strong capacity to acclimatize according to the increased sink demand that results in augmentation of the photosynthetic rate in leaf (McCormick et al., 2006). Several studies have indicated that sucrose acts as important signaling molecule that regulates the carbon partitioning between source and sink (Chiou and Bush, 1998; Gibson, 2005; Krapp et al., 1993) and therefore, presumably, sucrose content in the sink tissue regulates photosynthetic activity in leaf. Increased culm sucrose suppresses the synthesis of photoassimilates in leaf tissues, further limiting the phloem loading of sucrose in sink. This represents a close linkage between source photosynthetic activity, hexose, and sucrose concentrations in sink tissues (McCormick et al., 2006). Thus, uncoupling the negative feedback signals between supply and demand tissues may result in the extraordinary increase in sucrose yield (McCormick et al., 2009). Un-coupling or perturbed condition can be achieved by the application of the GA_3 (Gibberellic acid) on sugarcane, by restricting photosynthetic activity and improvising the sink strength (Figure 6.3).

Gibberellins are plant growth regulators belonging to large family of diterpenes. GA_3 (Gibberellic acid) is first broadly available active form of commercial GAs, best known for their influence on seed germination, leaf expansion, stem elongation, flower/trichome initiation, flower and fruit development (Yamaguchi, 2008). It also helps in improving plant photosynthetic efficiency by influencing photosynthetic enzymes, leaf area index, light interception and increased use of nutrients. Gibberellic acid is

(-)GA,   (+)GA,

FIGURE 6.3   Source-sink perturbation implying role of plant growth regulator in improving sink strength.

also reported to induce combined mechanisms that results in the increase of source potential and redistribution of photosynthate to enhance the sink strength (Khan et al., 2007), thereby improving the competitive ability of an organ to attract more assimilate (Farrar, 1993). GAs have also been reported to activate sucrose phosphosynthase which is involved in the synthesis of sucrose (Kozlowska et al., 2007), thus influencing phloem loading and regulation of sucrose synthesis (Aloni et al., 1986; Chen et al., 1994). Study of effect of GA_3 on the expression of genes associated with sucrose synthesis and accumulation, is limited in sugarcane, nevertheless, it has been utilized to visualize aspects such as bud sprouting, culm growth, sugar production, ripening, water relations, and weed control (Botha et al., 2013; Buren et al., 1979; Clowes, 1980; Moore and Buren, 1978; Moore and Ginoza, 1980;). At commercial level, GA_3 has been used to promote the sugarcane growth in Hawaii where some varieties were found more responsive than others (Moore, 1977; Moore and Buren, 1978).

Extracellular invertase is an important enzyme that plays significant role in the apoplasmic phloem unloading and initiates hydrolytic cleavage of sucrose. When released in apoplast it helps in assimilates translocation to long-distance (Iqbal et al., 2011). Exogenous application of GA causes increase in activity of acid invertase in many elongating plant tissues such as

Avena (Kaufman et al., 1973), *Phaseolus vulgaris* (Morris and Arthur, 1985) and elongating dwarf *Pisum sativum* shoot (Wu et al., 1993). An increase in expression of acid invertase was reported within 4 hrs of GA treatment in dwarf pea seedling (Wu et al., 1993). This shows that GAs enhance acid invertase activity that in turn increases the supply of hexoses (by the cleavage of sucrose), utilized by the elongating cells for its growth and metabolism (Ranwala and Miller, 2008). Thus increased acid invertase activity enhances phloem unloading in the sink resulting in increased sink demand (Zhang et al., 2007). This in turn up-regulates translocation activity and increases the capacity of phloem loading in sink tissue.

6.6   CONCLUSIONS

To improve sucrose content in sugarcane so as to bridge the existing gap between theoretical and experimental sugar yields, it is imperative to better understand the source-sink dynamics utilizing the latest molecular tools and techniques. Moreover results so far have convincingly demonstrated that sugarcane stalk has the potential to accommodate more sucrose than those of reported to date. To better understand the modulation of source-sink relationships in sugarcane it is imperative to look the scenario in totality. The complexity as exist in sugarcane regarding sucrose metabolism where continuous cycle of synthesis, breakdown and re-synthesis works in tandem needs thorough understanding at physiological, biochemical and genetical (molecular) levels. Transcriptome study will pave the way to identify genes and transcription factors regulating accumulation of sucrose especially in light of source-sink dynamics *vis-a-vis* genes associated with signal transduction regulating source-sink communication including feedback inhibition. Physio-biochemical changes and modulation of such changes through breeding or transgenic will also help in identifying crucial steps in source-sink dynamics.

The inverse relationship that exist between leaf rate of photosynthesis (source) and stalk sucrose content (sink), presumably assuage to address the nature of signaling mechanism directing requirements of the sink to be communicated to source and how the source sense these signals. Most importantly, how this communication or relationships is going to be affected among different varieties and also under different environmental conditions. In recent past the use of GA_3 that stimulates physiological responses which alters the source-sink metabolism leading to increased sink strength along with expression analysis of the pivotal genes that control the sucrose accumulation in both control and GA$_3$ perturbed condition provides a platform for better understanding of the source-sink relationship and identification of signal transduction components and transcription factors that will ultimately improve sugar accumulation and sugar yield in sugarcane.

KEYWORDS

- **gene expression**
- **genomics**
- **plant growth regulator**
- **source-sink**
- **source-sink perturbation**
- **sugarcane**
- **transgenic**

REFERENCES

Aloni, B., Daie, J., & Wyse, R. E., (1986). Enhancement of [^{14}C] sucrose export from source leaves of *Vicia faba* by gibberellic acid. *Plant Physiol, 82,* 962–966.

Batta, S. K., & Singh, R., (1986). Sucrose metabolism in sugarcane grown under varying climatic conditions: synthesis and storage of sucrose in relation to the activities of sucrose synthase, sucrose phosphate synthase and invertase. *Phytochemistry, 25,* 2431–2437.

Bosch, S., Grof C. P. L., & Botha, F. C., (2004). Expression of neutral invertase in sugarcane. *Plant Sci., 166,* 1125–1133.

Botha, F. C, Lakshmanan, P., O'Connell, A., & Moore, P. H., (2013). Hormones and Growth Regulators, in Sugarcane: *Physiology, Biochemistry, and Functional Biology,* P. H. Moore, & F. C. Botha (eds.), John Wiley & Sons Ltd, Chichester, UK. doi: 10.1002/9781118771280, pp. 331–377.

Buren, LL, Moore, PH, & Yamasaki, Y, (1979). Gibberellin studies with sugarcane. II. Hand-sampled field trials. *Crop Sci., 19,* 425–428.

Burg, S. P., & Bieleski, R. L., (1962). The physiology of sugarcane. V. Kinetics of sugar accumulation. *Australian Journal of Biological Sciences, 15,* 429–444.

Castleden, C. K., Aoki, N., Gillespie, V. J., MacRae, E. A., Quick, W. P., Buchner, P., Foyer, C. H., Furbank, R. T., & Lunn, J. E., (2004). Evolution and function of the sucrose phosphate synthase gene families in wheat and other grasses. *Plant Physiology, 135,* 1753–1764.

Chandra, A, Verma, P. K., Islam, M. N., Grisham, M. P., Jain, R., Sharma, A., Roopendra, K., Singh, K., Singh, P., Verma, I., & Solomon, S., (2015). Expression analysis of genes associated with sucrose accumulation in sugarcane (*Saccharum* spp. hybrids) varieties differing in content and time of peak sucrose storage. *Plant Biol., 17,* 608–617.

Chandra, A., (2011). Physio-biochemical and molecular approaches to enhance sucrose content in sugarcane: Indian initiatives. *Sugar Tech., 13,* 315–321.

Chandra, A., Jain, R., & Solomon, S., (2012). Complexities of invertases controlling sucrose accumulation and retention in sugarcane: The way forward. *Current Science, 102,* 857–866.

Chandra, A., Jain, R., Rai, R. K., & Solomon, S., (2011). Revisiting the source-sink paradigm in sugarcane. *Current Science, 100,* 978–980.

Chandra, A., Jain, R., Solomon, S., Shrivastava S., & Roy, A. K., (2013). Exploiting EST databases for the development and characterization of 3425 gene-tagged CISP markers in biofuel crop sugarcane and their transferability in cereals and orphan tropical grasses. *BMC Research Notes, 6,* 47.

Chen, W. S., Liu, Z. H., Yang, L., & Chen, W. H., (1994). Gibberellin and temperature influence carbohydrate content and flowering in *Phalaenopsis. Physiol Plant, 90,* 391–395.

Chiou, T. J., & Bush, D. R., (1998). Sucrose is a signal molecule in assimilate partitioning. *Proc Natl Acad Sci., 95,* 4784–4788.

De Souza, A. P., Gaspar, M., Da Silva, E. A., Ulian, E. C., Waclawovsky, A. J., Nishiyama, M. Y. Jr., Dos Santos, R. V., Teixeira, M. M., Souza, G. M., & Buckeridge, M. S., (2008). Elevated CO_2 increases photosynthesis, biomass and productivity, and modifies gene expression in sugarcane. *Plant Cell Environment, 31,* 1116–1127.

Doehlert, D. C., & Huber, S. C., (1985). The role of sulpha-hydryl groups in the regulation of spinach leaf sucrose phosphate synthase. *Bio-chimica et Biophysica Acta., 93,* 353–355.

Eschrich, W., (1980). Free space invertase, its possible role in phloem unloading. *Berichte der Deutschen Botanischen Gesellschaft, 93,* 363–378.

Farrar, J. F., (1993). Sink strength: what is it and how do we measure it? Introduction. *Plant Cell Environment, 16,* 10–15.

Farrar, J., (1996). Regulation of shoot-root ratio is mediated by sucrose. *Plant Soil, 185,* 13–19.

Gibson, S. I., (2005). Control of plant development and gene expression by sugar signaling. *Curr Opin Plant Biol., 8,* 93–102.

Grof C. P. L., So C. T. E., Perroux J. M., Bonnett G. D., & Forrester, R. I., (2006). The five families of sucrose phosphate synthase genes in *Saccharum* spp. are differentially expressed in leaves and stems. *Functional Plant Biology, 33,* 605–610.

Gutierrez-Miceli F. A., Rodriguez-Mendiola M. A., Ochoa-Alejo N., Mendez-Salas R., Arias-Castro C., & Dendooven L., (2005). Sucrose accumulation and enzyme activities in callus culture of sugarcane. *Biologia Plantarum, 49,* 475–479.

Hatch M. D., Sacher J. A., & Glasziou K. T., (1963). Sugar accumulation cycle in sugarcane. I. Studies on enzymes of the cycle. *Plant Physiology, 38,* 338–343.

Huber, S. C., & Huber, J. L., (1996). Role and regulation of sucrose phosphate synthase in higher plants. *Annual Review of Plant Physiology, 47,* 431–444.

Inman-Bamber, N. G., Bonnett, G. D., Spillman, M. F., Hewitt, M. L., & Jackson, J., (2008). Increasing sucrose accumulation in sugarcane by manipulating leaf extension and photosynthesis with irrigation. *Australian Journal of Agricultural Research, 59,* 13–26.

Iqbal, N., Nazar, R., Khan, M. I. R., Masood, A., & Khan, N. A., (2011). Role of gibberellins in regulation of source-sink relations under optimal and limiting environmental conditions. *Curr. Sci., 100,* 998–1007.

Jackson, P. A., (2005). Breeding for improved sugar content in sugarcane. *Field Crops Research, 92,* 277–290.

Jang, J. C., & Sheen, J., (1994). Sugar sensing in higher plants. *Plant Cell, 6,* 1665–1679.

Kaufman, P. B., Ghosheh, N. S., LaCroix, J. D., Soni, S. L., & Ikuma, H., (1973). Regulation of invertase levels in *Avena* stem segments by gibberellic acid, sucrose, glucose, and fructose. *Plant Physiol., 52,* 221–228.

Khan, N. A., Singh, S., Nazar, R, & Lone, P. M., (2007). The source–sink relationship in mustard. *Asian Aust. J. Plant Sci. Biotech., 1,* 10–18.

Koch K., (2004). Sucrose metabolism: Regulatory mechanisms and pivotal roles in sugar sensing and plant development. *Current Opinion in Plant Biology, 7,* 235–246.

Kozłowska, M., Rybus-Zając, M., Stachowiak, J., & Janowska, B., (2007). Changes in carbohydrate contents of *Zantedeschia* leaves under gibberellin-stimulated flowering. *Acta. Physiol. Plant 29,* 27–32.

Krapp, A., Hofman, B., Schafer, C., & Stitt, M., (1993). Regulation of the expression of rbcS and other photosynthetic genes by carbohydrates: mechanism for the sink regulation of photosynthesis? *Plant J., 3,* 817–828.

Lakshmanan, P., Geijskes, R. J., Aitken, K. S., Grof, C. L. P., Bonnet, G. D., & Smith, G. R., (2005). Sugarcane biotechnology: the challenges and opportunities. *In Vitro Cell Dev Biol Plant, 41,* 345–363.

McCormick, A. J., Cramer M. D., & Watt D. A., (2006). Sink strength regulates photosynthesis in sugarcane. *New Phytologist, 171,* 759–770.

McCormick, A. J., Watt D. A., & Cramer M. D., (2009). Supply and demand: sink regulation of sugar accumulation in sugarcane. *Journal of Experimental Botany, 60,* 357–364.

Moore, P. H., & Buren, L. L., (1978). Gibberellin studies with sugarcane. I. Cultivar differences in growth responses to gibberellic acid. *Crop Sci., 17,* 443–446.

Moore, P. H., & Ginoza, H., (1980). Gibberellin studies with sugarcane. III. Effects of rate and frequency of gibberellic acid applications on stalk length and fresh weight. *Crop Sci., 20,* 78–82.

Moore, P. H., (1995). Temporal and spatial regulation of sucrose metabolism in the sugarcane stem. *Aust J Plant Physiol., 22,* 661–679.

Morris, D. A., & Arthur, E. D., (1985). Effects of gibberellic acid on patterns of carbohydrate distribution and acid invertase activity in *Phaseolus vulgaris. Physiol Plant, 65,* 257–262.

Ranwala, A. P., & Miller, W. B., (2008). Gibberellin-mediated changes in carbohydrate metabolism during flower stalk elongation in tulips. *Plant Growth Regul., 55,* 241–248.

Roitsch, T., Balibrea, M. E., Hofmann, M., Proels, R., & Sinha, A. K., (2003). Extracellular invertase: Key metabolic enzyme and PR protein. *Journal of Experimental Botany, 54,* 513–524.

Scholes, J., Bundock, N., Wilde, R., & Rolfe, S., (1996). The impact of reduced vacuolar invertase on the photosynthetic and carbohydrate metabolism of tomato. *Planta, 200,* 265–272.

Walker, R. P., & Pollock, C. J., (1993). The purification and characterization of soluble acid invertase from coleoptiles of wheat (*Triticum aestivum* L. cv. Avalon). *Journal of Experimental Botany, 44,* 1029–1037.

Watt, D. A., McCormick, A. J., & Cramer, M. D., (2014). Source and sink physiology. In: *Sugarcane Physiology, Biochemistry and Functional Biology.* Moore, P. H., and Botha, F. C., (eds.). Wiley Blackwell, pp. 483–520.

Winter, H., Robinson, D. G., & Heldt, H. W., (1994). Subcellular volumes and metabolites in spinach leaves. *Planta, 193*, 530–535.

Wu, L., & Birch, R. G., (2007). Doubling sugar content in sugarcane plants modified to produce a sucrose isomer. *Plant Biotech J., 5*, 109–117.

Wu, L., Mitchell, J. P., Cohn, N. S., & Kaufman, P. B., (1993). Gibberellin (GA_3) enhances cell wall invertase activity and mRNA levels in elongating dwarf pea (*Pisum sativum*) shoots. *Int. J Plant Sci., 154*, 280–289.

Yamaguchi, S., (2008). Gibberellin metabolism and its regulation. *Annu Rev Plant Biol., 59*, 225–251.

Zhang, C., Tanabe, K., Tamura, F., Itai, A., & Yoshida, M., (2007). Roles of gibberellins in increasing sink demand in Japanese pear fruit during rapid fruit growth. *Plant Growth Regul., 52*, 161–172.

CHAPTER 7

RESIDUES USES AND ENVIRONMENT SUSTAINABILITY

R. ROSSETTO,[1] C. A. C. CRUSCIOL,[2] H. CANTARELLA,[3]
J. B. CARMO,[4] and C. A. C. NASCIMENTO[2]

[1]Agência Paulista deTecnologia do Agronegócio (APTA), Centro
de Cana-de-Açúcar do IAC, Rodovia SP 127 km 30, 13400–970
Piracicaba, SP, Brazil, E-mail: raffaella@apta.sp.gov.br

[2]São Paulo State University (UNESP), School of Agriculture,
Department of Crop Science, P.O. Box: 237, 18610-307 Botucatu,
State of São Paulo, Brazil

[3]Agronomic Institute of Campinas, Soils and Environmental Resources
Center, P.O. Box 28, 13001-970 Campinas, SP, Brazil

[4]Department of Environmental Sciences, Federal University of Sao
Carlos (UFSCar), Rod. João Leme dos Santos Km 110, 18052-780
Sorocaba, SP, Brazil

CONTENTS

7.1 INTRODUCTION

Sugarcane is composed of one-third of cane juice and two-thirds of biomass consisting of straw (dry and green leaves), tops, and bagasse. The juice provides the raw material for the manufacture of sugar and alcohol. In addition to its three main products—sugar, alcohol, and electricity—the sugarcane production chain generates various agro-industrial residues, all of which are recycled in sugarcane production.

Biomass has great potential for the production of energy for use in the sucrose and ethanol plant or cogeneration of electricity. The yield of straw dry mass varies from 8 to 15 tons (t) straw per hectare of sugarcane harvested.

In addition sugarcane also leaves an estimate of 5 to 8 t ha^{-1} of dry root matter inside the soil (Vasconcellos and Casagrande, 2008).

After milling, one tonne of sugarcane generates juice, which can produce sugar and ethanol, and some residues such as bagasse, filter cake, ashes, and vinasse, which is the residue that remains after ethanol production. Figure 7.1 shows the average quantities of products and residues generated in a sugar-ethanol plant.

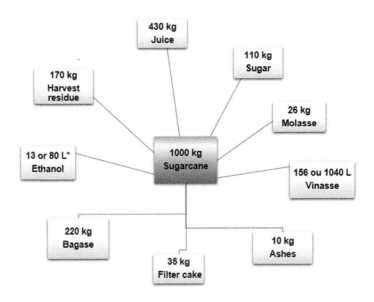

FIGURE 7.1   Amount of products and residues in kg per tonne of sugarcane (*in sugar-ethanol mill, or just distillery, respectively).

The quantities of residues generated in the State of São Paulo, in the harvest of 2012-13, and the production coefficients of such residues, are presented in Table 7.1.

Filter cake and vinasse contain organic matter and nutrients, which represents important options for recycling of nutrients. Table 7.2 presents the potential nutrient value of the straw, filter cake, and vinasse residues assuming that the entire production area, estimated at 9 million hectares, is not burned and 670 million tons of sugarcane, of which 350 million were used to produce sugar, which would generate 35 kg of cake per tonne of ground cane; cake with 70% moisture. Each hectare generate around 5 t ha^{-1} of dry straw, considering the entire sugarcane area of Brazil, as not burned and assuming the ethanol production of 30 billion L of ethanol that generates a vinasse volume equivalent to 13 L per L of ethanol produced. MAPA (2017).

The sugar and ethanol production chain has an important feature in comparison with most crops, which is its large nutrient recycling potential and the reduced use of fertilizer. In addition, unlike the N and P in soluble mineral fertilizers, great proportion of their nutrients in the residues are present in organic molecules and are slowly available to the crop.

TABLE 7.1 Estimated Amount of the Main Residues of the Sugarcane Agro-industry in the State of São Paulo and Their Yield Coefficients Per Unit of Sugarcane or Ethanol

Residue	Tons	Kg per tonne of sugarcane
Washing soil	5,767,370	14.2
Filter Cake	10,568,316	49.3
Bagasse	89,353,619	220
Boiler ash	836,675	2.06
Chimney soot	4,776,357	11.76
Trash of sweep	16,246	0.04
Lubricant	1,120,982	2.76
Vinasse	125,834,361	10.85

Source of residue production coefficient data: André Elia Neto. Tópicos sobre Impactos Ambientais. Resíduos da Indústria Canavieira. Centro de Tecnologia Canavieira. Presentation of the subject: Aspectos Ambientais da Unidade de Produção. MP Agro Professional Master in Agroenergy. Fundação Getúlio Vargas. FGV. São Paulo, 29 de maio de 2010.

Source of sugarcane production data: *SIDRA system. IBGE. In: http://www.sidra.ibge.gov.br/bda/tabela/protabl.asp?c=1612&z=t&o=11&i=P.
**Sugar Cane Industry Union. UNICA. In: http://www.unicadata.com.br/historico-de-producao-e-moagem.php?idMn=31&tipoHistorico=2&acao=visualizar&idTabela=1491&produto=acucar&safraIni=2011%2F2012&safraFim=2011%2F2012&estado=SP.

TABLE 7.2   Estimates of Nutrients Recycled Annually by the Sugar-Alcohol Agroindustry in Brazil

Residues	Nutrients			Volume of residues	Nutrients returned		
	N	P_2O_5	K_2O		N	P_2O_5	K_2O
	% in dry residue				t year^{-1}		
Filter Cake *	1.4	1.94	0.39	3.675 million tonne dry cake year^{-1}	51,450	712,900	143,000
Straw **	0.46	0.11	0.57	45 million tonne dry straw year^{-1}	207,000	49,500	256,000
	g m^{-3} vinasse						
Vinasse***	375	60	2035	390 billion L year^{-1}	150,000	24,000	814,000
Total					408,450	786,400	1,213,000

*Assuming: Production area in Brazil of 9 million hectares and 670 million tons of sugarcane, of which 350 million were used to produce sugar, which would generate 35 kg of cake per tonne of ground cane; cake with 70% moisture.

** Generation of 5 t ha^{-1} of dry straw, considering an entire sugarcane area of Brazil, as not burned.

*** Production 30 billion L of ethanol; vinasse equivalent to 13 L per L of ethanol produced MAPA (2017).

7.2   RESIDUES IN THE SUGAR-ETHANOL INDUSTRY

7.2.1   BAGASSE

In the sugarcane industry, the first by-product is the bagasse, which results from crushing the stalks to extract the cane juice; the yield of bagasse is approximately 250 kg (50% moisture) per tonne of cane. Due to its low nutrient and high-energy content, bagasse is seldom used as an organic fertilizer; instead, it is usually burned to produce heat and electricity in the mill. In this process, approximately 6 kg of ashes (dry basis) per tonne of cane is produced.

7.2.2   ASHES

The ashes of sugarcane bagasse contain oxides of most of the cations present in bagasse as well as portions of the N, S, and other elements that are partially lost as gases during combustion. The ashes are returned to the fields either singly or mixed with other by-products.

7.2.3 FILTER CAKE (PRESS MUD)

During sucrose production, small pieces of bagasse and sludge are vacuum-filtered in the process of cane juice clarification, yielding a mud known as filter cake. Currently, many distilleries also clarify the juice and generate filter cake. On average, 30 kg of filter cake (70% moisture) are produced per tonne of cane crushed. Filter cake contains high amounts of compostable organic matter (>50%) and variable amounts of nutrients, including those from the lime and phosphates added to facilitate the clarification process. Phosphorus (10 to 20 g P_2O_5/kg dry filter cake) is usually the nutrient of interest when filter cake is applied to fields; however, the filter cake contains other nutrients as well, and its easily mineralizable organic matter is highly beneficial as a soil conditioner. Sugar mills and distilleries generally have large areas and equipment for the preparation of filter cake compost. Other by-products such as vinasse, gypsum, and ashes may be added to enrich the compost. The compost is applied to the sugarcane fields in several ways; the most common is application to the planting furrows at rates that vary from 10 to 20 t/ha (wet basis).

The benefits of the use of filter cake compost usually exceed its nutrient value, especially in sandy, low-fertility soils; they include increased longevity and greater tolerance to soil pests such as nematodes and termites (Rossetto et al., 2010a).

In general, 50% of the P in filter cake may be readily available for sugarcane. The cake is used under various conditions, including:

a) broadcast and incorporation to soil at sugarcane reforming;
b) in the planting row;
c) surface application on the row or near the row, in ratoons, without incorporation or with slight incorporation.

Filter cake can be applied fresh or composted. Most commonly, wet filter cake is applied to the entire area at a rate between 80 and 100 t ha^{-1} during pre-planting, in the planting row or between the rows at a rate between 15 and 30 t ha^{-1}, and to the ratoon at a rate of between 40 and 50 t ha^{-1}. If necessary, it is supplemented with soluble phosphate fertilizer.

The main benefits of the use of filter cake are as follows:

a) source of organic matter, which increases cation exchange capacity (CEC) and retention of the cations provided by fertilization;
b) it aids in water retention;
c) in addition to P, filter cake is rich in calcium, N and micronutrients, although it contains low concentrations of K and Mg (Table 7.4). The nutrients present in the filter cake are resistant to leaching and will be available with mineralization of the organic matter.
d) during mineralization of the filter cake, microorganisms produce chelating and complexing substances that reduce P fixation and other substances that promote root growth;
e) most filter cake applied in the planting furrows helps sugarcane sprouting (Figure 7.2). Heat released during filter cake decomposition at the bottom of the furrow is also beneficial to sprouting, especially when planting is performed during the winter.

As with most residues, the nutrient composition of filter cake is not balanced or does not cover all the plants need, therefore, supplementation is often required. Table 7.3 shows the mean, maximum and minimum nutrient values of sugarcane filter cake according to Luz and Vitti (2008).

FIGURE 7.2 Effect of filter cake application on planting furrow compared to the control treatment (*Source*: (Rossetto et al., 2010a)).

TABLE 7.3   Chemical Composition of Filter Cake

Element	Maximum	Minimum	Mean
% in dry matter			
N	1.67	1.31	1.49
P_2O_5	2.12	1.32	1.72
K_2O	0.47	0.21	0.34
Ca	7.50	1.69	4.59
Mg	0.58	0.33	0.46
S	1.05	0.16	0.60
Relação C/N	23.96	20.46	21.96
pH CaCl$_2$	6.47	5.40	5.93
Density g cm^{-3}	0.66	0.53	0.6
Humidity %	70.20	59.92	65.06
OM %	65.36	49.89	57.62
C %	36.31	27.72	32.01
mg kg^{-1} dry matter			
Fe	23,504	13,821	22,189
Mn	458.7	224.0	576.9
Cu	115.1	10.9	119.1
Zn	148.3	26.3	142.9
Na	1,183.5	618.9	872.2
B	14.8	6.9	11.3

* *Source*: Luz and Vitti (2008).

Improvement of soil fertility is expected after the addition of filter cake. Rodella et al. (1999) demonstrated that 30 months after filter cake application, soil analyses still presented high Ca, P and CEC (Table 7.4). In addition, when applying the filter cake in the planting row there is a stimulus for the roots to deepen, which is quite desired for sugarcane (NARDIM, 2007).

7.2.3.1   Filter Cake Composting

The main objective of filter cake composting is the reduction of moisture content; this lowers transport costs and increase nutrients availability via the mineralization of organic matter, resulting in greater rationalization and economic gains. Fresh filter cake contains approximately 70% water. Under

TABLE 7.4 Soil Chemical Properties of the After 8 and 30 Months of Filter Cake Application

Treatments	pH	C	P	K	Ca	Mg	Al	CTC
		%	mg kg^{-1}	.------------------- mmol$_c$.dm^{-3}--------------------				
Control	5.0	1.07	32	1.2	10.7	5.4	11.1	27.5
Filter cake after 8 months	5.2	1.24	188	1.2	36.3	6.1	2.5	46.2
Filter cake after 30 months	5.0	1.21	109	1.0	31.5	5.7	4.5	42.9

Source: Rodella et al. (1999).

certain conditions, such as planting in the winter season when rainfall is scarce, the moisture of the filter cake may be essential for sugarcane sprouting, ensuring better sprout uniformity. For this use, composting is not so advantageous.

Composting may be performed by adding soot, gypsum, vinasse, or other sources of nutrients. The appropriateness and economic feasibility of mixing these materials with the filter cake should be confirmed. Freshly produced filter cake is transported from the plant to the composting yard, where it is distributed into 5- to 6-m-wide rows 1.5 to 2 m high, and spaced 6 to 10 m between rows to allow machines that transport and turn over the cake. The material is stabilized after 2 months of composting with 4 or 6 turnings. The filter cake pile can be turned by a self-propelled machine (Figure 7.3a) or by blades attached to a tractor (Figure 7.3b).

In certain processing plants, filter cake rows are mixed with chicken manure and natural rock phosphates and the resulting material is applied in the planting furrows, mapped by GPS, and the furrows are closed. At the time of planting, the furrows are again opened for planting the sugarcane seedlings. In this case, filter cake mineralization occurs inside the furrow, thus saving space in the composting yard.

7.2.4 HARVEST RESIDUE (STRAW)

Significant changes in the sugarcane agricultural sector have been observed in recent years, with increasing mechanization of planting and harvesting. In São Paulo state, mechanized harvesting was used in almost 100% of sugarcane area in the 2015/2016 season.

FIGURE 7.3A,B Self-propelled filter cake turner for composting, developed by UPP and CTC (a) and Filter cake turner coupled to tractor for composting (b). (Sources: J. Mangolini, R. Rossetto, respectively).

Although mechanized harvesting has brought clear benefits to the system, some studies show that the intensive use of agricultural machines can also cause damage to the soil, compromising sustainability. Braunbeck and Magalhães

(2010) showed how the current sugarcane mechanization management is characterized by yearly traffic involving over 15 wide and heavily operations of loaded harvesters, tractors, and wagons that cover nearly 60% of the field surface (inter rows), with negative effects on soil structure, mainly under wet conditions.

The sugarcane industry primarily exports sucrose and ethanol, both of which contain only C, H, and O; thus, all of the mineral nutrients that are brought from the field to the mill can theoretically be recycled. When the cane is not burned, the first waste is the straw, which recycles nutrients from the deep soil to the surface.

Despite the availability of cane straw (8 to 15 Mgha^{-1} of dry matter) resulting from green harvesting, it is just beginning to be considered as a fuel for the co-generation process and as feedstock for second-generation ethanol. Keeping the straw on the soil surface brings clear benefits to the production of sugarcane, such as protection against soil erosion, reduction in the variation of soil temperature, protection from direct radiation, increased biological activity, better water infiltration rate, increased availability of water due to lower evapotranspiration, and better control of weeds (Rossetto et al., 2008). Part of the residual straw left on the soil is incorporated into the soil as organic matter after the humification process, increasing the carbon stock. Several authors showed that maintaining the straw on the soil surface by the elimination of burning resulted in a higher concentration of organic matter that is composed mainly of about 50% organic carbon.

Soils can be either a source or a sink of atmospheric CO_2. Sourcing of CO_2 occurs primarily through the mineralization of organic matter, whereas sinking occurs when organic residues are incorporated into the stock of organic matter or into the inorganic phase of the soil C. The C stock is the result of the balance between the supply of organic material, its C:N ratio, and the climate. Cultivated soils generally have lower soil C stocks than the same soils in natural ecosystems – usually 50 to 70% of the original organic C – due to the effects of agricultural practices such as plowing, harrowing, cultivation, and other practices (Lal, 2008).

Sugarcane straw usually has a C:N ratio between 80:1 and 100:1. In principle, the low N level of the straw should make it difficult to incorporate C into the soil. The C stock may decline in areas of burnt sugarcane harvesting due to the low replenishment of organic C in the system, leading to soil degradation.

Results obtained in the sugarcane plantations of São Paulo state indicate accumulations of C in the soil of between 0.32 Mg ha^{-1} yr^{-1} and 0.8 Mg ha^{-1} yr^{-1} with the adoption of mechanical harvesting (Carvalho et al., 2010). The lower figure seems more representative because it refers to a period of 12 years. Moreover, the sugarcane crop is renewed every six to seven years, usually with great movement of the soil for planting; sugarcane management thus promotes the oxidation of part of the C accumulated in the soil (Cerri et al., 2009). Under the conditions used for sugarcane farming in São Paulo State (Brazil), Faroni et al. (2003) observed that 40 to 50% of the dry matter of straw remained in the soil after one year. The C:N ratio, however, decreased gradually from 85:1 in recently deposited straw to 34:1 after one year. In the medium and long terms, the soil can accumulate organic C and N when sugarcane is managed without burning; however, in the short term, the contribution of residues with high C:N ratios can increase the demand for mineral N. For sugarcane farming in Australia, a simulation model for 35 years shows that 86% of the straw N that would be lost by burning would be stored in SOM or exported with the harvested stems (Thorburn et al., 2001). Ramos et al. (2016) found the half-life of the residue decomposition to be 237 days. This high value confirmed that decomposition is slow. This was also confirmed by the long time (326 days) required to mineralize just 69% of the initial residue biomass. According to Trivelin et al. (2013), keeping crop residues over the soil will increase soil N stock and N recovery by sugarcane, reaching equilibrium after 40 years with recovery of approximately 40 kg ha^{-1} year^{-1} of N. Most of the nutrients are in the top leaves 75 % of the K$_2$O (81 kg ha^{-1} year^{-1}) and 50 % of the N (31 kg ha^{-1} year^{-1}) indicating the importance of maintaining tops in the soil to sustain soil fertility.

There are doubts as to whether retaining the sugarcane straw affects productivity compared to the burning system. The effects on productivity are complex and involve factors such as adaptation of varieties, ratoon sprouting under the straw, pressure from pests and diseases, water availability, and N nutrition. In the past, several authors reported that the presence of straw caused significant reduction in cane production, but the problem seemed to be related to the variety used. Many studies, including that of Urquiaga et al. (1995), reported higher yields of green sugarcane in most of the nine cycles evaluated.

All breeding programs currently operating in Brazil test the budding of specific sugarcane varieties under straw. Therefore, it is possible that the reduction in yield of sugarcane in areas due to problems with straw sprouting will be solved within a short time.

Current breeding programs also include research on obtaining energy from sugarcane; this includes research on sugarcane varieties that can be used for the production of second-generation ethanol from cellulose. To obtain plant material with high fiber and low sucrose content, *S. spontaneum* species are used in the crosses. The straw left in the field during the mechanized harvest process, can be used in part for the second-generation ethanol or to increase the electric energy production by boilers in the mill. Landell et al. (2013) provide information on the yield of straw, depending on the calculated leaf area index and the number of tillers per linear meter; primary energy production of several sugarcane genotypes; contribution of dry tops and leaves; biomass yield; and evaluation of fiber, cellulose, hemicellulose and lignin for several varieties in Brazil.

As discussed previously, burning of sugarcane is being increasingly restricted, resulting in large amounts of waste (8 to 20 t/ha of dry matter) remaining on the fields after harvesting. The amounts of nutrients in this material are approximately 64, 7, 66, 25, 13, and 9 kg/ha of N, P, K, Ca, Mg, and S, respectively (Oliveira et al., 1999a). Usually, more than 70% of the organic mass and of the N content of leaves and tops are lost after burning of a standing sugarcane field (Mitchell et al., 2000). When waste is burned, the losses of other nutrients are equally high. On the other hand, when waste is preserved, so are the nutrients. However, not all the elements are readily available for plant use. After almost one year, only 20% of the N contained in the waste was released; corresponding values for K were greater than 80%, and for other nutrients mineralization ranged from 50 to 68% (Oliveira et al., 1999b). Using waste marked with ^{15}N, Vitti et al. (2008) observed a much higher mineralization of waste N, 50 to 70% in 13 months. Therefore, waste preservation contributes to the recycling of nutrients and, in the long term, permits the reduction of fertilizer application.

7.2.5 VINASSE

Vinasse, an important liquid residue of bioethanol production, can supply several nutrients, particularly K, to cultivated sugarcane. When appropriately applied, these added nutrients can be deducted from conventional fertilization, which decreases production costs.

On the other hand, the large vinasse residue volumes produced by the Brazilian sugar-to-ethanol industry represent a potential environmental impact risk. Vinasse is typically generated at a proportion of 13 L vinasse per 1 L of bioethanol. In 2015/2016, Brazil produced 30 billion L of bio-ethanol, with the co-production of approximately 390 billion L of vinasse. In general, all the vinasse is distributed on the soil surface by spraying using irrigation equipment such as the semi-fixed moto-pump system, which captures the vinasse from the main field channels and directs it to irrigation pipes equipped with sprinklers. In practice, concentration processes can be used to reduce the volume by evaporation, with consequent changes in soil application.

Vinasse is very rich in organic matter (OM) and some mineral nutrients, particularly K. Several studies have been carried out on the influence of vinasse on soil properties such as pH, CEC, Ca, Mg, and K. The application of vinasse to soils increases soil fertility, especially when the supply of nutrients surpasses the nutritional requirements of sugarcane (Silva and Orlando Filho, 1981).

7.2.5.1 Vinasse Composition

The composition of vinasse is highly variable due to several factors. One is its origin. When cane juice is used for fermentation, the resulting vinasse is always less concentrated than the vinasse obtained from molasses must or mixed must. In addition, the concentration of vinasse varies from mill to mill, and within each mill, there are variations with harvest day and even within the same day due to the grinding of different varieties with different maturation rates and of material from different soils with different fertility levels, etc. Table 7.5 shows the maximum, mean, and minimum values for the chemical composition of vinasse and some of the physical characteristics that were determined in an evaluation conducted at sugarcane processing plants in the State of São Pauloby Neto and Nakahondo (1995).

7.2.5.2 Effect of Vinasse on Soils

In the 1950s, it was believed that the application of vinasse, which is a very acidic liquid residue, could increase soil acidity. In a pioneering study, Almeida et al. (1950) showed that soil acidity increased immediately after vinasse application; it then decreased gradually due to the microbial activity that was stimulated by the addition of organic matter present in the vinasse. The presence of basic cations, the growth of microorganisms and the transformation

TABLE 7.5 Physical-Chemical Characterization of Vinasse (Average of 64 Samples From 28 Sugar Mills in the State of Sao Paulo)

Description	Concentrations			Standard/L.
	Minimum	Average	Maximum	Alcohol
Process Data				
Must Brix (°B)	12.00	18.65	23.65	
Alcohol content of wine (°GL)	5.73	8.58	11.30	
Vinasse rate (L/L alcohol)	5.11	10.85	16.43	10.85 L
Reference Flow (m³ day⁻¹)	530.00	1908.86	4128.00	
Vinasse characterization				
pH	3.50	4.15	4.90	
Temperature (°C)	65.00	89.16	110.50	
Biochemical Oxygen Demand (BOD)(mg L⁻¹)	6680.00	16949.76	75330.00	175.13 g
Chemical Oxygen Demand (COD)(mg L⁻¹)	9200.00	28450.00	97400.00	297.60 g
Total Solids (TS) (mg L⁻¹)	10780.00	25154.61	38680.00	268.90 g
Total Suspended Solids (TSS) (mg L⁻¹)	260.00	3966.84	9500.00	45.71 g
Suspended Solids Fixed (SSF) (mg L⁻¹)	40.00	294.38	1500.00	2.69 g
Volatile Suspended Solids (VSS) (mg L⁻¹)	40.00	3632.16	9070.00	43.02 g
Total Dissolved Solids (SDT) (mg L⁻¹)	1509.00	18420.06	33680.00	223.19 g
Volatile Dissolved Solids (VDS) (mg L⁻¹)	588.00	6579.58	15000.00	77.98 g
Fixed Dissolved Solids (FDS) (mg L⁻¹)	921.00	11872.36	24020.00	145.21 g
Sedimentable waste (RS) 1 hora (mg L⁻¹)	0.20	2.29	20.00	24.81 mL
Calcium (mg L⁻¹ CaO)	71.00	515.25	1096.00	5.38 g
Chloride (mg L⁻¹ Cl)	480.00	1218.91	2300.00	12.91 g
Copper (mg L⁻¹ CuO)	0.50	1.20	3.00	0.01 g
Iron (mg L⁻¹ Fe₂O₃)	2.00	25.17	200.00	0.27 g
Total Phosphorus (mg L⁻¹ P₂O₄)	18.00	60.41	188.00	0.65 g
Magnesium (mg L⁻¹ MgO)	97.00	225.64	456.00	2.39 g
Manganese (mg L⁻¹ MnO)	1.00	4.82	12.00	0.05 g
Nitrogen (mg L⁻¹ N)	90.00	356.63	885.00	3.84 g

TABLE 7.5   (Continued)

Description	Concentrations			Standard/L.
	Minimum	Average	Maximum	Alcohol
Ammonia Nitrogen (mg L^{-1}N)	1.00	10.94	65.00	0.12 g
Total Potassium t (mg L^{-1}K2O)	814.00	2034.89	3852.00	21.21 g
Sodium (mg L^{-1}Na)	8.00	51.55	220.00	0.56 g
Sulfate (mg L^{-1}SO$_4$)	790.00	1537.66	2800.00	16.17 g
Sulfite (mg L^{-1}SO$_4$)	5.00	35.90	153.00	0.37 g
Zinc (mg L^{-1}ZnO)	0.70	1.70	4.60	0.02 g
Ethanol-CG (ml L^{-1})	0.10	0.88	119.00	9.1 mL
Glycerol (ml L^{-1})	2.60	5.89	25.00	62.1 mL
Yeast (dry base) (mg L^{-1})	114.01	403.56	1500.15	44.1 g

Source: Elia Neto and Nakahodo (1995).

of organic matter, increases the pH of the soil (Leal et al., 1983). The effects of increasing soil pH may be ephemeral (Rodella et al., 1983), and the soil pH may return to its original value after a certain period of time. Increased soil CEC may occur after years of vinasse use in soils due to the input of organic matter from vinasse addition. (Glória and Filho, 1983).

Many studies have investigated the effects of vinasse application on soils over time. In addition to the effect on CEC and soil pH, application of vinasse increases the availability of several nutrients, improves soil structure, increases water retention, and improves the biological activity of the soil by promoting the presence of a greater number of small animals (worms, beetles, etc.), bacteria, and fungi. The occasional harmful effects of vinasse application to soil or plants are usually due to the use of excessive doses (Camargo et al., 1983; Ferreira and Monteiro, 1987; Glória and Orlando Filho, 1983; Leme et al., 1980).

Although vinasse contains high concentrations of K, a monovalent cation whose characteristic of promoting the dispersion of clay particles is similar to that of Na, vinasse application may to reduce the amount of porous space and reduce the permeability of soil at doses of up to 300 m³ ha⁻¹. Uyedal et al. (2013) did not observe changes in the hydraulic conductivity of three soils but observed increases in the Ca and K content and in the CEC of the soils with vinasse application. Jiang et al. (2012) observed a decrease in soil bulk density and increased total porosity and capillarity porosity in the arable layer of the soil after 2–3 years of continuous vinasse application.

7.2.5.3 Effect of Vinasse on Sugarcane Yield

Positive effects of vinasse on sugarcane yield have been reported by many researchers in practically all varieties of cane under the most diverse soil and climatic conditions and are visible in commercial areas (Gloria, 1975; Gloria and Orlando Filho, 1983; Penatti et al., 1999a, b). Vinasse application has been shown to increase the yield level from many soils via the input of organic matter and the nutrients contained. Vinasse application also increases the number and distribution of sugarcane roots. In the study of Rossetto et al. (2013), root density increased by 34% in the top 0.1-m soil layer, by 122% in the 0.1 to 0.2-m soil layer, and by 36% in the 0.2 to 0.3-m soil layer relative to the control after vinasse fertigation. Linear correlations between the cationic saturation of the CEC base saturation with soil depth and between root number and density were found. These results indicate that successive fertigation with vinasse promotes soil fertility and the development of large and deep root systems.

The dose of vinasse used typically provides a sufficient amount of K for the ratoon cycle. Because sugarcane consumes large amounts of K, it is possible that in many areas, the vinasse dose may add greater amounts of K than necessary. Because vinasse is not a complete fertilizer that supplies all sugarcane needs, many researchers have studied how, when, and with what to supplement vinasse. Ratoons usually require that vinasse be complemented with N (Penatti, 1999a, b; Penatti and Forti, 1997). Vinasse can be an excellent fertilizer for other crops as well; for example, it is used to fertilize wheat in Egypt (Arafat and Yassen, 2002).

7.2.5.4 Vinasse application Dose

Vinasse doses are calculated by analyzing the soil K and the content of this nutrient in the vinasse in the same way that chemical fertilizers are analyzed. Amounts appropriate for the sugarcane needs in the crop year should be applied, taking into account what the soil can provide. Environmental legislation (CETESB, P4231) in São Paulo state, Brazil, restricts the rate of vinasse that can be applied to soils. The recommended vinasse dose is calculated according to the following formula:

$$\text{m}^3 \text{ of vinasse/ha} = [(0.05 \times x \text{ CEC} - \text{ks}) \times 3744 + 185]/\text{kvi},$$

where CEC = the cation exchange capacity expressed in cmolc/dm³ at pH 7.0, as determined by a soil fertility analysis performed in the laboratory of soil analysis using the soil analysis method of the Agronomic Institute of Campinas, signed by the technician responsible; 0.05 = 5% of the CEC; ks = the K concentration in the soil, expressed in cmolc/dm³, at the depth of 0.80 meters, as determined by a soil fertility analysis performed in the laboratory of soil analysis using the soil analysis method of the Agronomic Institute of Campinas, signed by the technician responsible; 3744 = a constant used to transform the results of the fertility analysis, expressed in cmolc/dm³ or meq/100cm³, for kg of K per volume in one hectare per 0.80 meters in depth; **185** = kg of K_2O extracted by the crop per ha per harvest; kvi = the K concentration in the vinasse, expressed in kg K_2O/m^3, presented in an analytical results report signed by the responsible technician.

7.2.5.5   Technology for Vinasse Application

Initially, vinasse was transported and applied using tanker trucks. Two early methods remain in use to this day: (i) application of vinasse in its original state; (ii) application of vinasse that has been diluted with water or with effluent from the mill (sugarcane washing water, condensation water, barometric column water, floor washing water, etc., otherwise known as wastewater).

Among the several alternatives presented for the use of vinasse, the method that was found to be the most economical and efficient from an agricultural point of view (and therefore became widespread and was adopted by most sugarcane mills) was the use of fertirrigation of sugarcane plantations. The term fertirrigation, although widely used in the sector, is not appropriate in this case because vinasse application is not exactly irrigation in the sense of controlling water depth and watering frequency. In the case of vinasse, the term refers to the application of liquid residue that also causes soil wetting (Freire and Cortez, 2000).

Technological innovations for the application of vinasse modified the rudimentary systems of application to include the use of flooding and infiltration furrows. A main pipe transported the vinasse to the main channels; these reached the planting rows by flooding or via PVC pipes with outlets to each row. A slope of 0.2% to 0.5% was required, and furrows were opened between the sugarcane rows (Rossetto, 1987). This system was complicated and inefficient, and the resulting application of the vinasse was very irregular. Subsequently, at the end of the 1970s, semi-fixed sprinkler systems were used in which the vinasse was pumped from the main channels by motor-driven pumps that fed lateral pipes to which the sprinklers were coupled. This system allowed greater control of the vinasse doses applied. In the 1980s, vinasse was applied by sprinkling with a gun sprinkler that moved on narrow towpaths next to the channels that transported the vinasse.

Introduction of the hose reel was an advance in the application of vinasse. Spraying was performed using self-propelled equipment with medium-density polyethylene tubing; this resulted in greater automation, higher operating yield, greater efficiency, less labor, fewer equipment changes, and less transport equipment (Leme et al., 1987).

Technological innovations continued with the use of irrigation bars 36 to 54 m long with sprinklers that delivered flow rates ranging from 25 to 150 m^3/h. In this system, the bar is attached to the end of the reel hose, permitting reduced application pressure and improving uniformity. Towable center pivot equipment can also be used for vinasse application, but its high corrosivity increases maintenance costs.

The spray application system involving a hose reel is still widely used today. Although it represents a lower initial investment compared to other irrigation systems, it has the highest operating cost (Hernandez, 2006).

7.2.5.6   Effect of Vinasse on the Quality of Sugarcane Raw Material

Despite its high agronomic value, which is reflected in the increase in sugarcane yield, a negative effect of high vinasse doses on the quality of sugarcane raw material for production has been observed. Silva et al., (1976) studied 16 sugarcane varieties planted with and without vinasse irrigation at a dose of 100 m^3ha^{-1}, which is equivalent to approximately 590 $kgha^{-1}$ of K_2O. They found that the addition of vinasse to the soil, especially of vinasse with high K content, delayed plant maturation, reduced the sucrose and fiber content of the plants, and promoted the accumulation of ashes in the cane juice, thereby decreasing the quality of the raw material, especially for the production of sugar.

7.2.5.7 Vinasse Use and Environmental Issues

The continuous use of vinasse in the same soils year after year, even in low amounts, can lead to cation saturation, especially K saturation, in the CEC of the soil and cause leaching of the soil constituents to groundwater.

Leaching of K to the subsurface is not an environmental problem because K is not a water pollutant, but K excess in the soil solution of soil cation exchange complex may stimulate nitrate leaching as KNO_3.

In the Brazilian state of São Paulo, Regulation P4.231 (CETESB) regulates the use of vinasse. There are also standards for the waterproofing of channels and reservoirs to decrease leaching losses. Vinasse can be applied at high dosages to the soil only if the saturation of K in the soil CEC is less than 5%. If this value has already been reached, regulation allows only the use of a K dose that will be utilized by sugarcane in the year in question, i.e., the vinasse dose equivalent of 185 kg ha^{-1} K_2O.

With the current regulation of vinasse use, many areas will have restrictions and the sugarcane industry is already moving toward transporting vinasse to greater distances. One of the solutions under study is vinasse concentration.

Although there are some indications that ions originating from vinasse, such as Cl$^-$ and SO_4^{-2} may be present in groundwater; contamination of the water table due to the use of vinasse at the doses currently used, such as in fertirrigation over sugarcane straw, may be a remote possibility that is restricted to certain conditions, such as those involving marginal, sandy, and shallow soils. If this occurs, the contamination process is likely to take many years, but the simple fact of the possibility of contamination demands caution.

In deep clay soils, there is a large soil layer for K adsorption. Cunha et al., (1986) studied leachate from a soil to which a dose of 800 m^3 ha^{-1} of vinasse was applied and did not find movement of K$^+$ in the soil below 40 cm of depth. Because sugarcane (especially ratoons) has a very deep root system, K can be utilized even if it leaches to deeper layers. Orlando Filho et al., (1995) studied leaching NO_3^- and NH_4^+ in sandy soils after the application of mineral N (60 kg/ha) and vinasse doses of 0, 150, 300, and 600 m^3/ha. The results obtained over four sampling periods at three soil depths (up to 2 meters) showed no N leaching up to 25 weeks after application.

Cruz et al. (1990) reported the results of a long-term study of the influence of vinasse applied at a dosage of 300 m^{3}ha^{-1}year^{-1} on soil properties (up to 1.5 m deep) and groundwater quality. The experiment covered periods of

0, 5, 10, and 15 years, and it was found that soil fertility improved over the years. The content of organic matter and P in the surface layer increased. The Ca, Mg, and S contents increased not only in the surface layer but also in the deeper layers. Nitrate was found in the groundwater but at a concentration below the limit harmful to human health.

Gloeden et al. (1990) evaluated the influence of vinasse application at doses of 150 and 300 m^3ha^{-1} to a sandy soil with low CEC on soil fertility and on the leaching of ions into groundwater. They found increased concentrations in the groundwater of: chloride, ammonia, organic C and forms of organic N. The K remaining in the soil was absorbed by the sugarcane and did not reach the water table during the study period. However, the authors caution that organic forms of N were found and that these could have generated nitrates.

Possible contamination of the water table by vinasse has been reported in the literature. Such contamination may occur under special conditions, such as when vinasse is applied in high doses, e.g., 12,000 m^3 of vinasse in the work of Hassuda (1989) or when it is applied to sandy or very shallow soils, as reported by Gunkel et al. (2007). Gloeden et al. (1991) reported that even with increasing doses of vinasse, the K concentration was maintained close to the natural level at depths of 2.90 and 4.50 m; the levels of dissolved organic carbon (DOC) indicated strong variation in the deeper layers after vinasse application, which the authors ascribe to its association with organic vinasse components, especially colloids. The soil is recognized as an efficient filtering element for organic components applied to the soil surface. Lyra et al. (2003) studied the chemical oxygen demand (COD) along the soil profile of soil fertigated with vinasse and observed a significant reduction in COD; the values in the soil in which vinasse was used were much higher than those in the water table.

Regarding greenhouse gas emissions, the production and use of ethanol in Brazil generates fewer greenhouse gas emissions (GHG) than does the production and use of gasoline (Egeskog et al., 2014) and is therefore considered a fuel with less environmental impact.

Greenhouse gas emissions are always higher when straw and/or N fertilizer is used than when vinasse is used in place of these materials. Thus, nitrous oxide (N_2O) emissions from N fertilizers may strongly affect the sustainability indicators of ethanol produced from sugarcane, and there is evidence that the application of vinasse could enhance the emission of GHGs from N fertilizers. Carmo et al. (2013) observed highest emissions for ratoon cane treated with vinasse especially as the amount of crop trash on the soil surface increased. Emissions of CO_2 and N_2O were 6.9 kg ha^{-1} yr^{-1} and 7.5 kg ha^{-1} yr^{-1}, respectively, totaling about 3000 kg in CO_2 equivalent ha^{-1} yr^{-1}.

The strategy of separating the application of N fertilizer and vinasse in time was tested in field experiments in Brazil by Cantarella et al. (2016). The strategy of anticipating or delaying the application of both regular and concentrated vinasse by approximately 30 days with respect to N fertilization in most cases resulted in lower N_2O emissions.

It should be noted that vinasse is originated from the cane juice organic origin and that it does not contain metals or other contaminants that prevent its agricultural use. In this sense, it is perfectly accepted by organic agriculture, and there are no restrictions on its use as a source of nutrients by certifiers. Vinasse represents an important source of K to be considered in organic agriculture activities. For the production of organic sugar, vinasse can supply all the N and K required by sugarcane.

7.2.5.8 Concentrated Vinasse

Concentration of vinasse is performed by evaporation, with the main objective of reducing transport costs. Several evaporation processes can be used; however, the high-energy cost of vinasse concentration, regardless of the process, is perhaps the main restriction. In the concentration process at the plant, water is removed from the vinasse and recycled for use in other processes. Because there is no loss of vinasse solids, the nutrients remain in the vinasse. Experiments in which the use of concentrated vinasse in sugarcane was investigated showed that the crop utilization of nutrients in concentrated vinasse is similar to that of the original vinasse (Barbosa, 2006; Rossetto et al., 2010). The N contained in concentrated vinasse is mineralized slowly compared to that in non-concentrated vinasse, allowing better utilization by sugarcane (Silva et al., 2013).

In Brazil, an increase in the number of vinasse concentrators is expected in the coming years to meet the environmental standards of vinasse distribution in soils with K saturation below 5% CEC.

In the coming years, greater efficiency in the ethanol production process is expected as a result of genetic improvement of yeast, leading to lower production of vinasse per liter of ethanol produced and reducing the impact generated by this residue.

7.2.5.9   Other Uses of Vinasse

In addition to the use of vinasse as a fertilizer, other options for the use of vinasse may have economic importance, as suggested by Camhi (1979). These options include its use in (a) animal feed; (b) aerobic fermentation by microorganisms for the production of specific proteins; and (c) anaerobic fermentation using methanogenic bacteria to produce methane (biogas). Vinasse was also used at the Campinas State University by Rolim (1996) to obtain so-called soil-vinasse, a material that can be used to make bricks for construction. In addition, Freire and Cortez (2000) studied the possibility of combustion or incineration of vinasse for recovery of K salts and energy utilization.

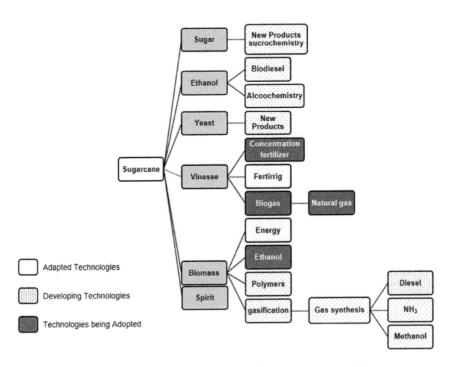

FIGURE 7.4   In use and promising technologies from the products of the sugar-ethanol industry (Adapted from Dedini, 2013).

7.3 NEW TECHNOLOGIES FOR THE USE OF RESIDUES

In the coming years, it is expected that residues of the agro-ethanol industry will be increasingly utilized in the production of novel products such as enzymes, proteins, and amino acids and also energy. The biodigestion of vinasse for the production of energy is a technology that is also expected to increase in the coming years.

7.4 CONCLUSIONS

The agricultural use of residues from the sugarcane agroindustry has undergone several changes over the years. In addition to concern with greater agronomic efficiency and management optimization the environmental awareness after the 1990s required a self-organization of the sugarcane industry, to address new problems arising from the use of large amounts of residues in soils, such as the leaching of chemical elements, the consequent transport of these elements to groundwater, and aquifer pollution. New technologies, such as vinasse concentration and biodigestion, have emerged to make the sector more competitive and sustainable. Additional new technologies must be developed, and these must always be accompanied by relevant legislation and environmental monitoring to help preserve the environment.

The sugarcane crop is becoming more sustainable due to advances in technology, which have resulted increased sugarcane yields even after more than 100 years of sugarcane farming. Soil quality and function have been maintained and even improved over the years. The addition of organic matter, mainly residues from sugarcane production, represents a great potential for reducing the use of fertilizer by nutrient recycling and for the maintenance of soil quality and sustainability.

Maintaining sugarcane straw in the field presents several advantages, including greater soil protecting the soil, reducing water loss through evaporation, and controlling erosion including greater soil protection. Straw also increases the organic matter content of the soil, increases the soil's nutrient retention capacity, and recycles nutrients by bringing nutrients to the soil surface that had been removed by roots from deeper soil layers. Furthermore, straw can be recovered from the field to produce energy or more ethanol in the second-generation process. It is expected for the coming years that residues of the sucro-ethanol industry will be increasingly utilized in the production of novel products such as enzymes, proteins, and amino acids and also energy diversifying the chain and bringing more opportunities to the sector.

KEYWORDS

- biogas
- biomass
- composting
- fertirrigation
- harvest residue
- vinasse

REFERENCES

Arafat, S., & Yassen, A. E., (2002). Agronomic evaluation of fertilizing efficiency of vinasse. *17th World Congress Soil Science Proc.*, 14–21 August, Thailand, pp. 1991–1996.

Barbosa, V., (2006). New technologies in the use of vinasse and legal aspects. In: MARQUES, MO et al. (Ed.) Topics in sugarcane technology. Jaboticabal, Unesp, pp. 141–150,

Braunbeck, O. A., & Andmagalhaes, P. S. G., (2010). *"Technological Evaluation of Sugarcane Mechanization",* "Sugarcane Bioethanol: R & D for Productivity and Sustainability", 05/2010, ed. 1, Chap. 12, Blucher Publishing, pp. 14, pp. 411–424.

Camargo, de, O. A., et al., (1983). Chemical and physical characteristics of soil that has been vinasse for a long time. IAC Technical Bulletin. Institute of Agronomy of Campinas, Campinas, SP, (76), 1983, 30 p.

Camhi, J. D., (1979). Treatment of vinhoto, by-product of alcohol distillation. Brazil Açucareiro, Vol. *94*, n. 1, p. 18–23.

Cantarella, H., et al., (2016). Nitrous oxide emission from N fertilizer and vinasse in sugarcane. Proceedings of the 2016 International Nitrogen Initiative Conference, "Solutions to improve nitrogen use efficiency for the world," 4–8 December, Melbourne, Australia. *www.ini2016.com.*

Carmo, J. B., et al., (2012). Infield greenhouse gas emissions from sugarcane soils in Brazil: effects from synthetic and organic fertilizer application and crop trash accumulation. *Glob. Change Biol. Bioenergy*, *5*, 267–280.

Carvalho, J. L. N., et al., (2010). Potential for carbon sequestration in different Brazilian biomes *Rev. Bras. Cienc. Solo.*, *34*(2), 277–289.

Cerri, C. C., et al., (2009). Brazilian greenhouse gas emissions : the importance of agriculture and livestock. *Sci. Agric. (Piracicaba, Braz.)*, *66*(6), 831–843.

Christofoletti, C. A., et al., (2013). Sugarcane vinasse: Environmental implications of its use. *WasteManage.*, *33*, 2752–2761.

Cruz, R. L., Righetto, A. M., & Nogueira, M. A., (1990). Experimental investigation of soil and groundwater impacts caused by vinasse disposal. In: International Seminar of Pollution, Protection and Control of Ground Water, Porto Alegre, RS, Brazil, pp. 66–74.

Cunha, A. R. C., et al., (1986). Effect of irrigation with vinasse and the dynamic of its constituints in the soil. I: Physical and chemical aspects. In use of soil treatment and final disposal of effluent and sludge. 13–15 August. Salvador, Bahia, Brazil. Ed. P. R. C. Oliveira & S. A. S. Almeida.

Dedini, (2013). http://www.codistil.com.br/index.php?option=com_docman&task=cat_view&gid=51&Itemid=40&lang=pt.

Egeskog, A., et al., (2014). Greenhouse gas balances and use changes associated with the planned expansion (to 2020) of the sugarcane ethanol industry in Sao Paulo, Brazil. *Biomass Bioenergy, 63,* 280–290.

Elia Neto, A., & Nakahodo, T., (1995). Physical-chemical characterization of vinasse - project n. 9500278. Technical Report of the Water Treatment Technology Section of the Copersucar Technology Center, Piracicaba, 1995. 26 p.

Faroni, C. E., et al., (2003). Degradation of sugarcane straw (15N) in two consecutive years (Compact disc). In: Brazilian Congress of Soil Science, 29, 2003, Ribeirão Preto. Anais – Ribeirão Preto: UNESP / SBCS, 2003.

Freire, W. J., & Andcortez, L. A. B., (2000). Vinasse of sugarcane. Guaíba: Agropecuaria Publishing House. 203 p. 2000

Garcia, S. S., et al., (2003). The Sustainable Management Group of Sugar Cane from the School of Postgraduates, *México,* 395 p.

Gloeden, E., Cunha, R. C., Fraccaroli, M. J. B., & Cleary, R. W., (1990). The behavior of vinasse constituents in the unsaturated and saturated zones in the Botucatu Aquifer recharge area. São Paulo: Cetesb, 11 p.

Gloeden, E., et al., (1991). The behavior of vinasse constituents in the unsaturated and saturated zones in the Botucatu aquifer recharge area. *Water Sci Technol., 24*(11), 147–157.

Gloria N. A., (1975). Agricultural use of vinasse. *Brasil Acucareiro, Rio de Janeiro., 86*(5), 11–17.

Glória, N. A., & Magro, J. A., (1977). Agricultural use of waste from sugar and distillery at Usina da Pedra. In: Copersucar Seminar of the Sugar Agroindustry, 4, Aguas de Lindóia, 1976. *Anais. São Paulo, Copersucar,* pp. 163–180.

Gloria, N. A., & Orlando Filho, J., (1983). Application of vinasse as fertilizer. Technical Bulletin, *5,* 38.

Gunkel, G., et al., (2007). Sugarcane industry as a source of water pollution case study on the situation in Ipojuca river Pernambuco Brazil. *Water Air Soil Pollu., 180,* 261–269.

Hassuda, S., (1989). Impacts cane vinasse infiltration in Bauru aquifer. Masters dissertation. USP-ICG. 1989. 92 pp.

Hernandez, F. B. T., (2006). Management of irrigation in sugarcane, http://www.agr.feis.unesp.br/rcn_abril, acesso em 27/01/2007.

Jiang, Z. P., et al., (2012). Effect of long-term vinasse application on physico-chemical properties of sugarcane field soils. *Sugar Tech., 14,* 412–417.

Lal, R., (2008). Carbon sequestration. *Philos. Trans. R. Soc. Lond., 363,* 815–830.

Landell, M. G. A., et al., (2013). Residual biomass potential of commercial and pre-commercial sugarcane cultivars. *Sci. Agric. (Piracicaba, Braz.), 70*(5), Piracicaba Sept/Oct. http://dx.doi.org/10.1590/S0103--90162013000500003.

Leal, J. R., Amaral Sobrinho, N. M. B., Velloso, A. C. X., & Rossiello, R. O. P., (1983). ROP Redox potential and pH: variations in a treated soil with vinasse. R. Bras. Ci. Solo, 7, 257–261.

Leme, E. J. A., et al., (1987). The use of self-propelled in the application of vinasse by sprinkling: technical-economical viability. Piracicaba, IAA Planalsucar, 1987. 65 p.

Luz, P. H. C., & Vitti, G. C., (2008). Management and use of fertilizer for sugarcane. In: Marques, MO et al. Technologies in the sugarcane farming. Jaboticabal: FCAV,2008. pp. 141–167.

Lyra, M. R. C. C., Rolim, M. M., & Silva, J. A. A., (2003). Toposequence fertirrigated of soils with vinasse: contribution to the quality of ground sheet. Journal of engineering Agriculture and Environment, v.7, n.3, pp. 525–532.

MAPA (2017). *http://www.udop.com.br/download/estatistica/area_cultivada/23fev17_area_colhida_producao_produtividade_cana.pdf.*

Mitchell, R. D. J., Thorburn, P. J., & Larsen, P., (2000). Quantifying the loss of nutrients from the immediate area when sugarcane residues are burnt. *Proc. Aust. Soc. Sugar Cane Technol.*, Bundaberg, Queensland, Australia, 2–5 May. pp. 206–211.

Oliveira, M. W., et al. (1999a). Sugarcane trash degradation. *Sci. Agric. (Piracicaba, Braz.)*, *56*, 803–809.

Oliveira, M. W., et al. (1999b). Field decomposition and release of sugar cane trash nutrients. *Pesq. Agropec. Bras.*, *34*, 2359–2362.

Orlando Filho, J., Bittencourt, V. C., De, & Alves, M. C., (1995). Aplicação de vinhaça em solo arenoso do Brasil e poluição do lençol freático com nitrogênio. In: *Congresso Nacional da Sociedade de Técnicos Açucareiros e Alcooleiros do Brasil.*, *13*(6), 14–17.

Penatti, C. P. (1999a). Doses of vinasse versus doses of nitrogen in cane soca during four harvests. Internal Report Copersucar, Usina São Luiz SA, clay soil (LR-2).

Penatti, C. P. (1999b). Doses of vinasse versus doses of nitrogen in cane soca during four harvests. Internal Report Copersucar (RT928), São José da Estiva Plant, sandy soil (LVA-9).

Penatti, C. P., & Forti, J. A., (1997). Doses of vinasse versus doses of nitrogen in sugar cane. In: VII Seminar on Agronomic Technology. Piracicaba, 1997. Anais ..., Copersucar, Nov, 1997. pp. 328–339.

Ramos, N. P., et al., (2016). Carbon dioxide enrichment effects on the decomposition of sugarcane residues. *International Society of Sugarcane Tech. Chiang Mai*, Thailand, Vol. *26*.

Rodella, A. A., Silva, L. C. F., Orlando Filho, J. O., (1999). Effect of filter cake application on sugarcane yields. *Turrialba*, *40*(3), 323–326.

Rodella, A. A., Zambello Jr., E., & Orlando Filho, J. O., (1983). Calibration of soil phosphorus and potassium analysis in sugarcane - 2nd approach. Saccharum, 28: 39-42, 1983.

Rolim, M. M., (1996). Measure the single-phasic the concentrated ground material and its vinase utilization for the purposes of fabrication of the bricks. Dissertation (Master of Engineering Used Farm) Used Farm School of Economics, University of Campinas, Campinas, 1996. 90 p.

Rossetto, A. J., (1987). Products and waste from the sugar and alcohol industry. In: Paranhos, SB (ed.). Sugar cane: cultivation and use. Campinas: Cargill Foundation, 1987, Vol. *2*, pp. 435–504.

Rossetto, R., et al. (2010a). Fertility maintenance and soil recovery in sugarcane crop. In: Cortez, L. A. B. (Ed.). *Sugarcane Bioethanol. R&D for Productivity and Sustainability*, São Paulo: Blucher. *1*, 1381–403.

Rossetto, R., et al., (2013). Vinasse Enhances Sugarcane Roots in a Brazilian Sandy Brazilian Soil. *International Society of Sugarcane Technologists Proc. XXVIII*, São Paulo, Brazil. CD, pp. 58.

Silva, A., et al., (2013). Net and Potential Nitrogen Mineralization in Soil with Sugarcane Vinasse. *Sugar Tech.*, *15*(2), 159–164.

Silva, G. M., de A., Pozzi D. E., Castro, L. J., & Magro, J. A., (1976). Agroindustrial behavior of sugarcane in irrigated and non-irrigated soil with vinasse. *Anais., Águas de Lindoia*, pp. 107–122.

Thorburn, P. J., Probert, M. E., & Robertson, F. A., (2001). Modeling decomposition of sugar cane surface residues with APSIM-Residue. *Field Crops Res.*, *70,* pp. 223–232.

Trivelin, P. C. O., et al., (2013). Impact of sugarcane trash on fertilizer requirements for São Paulo, Brazil. *Sci. Agric. (Piracicaba, Braz.) 70*(5) Piracicaba Sept./Oct. http://dx.doi.org/10.1590/S0103-90162013000500009.

Urquiaga, S., et al., (2011). Evidence from field nitrogen balance and 15N natural abundance data for the contribution of biological N2 fixation to Brazilian sugarcane varieties. *Plant Soil, 356,* 5–21.

Uyedai, C. A., et al., (2013). Influence of vinasse application in hydraulic conductivity of three soils. *Eng. Agríc.*, *33*(4), 689–698.

Vasconcelos, Acm of; Casagrande, AA Physiology of the root system. In: Dinardo-Miranda, Ll; Vasconcelos, Acm Of; Landell, Mg, A. (Ed.). Sugar cane. Campinas: Instituto Agronômico, 2008. 882 p.

Vitti, A. C., et al (2008). Straw mineralization and root growth of sugar cane as related to nitrogen fertilization at planting. Rev. Bras. Ciênc. Solo [online]. 2008, vol. *32*, n.spe, pp. 2757–2762.

CHAPTER 8

WATER INDUCED STRESSES IN SUGARCANE: RESPONSES AND MANAGEMENT

R. GOMATHI

Plant Physiology, ICAR-Sugarcane Breeding Institute, Coimbatore–641007, Tamil Nadu, India, E-mail: gomathi_sbi@yahoo.co.in

CONTENTS

8.1   INTRODUCTION

Abiotic stresses are the most important limiting factors for cane productivity. These stresses include drought, flooding and temperature extremes, which cause detrimental effects on plant growth and yield. These negative factors affect the root function, growth rates, metabolism and in extreme cases lead to dehydration and death. Also, the expected rise in global temperatures indicates that there is an urgent need to understand and improve plant tolerance to these stresses. In India, the productivity losses due to various abiotic stresses vary from 20 to 50% (Dwivedi, 2000). In Maharashtra, a high recovery zone, large areas have gone out of cultivation due to salinity, alkalinity and waterlogging (Zende, 2002; Zende and Hapse, 1986). Drought is the primary abiotic stress causing not only differences between the mean yield and the potential yield but also causing yield instability. Drought stress associated with high day temperature causes poor growth and high tiller mortality particularly during primary growth stage, which normally coincides with summer months in tropics. High temperatures have deleterious effects on plant photosynthesis, respiration and reproduction. A small increase in temperature results in conspicuous effect on growth and survival. Elevated temperatures cause rapid loss of water resulting in dehydration. In addition, drought coupled with water logging, i.e., early drought and subsequent water logging in Bihar, U.P., and Orissa is becoming a serious productivity constraint affecting considerable area under sugarcane cultivation. Sugarcane is moderately tolerant to flooding and water logging. However duration of water logging and the physiological stage at which the problem occurs determines the final yield and quality. Higher water table during active growth phase adversely affects stalk weight and plant population resulting yield loss at the rate of about one ton per acre for one inch increase in excess water (Carter, 1976; Carter and Floyed, 1974), although sugarcane is relatively tolerant to high water tables (Deren et al, 1991a, 1993; Roach and Mullins, 1985). It is reported that well-established cane can survive few months into flood, while less established cane appears to be much more vulnerable to flooding (Deren et al., 1991b). These abiotic stresses are location specific, exhibiting variation in frequency, intensity and duration and might occur at any stage of plant growth and development. This chapter provides an overview of recent research on sugarcane response in terms of growth and development, yield and quality to water induced stresses and high temperature and their management to cope up or sustain the sugarcane production for future challenges.

8.2 SUGARCANE UNDER MOISTURE STRESS

Water stress remains an ever-growing problem and it is the major limiting factor in crop production worldwide (Jones and Corlett, 1992). In India, nearly 60% of the total sugarcane agriculture suffers from lack of adequate water supply mainly because of limited availability of water for irrigation in lift irrigated areas, canal closure during summer in many of canal irrigated tracts, and drought which occur in a cyclic manner (Sundara, 1998). Therefore water stress of varying degrees is experienced at one stage or the other of the crop growth in all most all the sugarcane growing regions of the country.

8.2.1 WATER REQUIREMENT AND EVAPOTRANSPIRATION:

Total water requirement of annual sugarcane crop varies from 1850 mm to 2500 mm. It is estimated that 250 tonnes of water is required for production of a tonne of sugarcane. Daily evaporation in sugarcane fields varies from 8–10 mm. Solar energy, wind velocity, temperature, and humidity affect the evapotranspiration. Earlier trials on response of sugarcane to irrigation suggested that maximum tonnage was obtained at Et/Ep of 0.8. Sheath moisture and moisture content of immature nodes also served as useful indices for determining the water requirement of sugarcane crop. For high yield, sheath moisture index at 5th month stage should be high enough (83–85%), and for higher CCS%, proper drying off with sheath moisture index of about 72% at 12th month was found to be desirable.

8.2.2 CRITICAL WATER DEMAND PERIOD

Formative growth stage (60–150 days) has been identified as the critical water demand period and stress at this early growth phase had a direct influence on the cane yield and juice quality. Yield reduction up to 60% has been recorded in a typical drought year. Water stress especially during summer months coincides with the formative phase of the crop which affects the final yield through reduction in tiller productivity, number of millable canes, individual cane weight, and finally the cane yield and juice quality (Naidu, 1987).

8.2. 3 PLANT RESPONSES TO DROUGHT STRESS

8.2.3. 1 Root System

Extensive root investigations revealed that the sett roots emerge from the root band (present at nodal region of sugarcane sett), and start growing within 24 hr of planting. At the third day, some roots extend at a rate of 10 mm/day and by day 5, the elongation reaches to 20 mm/day. These thin and branched sett roots are replaced by thick; fleshier and less branched shoot roots by 90 days age. Rooting depth, distribution and activity are generally affected by soil water relationships (Naidu and Venkataramana, 1993). Generally more root mass occur at less than 50 cm depth in normally irrigated condition while under stress, roots penetrate vertically downwards in the form of a rope. The root system also shows penetrating type roots which reach out for water source and hence longer and thicker roots are seen under drought (Venkataramana and Naidu, 1989). The varieties selected for greater rooting depth suffered the least water deficits as compared to the normally irrigated plants. However, reports of diminished root development under moisture stress have been reported by Mongelrad (1968) and Rao (2000). Differences in root growth were related to differences in growth of susceptible and toler- ant varieties (Mongelrad, 1968a).

8.2.3.2 Shoot System

The maximum LAI is generally achieved by about 6 months from planting and then slowly declines. A high LAI produces large structural apparatus for the production of photosynthate and a higher yield. Leaf expansion is very sensitive to stress. Large differences occur in the density of stomata of crop plants. The activity of stomata is greatly affected by external factors such as light, temperature, and humidity. Direct sunlight makes stomata to open, while weak and diffusive light result in closure. This explains the beneficial effect of early morning sunshine on sugarcane. Since drought is common in many sugarcane growing areas, it is important to consider reducing the tran- spiration and thereby reducing consumptive water use. Transpiration occurs predominantly (>90%) through the leaves while nodal region, which is free from wax deposition. Significant reduction in water loss (10 to 20%) was demonstrated due to passive curling of leaves, which reduce the radiation receipt by leaves thereby reducing water loss and increasing water use effi- ciency to a greater extent. Cell growth is retarded under mild stress which in turn results in reduced leaf area, followed by reduced sink growth and reduced stem elongation. The major attribute is the drying off of older leaves and stunted growth of stem resulting in a dwarf canopy. The young leaves however remain green, but when the stress intensity becomes severe, the entire crop loses its turgidity and drying will be hastened. Characters like leaf thickness, leaf dry weight and leaf area ratio are highly sensitive to drought. Deposition of wax, which is a protective mechanism, is also seen on the upper surfaces of the sugarcane leaves and stem.

8.2.3.3   Light Interception and Photosynthesis

Sugarcane is one of the most efficient crops capable of converting a maximum of 2–3% of solar energy into organic matter through an efficient photosynthetic system (Bull and Glasziou, 1975). It has been estimated that one hectare of sugarcane can produce 100 tonnes of green matter which is more than twice the yield of most other commercial crops (Almazan et al., 2001). Majority of the clones intercepted 60–80% of the radiation at the completion of formative phase. The light falling on the crop surface varied from 1275 to 1950 μmol m^2 s^{-1}. In the initial stage of the stress, stomatal closure occurs which reduces transpiration rates and a decrease in leaf water potential which collectively influence the photosynthesis and productivity. The chlorophyll content also decreases resulting in low CO_2 fixation. Drought during the vegetative period tends to slow down the leaf development and canopy expansion. Chlorophyll fluorescence kinetics changed significantly during moisture stress indicating that photosynthetic electron transfer system (PETS), especially PSII and carbon assimilation were inhibited (Luo et al., 2000). The decrease in chlorophyll fluorescence was related to drought tolerance of varieties (Luo et al., 1999). Leaf water potential and stomatal diffusive resistance are the measure of stress intensity and were found to be related to the yield of a variety. These two parameters which were identified as water stress indicators were found useful for screening varieties for drought resistance (Naidu et al., 1983; Venkataramana et al., 1986). The carbon isotope discrimination at 240 days was negatively and significantly associated with leaf area and total dry matter, but with photosynthesis and transpiration, the relationship was not significant (Gururaja Rao, et al., 2008).

8.2.3.4   Dry Mass Accumulation and Distribution

Sugarcane has the capability of producing 65 MT of above ground dry mass per year. The dry mass production rate ranged from 20 to 35 g/day during active growth phase and the energy conversion efficiency was estimated to reach a maximum of about 1.8% (Ramanujam and Venkataramana, 1999). The increase in dry matter was low during periods of incomplete canopy development. The average dry matter produced was 16.83, 41.23, or 49.41 tonnes/ha or 4.81, 22.41 and 47.48 tonnes/ha under drought at the completion of formative (150 days), grand growth (240 days) and maturity (360 days), respectively (Venkataramana and Naidu, 1989). The growth analysis studies indicated that net assimilation rate (NAR) and relative growth rate (RGR) were high during early growth phase, but declined with the age of the crop. Leaf area ratio (LAR) and leaf area index (LAI) increased with crop growth under normal irrigation while drought caused 34.62% reduction in LAI (Venkataramana et al., 1984). Harvest index was significantly associated with cane yield, sugar yield and CCS% (Naidu and Venkataramana, 1989).

8.2.3.5   Biochemical Responses

Sugarcane plant responds to the stresses at the biochemical level. The cellular water deficits results in the concentration of solutes, loss of turgor, change in cell volume, disruption of water potential gradients, change in membrane integrity, denaturation of proteins and several other physiological and molecular components. The concentration of malondialdehyde, a lipid peroxidation product doubled as the leaf water potential declined (Venkataramana et al., 1987). Epicuticular wax content was significantly high in drought resistant varieties when compared to drought susceptible types. Cellular membrane thermo stability and electrolyte leakage decreased due to water stress thereby increasing the membrane injury to as high as 85% in susceptible types. Drought tolerant varieties recovered effectively during rehydration (>60%). The capacity to maintain high membrane themostability is an important feature of tolerance to water stress (Venkataramana et al., 1983).

8.2.3.5.1   Osmoprotection

Recently, interest has been generated on osmotic adjustment, turgor maintenance and growth. Turgor can be maintained by increasing various osmolytes. Accumulation of osmolytes (proline, glycine-betaine, polyamines, sugars, etc.), which maintain the turgor and reduce the osmotic potential, help the plant to cope with the drought effect, the phenomenon called as osmoregulation. Osmotic adjustment occurs through increased accumulation of various osmolytes such as soluble sugars, soluble carbohydrates, proline, potassium, sugar alcohols and organic acids. Concomitant with 70% reduction in leaf water potential, the osmotic potential increased in many varieties suggesting an increased accumulation of osmolytes. Under water deficit conditions, the proline accumulation increased several folds in sugarcane and a significant varietal variation was noticed by Rao and Asokan (1978). Drought stress leads to the generation of reactive oxygen species (ROS), which include superoxide anion radicals (O_2), hydroxyl radicals (OH), hydrogen peroxide (H_2O_2) and singlet oxygen (O·) which cause damage to the cellular system. Drought enhanced activities of peroxidase and polyphenol oxidase have been reported in popular cultivars of sugarcane (Vasantha and GururajaRao, 2003). The process of osmoprotection prevents protein denaturation, helps preserve enzyme structures and protects membranes from damage by ROS. Increase in solute concentration or accumulation of solutes causes osmotic adjustment and the compounds that are accumulated during stress are soluble sugars, soluble carbohydrates, proline, potassium, sugar alcohols and organic acids. Osmotic adjustment has a few advantages such as maintenance of cell turgor, continued cell elongation, maintenance of stomatal opening, and photosynthesis and survival under dehydration.

Among higher plants, proline and glycine betaine are the most commonly reported osmolytes and accumulation of these compounds sustains the viability of plants and their yields under moderate and severe drought. Osmotic adjustment helps plants to maintain turgor by increasing various osmolytes thus resulting in good growth. Free proline accumulates in water stressed leaf tissue. Oxidation of proline (to glutamate) in turgid tissues generally prevents accumulation while in stressed tissue proline accumulates only to serve as buffer of nitrogenous substances.

In several studies proline accumulation was used as a screening test for drought resistance. Another metabolically inert compound called betaine also accumulates under stress. Carlin and Santos (2009) evaluated the sugarcane variety IAC91-5155 under water stress and observed a trehalose accumulation of 25.9% (increase of 0.54 µmol g^{-1} fresh mass weight) at the 60th day under stress, reaching concentrations of trehalose of 2.54 µmol g^{-1} of the fresh weight. Queiroz1 et al. (2011), reported the increase in levels of trehalose and free proline found to confirm what many others have reported: the importance of the osmotic adjustment of plant species, genotypes and cultivars to water deficiency in the soil. These authors declare that this mechanism is related to the accumulation of compatible osmolytes to maintain cell turgidity and facilitate physiological and biochemical processes under drought conditions.

8.2.3.5.2 Nutrients

Drought imposed during formative phase significantly reduced P content while N and K did not decrease (Ramesh, 2000) contrary to the earlier report of decreasing N and K content by Samuels (1971).

8.2.3.5.3 Abscisic Acid

Abscisic acid (ABA) accumulates in drought-affected leaves. ABA content enhances the leaf water potential by 1 to 2 bars and thus helps in dehydration postponement. The ABA was also found to possess a direct and stabilizing effect on protoplasm, and drought induced senescence of leaves. Dry matter production by ABA treated plants was greater than that of control. This was due to a greater development of shoot at the expense of roots. External application of abscisic acid (1×10^{-5} M) exerted a regulatory role on stomatal diffusive resistance and helped in maintaining relatively high water potential (Venkataramana and Naidu, 1993). ABA content enhanced the leaf water potential by 1 to 2 bars and thus helped in dehydration postponement and drought induced senescence of leaves.

8.2.3.5.4 Enzymes

Enzymes such as nitrate reductase (NRase), sucrose phosphate synthase (SPS), invertase, etc. have been found to be regulated by the tissue water status. Nitrate reductase activity is reversible and the extent of loss under stress is to an extent of 30% and the regulation of nitrogen metabolism and the constituent end products are affected in the rate limiting way. Moisture stress induced reduction in the activity of SPS and sucrose synthase was reported in popular cultivars of sugarcane, which on rehydration resumed to normal level (Vasantha et al., 2003).

8.2.3.5.5 Cane Elongation

Cane elongation is positively correlated with amount of irrigation water (Chang et al., 1968). The rapid cane elongation (60 to 70%) takes place during grand growth during which the seasonal available water will be utilized. Large reduction in stalk number, height, cane yield and sucrose yield were noticed due to drought. Shih and Gascho (1980) reported that stalk elongation was positively and strongly correlated with water content of the elongating and meristematic tissues and cumulative soil water depletion.

8.2.3.5.6 Sucrose Accumulation

The sucrose content in cane will be high during maturity period in the normal crop as compared to the stressed cane. Sucrose accumulation increases by about 100% while cane tonnage increases by only 20% during the maturity phase (Gascho and Shih, 1983). Sucrose accumulation begins at the bottom of the stem and progresses upward to the top internodes. After about 11 months age, the sucrose% in the normal crop remains constant while the percentage in the stressed crop continue to increase until 14 months.

8.3 SUGARCANE UNDER WATERLOGGING

Water logging drastically reduces the growth and survival of sugarcane worldwide and cane yield reduction is estimated between 15–45%. A considerable area under sugarcane crop in several parts of India (Assam, Bihar, and West Bengal, and eastern Uttar Pradesh, coastal region of Andhra Pradesh, Tamil Nadu, Kerala and Karnataka) are exposed to stagnant water for two to three months during monsoon season. Sugarcane is fairly tolerant to flooding and water logging. It was observed that sugarcane crop was susceptible to water logging in the first 3–4 months, somewhat tolerant at 4–9 months age and helped by it in maturity beyond that age. Higher water table during active growth phase adversely affects stalk weight and plant population resulting yield loss at the rate of about one ton per acre for one inch increase in excess water (Carter, 1976; Carter and Floyed, 1974).

Some physiological effects of cane are found due to water-logging are: (i) transpiration rates are reduced due to stomatal closer, (ii) rate of photosynthesis is considerably reduced presumably that cause the reduction of effective leaf area, (iii) growth rates are drastically reduced during waterlogging, and (iv) higher respiration rate of submerged organs compared to leaves. A shift in respiratory metabolism from aerobic to anaerobic pathways is one of the main effects of oxygen deficiency causing from waterlogging. The effects of water logging on respiration rate depend on the varieties and its physiological age. It is also reported that under waterlogging condition some morphological, anatomical, physiological and biochemical changes take place in plant for sack of adaptation/survival (Barcly and Crawford, 1982; Gomathi and Chandran, 2009; Gomathi et al., 2010d, 2012a, 2015).

8.3.1   GERMINATION

Studies conducted in Australia indicated that waterlogging decreased germination if soil was saturated for more than 3 days (McMohan et al., 1993). Flooding at planting affects emergence. A sugarcane variety Cp 89–2376, less than 6 days of flooding has more emergences and CP 72-2086 had externally low emergence.

8.3.2   ROOT SYSTEM

In the absence of oxygen, root hairs die and eventually the roots blacken and rot with the results entire underground root system gets choked and root respiration is also impaired. Because of the insufficient and inadequate root system absorption of nutrients and water is seriously affected. Nutrient absorption is further affected by their unavailability. During natural water logging of the soil, roots exposed to hypoxia condition under such situation roots able survive either by inducing biochemical acclimation or by anatomical acclimation. Following sensing of partial oxygen deficiency, genes coding for so called anaerobic proteins (HIPs protein) are up –regulated at transcriptional and post- transcriptional levels and the HIPs are necessary for the acclimation (Jackson and Richard, 2003).

8.3.3.1   Anatomical Acclimation Through Aerenchyma Formation

Sugarcane is supported by adventitious roots in waterlogging conditions, which develop possibly as result of hormonal imbalance induced by hypoxia and decreased supply of oxygen to the submerged tissue. These roots remain in the upper layers of the water, which are presumably richer in oxygen content. These are adapted to water logging conditions than the original roots because they have much larger intercellular spaces (Van der Heyden et al., 1998). Study examining genetic correlation of sugarcane traits under flood, found that selection for adventitious root development may not increase sugarcane yield under flooding (Sukchain and Dhaliwal, 2005). All these supportive roots developed during flooded situations, helped in maintenance of root activity by supplying necessary oxygen (Drew, 1997).

8.3.3.2   Role of Ethylene

Ethylene involvement in adventitious root formation and aerenchyma formation in several crops was summarized by Jackson and Richard (2003). Ethylene synthesis increases under flooded conditions when ACC synthase concentrations increase, which stimulates ACC synthesis. ACC then diffuses to aerated parts of the root and is converted into ethylene by ACC oxidase. Ethylene is far less soluble in water than in air. Therefore, more ethylene is retained inside plant tissues when flooding occurs and ethylene concentrations increase (Gomathi et al., 2010d, 2015).

8.3.3.3   Shoot Growth

It is inevitable that, because of the close functional inter dependence between roots and shoots, stress on roots from water logging also threatens the shoot system. One example of this is the arrest of nitrate uptake that arises from microbial de nitrification and damage to uptake mechanism from an absence of oxygen. Young leaves remobilize the nutrients from older leaves leading to premature senescence of the later (Drew and Sisworo, 1979). The impact of water logging on shoot growth can be observed on changes in growth habit, visual health, internal anatomy, water relations, hormonal and nutritional composition. Water logging can inhibit leaf and stem expansion and tiller production; and cause epinastic curvature of petioles, orientation of shoot extension (Jackson, 1990b). Sarkar et al. (1999) was observed that plant height after 12 days of submergence showed significant positive association with survival percentage. In sugarcane, varieties which maintained better shoot height and internodal length yielded better under flooding condition (Gomathi, 2009; Gomathi and Chandran, 2009).

8.3.3.4   Tiller Production

Flooding during tillering, resulted in greater tiller mortality and reduced stalk population. Flooding at any stage reduced production of new tillers and rate of elongation of the established tillers (Webster and Eavis, 1971) and decrease was more with longer duration of flooding. Varieties also differed in the regard (Rahman et al., 1989). Studies conducted at SBI, Coimbatore, indicated that the waterlogging stress during formative phase of the (90–170 DAP) caused 13.00, 21.63 and 26.52% reductions in plant height, tiller production and leaf area respectively. However, the reduction was less in the resistant clones (Gomathi and Chandran, 2009).

8.3.3.5   Leaf Development and Growth Parameters

The expression of yellowing symptom in leaf, higher stalk mortality, faster drying of lower leaves, reduction in leaf number and size, are the morphological changes due to waterlogging stress. Further, waterlogging resulted in 26.5%, 25.2% and 24.0% mean reductions in, leaf area index (LAI), leaf nitrogen content and total chlorophyll content, respectively (Gomathi and Chandran, 2009; Gomathi et al., 2010b,c; Pandey et al., 2001).

8.3.4   YIELD AND QUALITY

Cane yield losses depend upon the duration of water logging, stage of crop growth and management practices before, during and after water logging. Yield loss occurs due to stalk mortality, reduced crop growth due to lack of nutrition and water uptake, lodging, cane breakage, etc. About 5–30% loss in yield was reported for 15–60 days of water logging condition created artificially during the late grand growth phase (7.5–9.50 months). A study conducted in tropical India (Tamil Nadu) indicated that in a water stagnation period of 2 months the reduction in cane yield was to the tune of 26–36% in various varieties (Monoharan et al., 1990). A study conducted at SBI, Coimbatore showed that the waterlogging caused 22.4%, reduction in NMC, 45.6% reduction in single cane weight, 30.0% reduction in cane length, 15.9% reduction in internodal length, 17.8% reduction in cane thickness and 40.1% reduction in cane yield (Gomathi et al., 2010b,c).

8.3.5 PHOTOSYNTHESIS AND PARTITIONING OF ASSIMILATES

Under anaerobic condition, photosynthesis declined due to slow diffusion of CO_2 in water and reduced availability of light as result flow rate of assimilates to the roots also decreased. In sugarcane, chlorophyll content reduced under submergence and the reduction was more pronounced in susceptible varieties results in reduction photosynthetic rate and leaf dry matter accumulation (Gomathi et al., 2010b). Reports available in sugarcane that the reduced photosynthesis after lowering of the water table to end saturation caused a reduction in biomass by harvest date in the range 40 to 50% in saturated treatments as compared with the control (Giaz and Gilbert, 2006; Giaz et al., 2004).

8.3.6 ANAEROBIC PROTEINS (ANP'S) IN RESPONSE TO FLOODING STRESS

Plants also respond to anoxia by altering the pattern of root protein synthesis. The proteins which are synthesized as a specific response to anaerobiosis are called the anaerobic polypeptides (ANPs) (Sachs et al., 1980). Flooding stimulated the synthesis of a small group of proteins known as anaerobic polypetides (ANP's) appear to play an essential role for anoxia survival. All the characterized polypeptides are glycolytic enzymes (Mujer et al., 1993). Among the ANPs, ADH is predominating one and has been extensively studied (Sachs et al., 1980). New synthesized ADH isozymes emerge during flooding in many plants (Liao and Lin, 1995; Lin and Lin, 1992) and with different biochemical properties. Both in leaf and root, specific expression of ANP's viz., 66 kDa, 98 kDa and 132 kDa proteins in response to short term flooding stress was recently reported in sugarcane especially in tolerant genotypes (Co 99006 and Co 8371), indicating their possible role in tolerant behavior (Gomathi et al., 2010a).

8.3.7 NITRATE REDUCTASE ACTIVITY

Reduction of NRase in leaves of waterlogged plants results in the rapid depletion of the nitrate and oxygen is consumed by soil biota and then anaerobic conditions develop. Gomathi and Chandran (2009) found a positive association between Nitrogen content of index leaf and nitrate reductase activity (NRase) with flooding tolerance of sugarcane clones exposed to long term flooding stress.

8.3.8 ALCOHOL DEHYDROGENASE (ADH)

Alcohol Dehydrogenase (ADH) is responsible for the synthesis of alcohol and regeneration of NAD in alcoholic fermentation (Kennedy et al., 1992; Russel et al., 1990). This regenerated NAD enables glycolysis to continue under anoxia, thus producing a net 2 moles of ATP per mole of glucose relative to the 38 moles of ATP produced under aerobic conditions through respiration (Davies, 1980). A significant increase in ADH activity was recently reported in sugarcane due to short term flooding (Gomathi et al., 2010a, 2012a).

8.3.9 ANTIOXIDANT SYSTEM IN RESPONSE TO WATERLOGGING

Tolerance to wide varieties of environmental stress conditions has been correlated with increased activity of antioxidant enzymes and levels of antioxidant metabolites (Davies, 1987). A short-term waterlogging treatment led to an increase in the activities of anti-oxidant enzymes viz., APX, CAT and SOD in sugarcane (Gomathi et al., 2012a). Weijun Zhou and Xianqing Lin (1995) concluded that Leaf chlorophyll content and SOD and CAT activities were markedly reduced after plants were waterlogged for 30 days at various stages of growth and results indicated that waterlogging could promote the degradation of chlorophyll, reduce the activities of SOD and CAT, and therefore accelerate leaf senescence.

8.4 MOLECULAR INTERVENTIONS FOR MITIGATION OF ABIOTIC STRESSES

A number of genes have been reported to be induced by drought, high salinity and low temperature stresses and their products are thought to function in stress tolerance and response (Bray, 1997; Shinozaki et al., 2003; Thomashow, 1999). The direct introduction of small number of genes by genetic engineering seems to be a more attractive and rapid approaches for improving stress tolerance (Cushman and Bohnert, 2000). Present day biotechnological strategies rely on the transfer of one or several genes that encodes either biochemical pathways or endpoints of signaling pathways. These genes/products protect either directly or indirectly against environmental stresses.

8.4.1 DROUGHT STRESS

Studies were conducted to detect genes expressed under drought by macroarray, identified 165 genes responded to drought. Important stress related pathways were repressed in sensitive cultivar. A great number of genes with unknown function were reported which may provide new insight into the tolerance mechanism (Rodrigues et al., 2009). Up regulation of genes encoding for polyamine oxidase, cytochrome-c-oxidase, s-adenosyl methionine (SAM) decarboxylase and thioredoxins, which directly or indirectly involved in the regulation of redox status was reported in sugarcane under drought stress (Prabhu et al., 2011). Rodrigues et al (2009) demonstrated increased expression of a gene encoding a peroxidase in a drought tolerant sugarcane cultivar. This enzyme is responsible for the reduction of H_2O_2 to H_2O and O_2 and a decline in peroxidase activity is considered a limiting step to ROS neutralization in sugarcane (Chagas et al., 2008). The accumulation of the osmolytes trehalose and proline also contributes to the reduction in the damage caused by the accumulation of ROS and is associated with drought tolerance in sugarcane (Guimaraes et al., 2008; Molinari et al., 2007; Zhang et al., 2006).

Area that deserves attention is the response mediated by the phytohormone abscisic acid (ABA). ABA is the major plant hormone related to water stress signaling and regulates plant water balance (Riera et al., 2005; Wilkinson and Davies, 2010). Rocha et al., (2007) found drought responses in sugarcane analogous to those induced by exogenous ABA application. Both drought and ABA induced the expression of genes encoding a PP2C-like protein phosphatase, a S-adenosylmethionine decarboxylase and two deltas – 12 oleate desaturases. Trujillo et al. (2008) showed that SodERF3, a sugarcane ethylene responsive factor (ERF), is induced by ABA and drought stress and may be involved in salt and drought tolerance. However, plant response to drought is a complex phenomenon, especially with a polyploid genome like sugarcane (Givet and Arruda, 2001). Besides, the fact that the drought stress involves biochemical networks which are still being elucidated. For example, phosphorus supply improved sugarcane acclimation capacity by affecting plant characteristics related to water status and photosynthetic performance and causing network modulation under water deficit (Sato et al., 2010).

8.4.2   WATERLOGGING STRESS

Molecular mechanisms of waterlogging tolerance include changes in expression of a suite of genes under waterlogging conditions. Studies revealed that genes involved in complex pathways, such as signal transduction, protein degradation, ion transport, carbon and amino acid metabolism, and transcriptional and translational regulation play important roles in response to waterlogging. Transcriptional factors (TFs) that constitute the signal transduction components play an extremely important role in waterlogging tolerance. In rice, two TFs, Snorkel and Submergence-1A, have been cloned by mapped based cloning, and both of them encode ethylene-responsive factor-type transcription factors that helps in adaptation to different types of flood. The understanding of the regulatory mechanisms and signaling events responsible for triggering responses to hypoxia or anoxia in sugarcane plants is a prospective area of research. Although many studies of the molecular mechanism of tolerance to waterlogging have been reported in several crops, the limited knowledge of stress-associated metabolism in sugarcane remains a major gap. Exploring these phenomena in sugarcane will be of relevance to identify candidate genes for waterlogging tolerance and will provide fundamental knowledge on the molecular mechanisms operating under anaerobiosis. The candidates can further be used for manipulating the sugarcane genome for improved stress tolerance or can be used to generate gene-targeted markers for marker assisted breeding programs for flooding tolerance. The understanding of the regulatory mechanisms of signaling perception and transduction events in responses to hypoxia is also a major future prospective research area of sugarcane.

8.5 TECHNOLOGY INTERVENTION TO OVERCOME ABIOTIC STRESSES

Since climate change is projected to reduce sugarcane yields in the next century, it is vital to come up with mitigation strategies that can lower the effects. A number of mitigation measures can be drawn from understanding the potential effects of climate change relying much on climate models. However, the projections of climate change using models are uncertain because of errors that may be encountered in these models (Mall et al., 2004). Water stress generated by high temperatures and low rainfall can be mitigated by growing varieties that are tolerant or resistant to drought. Inman-Bamber et al. (2012) reported that sugarcane cultivar differences in drought adaptation exist. Therefore, evolution of sugarcane varieties with drought resistance and search for genotypes, which possess inherent capabilities of drought tolerance has been on the forefront of sugarcane research. Researchers should therefore continue to breed sugarcane varieties or cultivars that adapt to drought conditions or greater water use efficiency (Matthieson, 2007). Hence, each year genetic potential of advanced breeding material (AVT clones) are being exploited for adverse environments like drought and salinity wherein the competitive advantage for one cultivar over another is likely to be greater. Traits such as higher NMC, maintenance of better leaf production, higher single cane weight and rapid stem elongation contribute to cane yield under drought (Gomathi et al., 2011).

- Screening Potential Clones for Drought Tolerance – Involvement in the Varietal Development Programme: Totally 39 IVT and 88 AVT clones were screened for drought tolerance under field condition, out of which 46 clones were identified as a drought tolerant (Gomathi and Vasantha, 2006; Gomathi et al., 2010, 2012). Among these, varieties Co 99004 (Damodar), Co 2001–13 (Sulabh), Co 2001–15 (Mangal), Co 0218, Co 0403, and Co 06027, Co 06030 were notified by CVRC for cultivation was released in peninsular zone. The variety Co 0212 was released during 2015 as a state release at Tamil Nadu by TNAU. Presently the crop occupies around 3 lakh hectares with a cane production of 350 lakh tonne annually. Totally 38 potential genetic stock sources were identified for drought tolerance. Currently, the clone Co 06015 is proposed for registering genetic stock for improving tolerance for drought, salinity and high temperatures stresses.

- Rapid and Non Destructive Screening Technique for Screening Drought Tolerance: The ability of four interrelated physiological parameters (chlorophyll fluorescence ratio: Fv/Fm; SPAD index, leaf relative water content (RWC), cell membrane injury (%) were evaluated at two drought cycle during formative phase of the crop in known resistant and susceptible varieties. Result suggested that the Chlorophyll fluorescence ratio and SPAD index seems to be the most promising for rapid and non-destructive screening for drought tolerance. Hence, these parameters could be effectively used in screening larger genotypes in the early segregating population in breeding program as they are high throughput and non destructive means of estimation (Gomathi et al., 2013).

8.5.1 DROUGHT MANAGEMENT

- Early Planting: In the tropical belt, November to January planting is better than March–April planting to overcome the problem of moisture stress.
- Seed rate and spacing: Higher seed rate (or) closer spacing is to establish a higher stalk population to make up the greater less of individual stalks, row spacing can be narrowed down to 60 (or) 75 cm to give 15—20% higher cane yield over normal spacing.
- Seed treatment: Soaking the setts in a saturated lime solution for one hour before planting (dissolve 80 kg of lime (calcium carbonate $CaCO_3$) in 400 lit of water.
- Trash mulching: Trash mulching (5–7 t/ha) helps in conserving soil moisture, checks the weed growth and reduces the soil temperature by 2°C.
- Deep trench system of planting: Deep trench system of planting helps deep root development and efficient use of nutrients and moisture.
- Foliar application of urea and potassium: Foliar application of urea and KCl each at 2.5% (2.5 kg urea + 2.5 kg MOP in 100 liters of water) at

15–20 days interval maintains the crop turgidity.

- Protective irrigation: During drought available water can be given in alternate furrows alternatively.
- Use of anti-transpirants: Kaolin acts as a reflectant and reduces the transpiration loss.
- Rain gun/mobile sprinklers may be used to give protective irrigation at critical period.
- Tolerant varieties: Varieties CoC 671, Co 8208, Co 85007, Co 85004, Co 86032, Co 85019 and Co 87263, Co 99004 (Damodar), Co 2001–13 (Sulabh) and Co 2001–15 (Mangal), Co 0218, Co 0325 and Co 0328, Co 0403, Co 06015 and Co 06022 are suitable for water limited condition.

8.5.2 WATERLOGGING MANAGEMENT

Climate change is projected to result in floods in some areas or years. Since floods result in waterlogging conditions, salinity and raised water table, reducing yields significantly (Glaz et al., 2004), it is therefore important to adapt sugarcane production to such conditions. Drainage systems in the fields that are likely to be affected (flat) areas may need to be installed. Once the drainage improves, excessive salts causing salinity can be leached by irrigation. Varieties that adapt to waterlogging and saline conditions may be grown.

1. Proper drainage system should be provided.
2. Early and deep planting is beneficial.
3. Seed rate and row spacing: Normally 38 to 40 thousand healthy three-bud setts/ha are used for planting.
4. The row-to-row distances should be widened and deepened to 135 cm to make drainage channels in between them at the time of water logging.
5. Split application of nitrogen (2–3 times) helps in minimizing nitrate leaching, the chances of which are more under waterlogging.
6. Foliar application of 5 per cent urea during water logging increases the yield of cane.
7. Foliar application of potassium and phosphorus along with nitrogen causes greater root proliferation and stiffness of cane.

8. Sugarcane matures earlier in waterlogged areas, early harvesting facilitates to get maximum amount of sucrose.
9. Waterlogging tolerant varieties:

- Co 8231, Co 8232, Co 8145, CoSi 86071 and Co Si 776, Co 8371 & Co 99006 by SBI.
- At Anakapalle, promising clones 93A4, 93A11, 93A145, and 93A21 have been identified under water logging conditions.
- In the Kolhapur region of Maharashtra, Co 8371 has been found to perform well under river flood.
- Some of the Bo varieties like Bo 91 and varieties Co 87263 and Co 87268 are suitable for flooded conditions of Bihar, while CoT1 8201 and CoT1 88322 are grown in Kerala in Tiruvalla are water logging is very common (Gomathi et al., 2010c).

8.6 CONCLUSIONS AND FUTURE RESEARCH NEEDS

Recurring flooding and waterlogging are significant abiotic production constraints in low lying, high rainfall areas of sugarcane. Most of the field studies reported were conducted with long – term waterlogging, but in practice, short-term flooding is also more common, and some sugarcane clones that tolerate such episodes quit well (Glaz, 2004b). Plant adaptation to waterlogging stress is dependents upon the activation of cascades of molecular networks involved in stress perception, signal transduction, and expression of specific stress-related genes and metabolites (Gomathi et al., 2015; Figure 8.1). Short term or long term flooding significantly down regulated the various physiological and metabolic processes viz., light interception, degradation of chlorophyll pigment, photosynthetic rate, reduction of key enzyme activity (NRase) and soluble protein content and nutrient uptake and finally shoot and root growth behavior (Gomathi et al., 2015; Figure 8.1). However, the degree of tolerance depends upon the varieties used and duration as well as intensity of flood. For example, leaf senescence in sensitive clones was associated with the degradation of chlorophyll, the accumulation of MDA and ethylene production, the decrease of SOD and CAT activities under long term flooding stress.

Plants have evolved the mechanism of adaption, avoidance, acclimation, or a combination thereof to survive and grow under waterlogging condition. Adaptation involves permanent genetic modification of the plant's structure

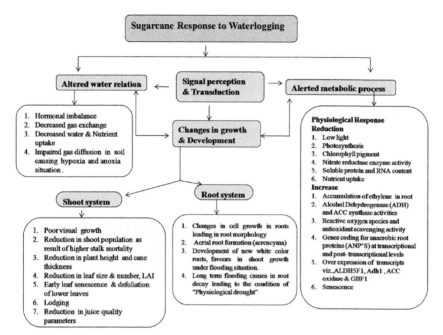

FIGURE 8.1   A schematic representation of sugarcane response and adaptation strategies to waterlogging stress. Note that the extent of variation of these responses varies greatly between the species and varieties. (Reprinted with permission from Gomathi, R., Gururaja Rao, P. N., Chandran, K., & Selvi, A., (2015). Adaptive responses of Sugarcane to waterlogging stress – An over view. Sugar Tech., 15(1), 17–26. © 2014, Springer Nature.)

and function. For example, flood tolerant plants could able to survive in high water table, through formation of adaptive traits like aerial roots, which are temporary source of oxygen for the respiratory event. The tolerant species are able to form aerenchyma, which helps for functioning of the plant processes under anoxia conditions. Ethylene involvement in adventitious root formation and aerenchyma formation was well reported in sugarcane and constitutive formation of stalk aerenchyma possibly enables sugarcane to tolerate periodic floods. Many of the adaptive growth responses occur in response to ethylene. The key enzymes in ethylene biosynthesis like ACC synthase and ACC oxidase were accumulated to elevated levels during submergence (Figure 8.1). In contrast to adaptation, acclimation is a plastic response, often short lived and reversible because it does not involve permanent genetic changes. It leads to transient changes in physiological and molecular process which import tolerance behavior of the crop under waterlogging situation. Flooding stimulated the synthesis of a small group of proteins known as anaerobic polypeptides (ANP's) appear play major role ethanol fermentation that necessary to maintain energy production under anaerobic conditions. Both in leaf and root, specific expression of ANP's viz., 66 kDa, 98 kDa and 132 kDa proteins in response to short term flooding stress was recently reported in sugarcane especially in tolerant genotypes, indicating their possible role in tolerant behaviors.

Extraordinary progress has been made in recent years in diverse areas ranging from meteorological analysis, water balance models which have provided new avenues to achieve improved yields under water limited environments. Using such selection methods, it has been difficult to manipulate precisely the individual components of stress tolerance, and thus development of stress tolerant cultivars has not made much headway. Improvement of qualitative traits such as photosynthesis, transpiration, roots and water use efficiency is difficult through conventional breeding because of the complexity of quantifying these traits in a large number of germplasm lines. Therefore, the immediate task ahead is to look for new options to exploit the variability present in our rich germplasm through new evaluation techniques. Several stress related genes that alter the expression of important stress traits for improving the plant adaptation have to be fully characterized. The donor sources for various stress resistant features have to be identified and the genes responsible for these traits need to be transferred through molecular breeding. In recent years, many genes and gene products have been identified which get induced upon exposure to various abiotic stresses (Chaves et al., 2003). Genes encoding enzymes of the biosynthetic pathways of different osmolytes such as proline, glycine betaine, sorbitol, pinitol, have been cloned and exploited in improving abiotic stress tolerance. Response of crop plants to abiotic stresses involves several gene alterations. Proteins which are up regulated are synthesized in response to high temperature, drought, salinity and several other abiotic stresses. New research programme should be aimed at identifying or developing cultivars and management practices appropriate for altered climates.

KEYWORDS

- abiotic stress
- abscisic acid
- evapotranspiration

- **sugarcane**
- **water stress**
- **waterlogging**

REFERENCES

Almazan, O., Gonzalez, L., & Galvez, L., (2001). Sugar Cane International, 7, 3–8. *Journal of Agricultural Science*, Cambridge, (1992), pp. *119*, 291–296.

Barcly, A. M., Crawford, R. M. M., (1982). Plant growth and survival under strict anaerobiosis. *J. Expt. Bot., 33,* 541–549.

Bray, E. A., (1997). Plant responses to water deficit. *Trends in Plant Science, 2*, 48–54.

Bull, T. A., Glasziou, K. T., (1975). Sugarcane. In: *Crop Physiology, Case Histories*, Ed. LT Evans, Cambridge Univ. Press, Cambridge, pp. 51–72.

Carlin, S. D., Santos, D. M. M., (2009). Physiological indicators of the interaction between water deficit and soil acidity in sugarcane. *Pesquisa Agropecuária Brasileira, 44,* 1106–1113.

Carter, C. E., (1976). Excess water decreases cane and sugar yields. *Proc. American Soc. Sugar cane Technol., 6,* 44–51.

Carter, C. E., Floyed, J. M., (1974). Inhibition of sugarcane yields by high water level during dormant season. *Proc. Soc. Sugarcane Technol., 4,* 14–18.

Chagas, R. M., Silveira, J. A. G., Ribeiro, R. V., Vitorello, V. A., Carrer, H., (2008). Photochemical damage and comparative performance of super oxide dismutase and ascorbate peroxidase in sugarcane leaves exposed to paraquat-induced oxidative stress. *Pestic. Biochem. Physiol., 90,* 181–188.

Chang, H., Wang, J. S., Ho, F. W., (1968). The effect of different pan ratio for controlling irrigation of sugarcane in Taiwan. *Proc. Int. Soc. Sugarcane Technol., 13,* 652–663.

Chaves, M. M., Maroco, J. P., Pereira, J. S., (2003). Understanding plant responses to drought: from genes to the whole plant. *Functional Pl. Biol., 30,* 239–264.

Cushman, J. C., Bohnert, H. J., (2000). Genomic approaches to plant stress tolerance. *Curr. Opin. Plant Biol.,* 3117–124.

Davies, D. D., (1980). Anaerobic metabolism and the production of organic acids. In: The biochemistry of plants. Davies, D. D. (ed.). Acad. Press, New York, USA, Vol. *2*, pp. 581–611.

Davies, K. J. A., (1987). Protein damage and degradations by oxygen radicals. Journal of *Biological Chemistry, 262*, 9895–9901.

Deren, C. W., Cherry, R. H., & Snyder, G. H., (1993). Effects of flooding on selected sugarcane clones and soil insect pests. *J. Am Soc. Sug. Cane Technol. 13,* 22–27.

Deren, C. W., Snyder, G. H., Miller, J. D., & Porter, P. S., (1991a). Screening for and heritability of flood tolerance in the Florida (CP) sugarcane breeding population. *Euphytica, 56,* 155–160.

Deren, C. W., Snyder, G. H., Tai, P, Y. P., Turick, C. E., & Chynoweth, D. P., (1991b). Biomass production and biochemical methane potential of seasonally flooded intergeneric and *interspecific Saccharum hybrids. Bioresource Tech., 36,* 179–184.

Drew, M. C., (1997). Oxygen deficiency and root metabolism: Injury and acclimation under hypoxia and anoxia. Annual Review of *Plant Physiology Plant Molecular Biology, 48,* 223–250.

Drew, M. C., & Sisworo, E. T., (1979). The development of waterlogging damage in young barley plants in relation to plant nutrient status and changes in soil properties. *New Phytologist, 82,* 301–314.

Dwivedi, R. S., (2000). Adaptability mechanisms of sugarcane cultivars to abiotic stresses. In: *Sugarcane Production: Strategies and Technology* (Shahi, H. N., Lal, M., Sinha, O. K., Srivastava, T. K., Eds), Technical Bulletin No. *40,* Indian Institute of Sugarcane Research, Lucknow, pp. 42–45.

Gascho, G. J., & Shih, S. F., (1983). Sugarcane. In: *Crop Water Relations*, Ed Teare, I. D., and, M. M., Peet, Wiley *Inter science*, New York, 445–479.

Glaz, B., & Gilbert, R. A., (2006). Sugarcane response to water table, periodic flood, and foliar nitrogen on organic soil. *Agronomy Journal, 98,* 616–621.

Glaz, B., Morris, D. R., & Daroub, S. H., (2004b). Periodic flooding and water table effects on two sugarcane genotypes. *Agronomy Journal, 96,* 832–838.

Glaz, B., Morris, D. R., & Daroub, S. H., (2004a). Sugarcane photosynthesis, transpiration and stomatal conductance due to flooding and water table. *Crop Sci., 44,* 1633–1641.

Gomathi, R., & Chandran, K., (2008). Annual Report, SBI, Coimbatore 39–40.

Gomathi, R., & Chandran, K., (2010b). Physiological and growth response of sugarcane clones to waterlogging. *Proceedings of ISPP Zonal Conference on "Recent Trends in Plant Physiology and Crop Improvement"* at VIT University, Vellore, Tamil Nadu. Abstract, No. *17.*

Gomathi, R., & Gowri Manohari, N., (2010a). Anaerobic proteins and enzymes in relation to flooding tolerance of sugarcane varieties. *Proceedings of National Plant Physiology Conference on "Physiological and Molecular Approaches for Crop Improvement under Changing Environment."* At Banaras Hindu University, p. 120.

Gomathi, R, Sindhuja, S., & Kohila, S., (2014). Metabolic and Molecular Response of Tolerant and Sensitive Sugarcane Genotypes Subjected to Short Term Dehydration Stress. Proceedings in the International conference on *"Nanobio, Biomimetic Materials and Its Applications: ICNBM-14."* Hindusthan College of Arts and Science & Institute of Technology, Coimbatore, pp. 8–9.

Gomathi, R. (2009a). Physiological Basis of Waterlogging Resistance in Sugarcane. In: *Sugarcane Physiology*. Ed. By. P. N. Gururaja Rao, Published by Sugarcane Breeding Institute (ICAR) and Directorate of open and distance learning, TNAU, Coimbatore, Tamil Nadu, pp. 53–61.

Gomathi, R. G., Hemaprabha, R. M., Shanthi, S., & Alarmelu, (2010). Evaluation of elite sugarcane clones for drought tolerance. Presented paper in ISPP Zonal Conference on "Recent trends in Plant Physiology & Crop Improvement" at VIT University, Vellore, Tamil Nadu. Abstract No. *52.*

Gomathi, R., & Chandran, K., (2012b). Physiological markers for screening waterlogging resistance in sugarcane. *Proceedings of International symposium on "New Paradigms in Sugarcane Research" ISNPSR 2012* organised by SSRD & SBI at Coimbatore. Abstract No. *129.*

Gomathi, R., & Chandran, K., (2013). Juice quality as influenced by water-logging stress in sugarcane. *Proceedings in National Conference of Plant Physiology on "Current Trends in Plant Biology Research "NCPP-13."* Directorate of Groundnut Research (DGR), Junagath, Gujarat. 13 to *16,* December 2013, pp. 410–411.

Gomathi, R., & S. Vasantha., (2006). Field evaluation of sugarcane clones (*Saccharum Officinarum L.*) for drought tolerance. Presented in National seminar on" Physiological and Molecular approaches for the improvement of Agricultural, Horticultural and Forestry crops," Department of Plant Physiology, KAU, Vellanikkara, *Thrissur & Indian Society of Plant Physiology (ISPP)*, New Delhi, pp. 69.

Gomathi, R., & Chandran, k., (2009b). "Effect of waterlogging on growth and yield of sugarcane clones." Sugarcane Breeding Institute (SBI-ICAR), *Quarterly News letter* Vol. *29*(4). pp. 1–2.

Gomathi, R., Chandran, K., Gururaja Rao, P. N., & Rakkiyappan, P., (2010c). "Effect of Waterlogging in Sugarcane and its Management" Published by The Director, Sugarcane Breeding Institute (SBI-ICAR), Coimbatore. *Extension Pub., No. 185.*

Gomathi, R., Gowri Manohari, N., & Rakkiyappan, P., (2012a). Antioxidant enzymes on cell membrane integrity of sugarcane varieties differing in flooding Tolerance. *Sugar Tech., 14, 261–265.*

Gomathi, R., Gowri Manohari, N., Vasantha, S., & Rakkiappan, P., (2010d). Anoxia Induced enzymes and polypeptides (ANP'S) in relation to flooding tolerance of sugarcane genotypes. *Proceedings of International Conference on "Computational Biotechnology and Nano technology" (ICCBN – 2010)* at Vivekanandha College of Engineering for Women, Elayampalayam, Tamil Nadu Abstract No. BTBI02.

Gomathi, R., Gururaja Rao, P. N., Chandran, K., & Selvi, A., (2015). Adaptive responses of Sugarcane to waterlogging stress – An over view. *Sugar Tech., 15*(1), 17–26**.**

Gomathi, R., Vasantha, S., Alarmelu, S., & Anna Durai, A., (2012). Screening of promising sugarcane (AVT) clones for drought tolerance. Presented paper in the "Regional Science congress to be held at Kongunadu Arts and Science College, Coimbatore, pp. 34.

Gomathi, R., Vasantha, S., Govindaraj, P., & Manjunatha, T., (2014)."Evaluation of promising sugarcane (AVT) clones for drought tolerance" Proced. In National Seminar on "Bio-Energy for Sustainable Role of Sugar Crops." *Organized by SSRD and Sugarcane Breeding Institute,* Coimbatore, pp. 293–294.

Gomathi, R., Vasantha, S., Govindaraj, P., Alarmelu, S., & Anna Durai, A., (2013). Rapid physiological tools for screening drought tolerance in sugarcane. *SBI NEWS, 33*(3).

Gomathi, R., Vasantha, S., Hemaprabha, G., Alarmelu, S., & Shanthi, R. M., (2011). Evaluations of elite sugarcane clones for drought tolerance. *J. Sugarcane Res., 1*(1), 55–62.

Grivet, L., & Arruda, P., (2001). Sugarcane genomes: depicting the complex genome of an important tropical crop. *Curr. Opin. Plant Biol., 5*, 122–127.

Guimaraes, E. R., Mutton, M. A., Mutton, M. J. R., Ferro, M. I. T., Ravaneli, G. C., & Silva, J. A., (2008). Free proline accumulation in sugarcane under water restriction and spittlebug infestation. *Sci. Agric., 65*, 628–633.

Gururaja Rao, PadmajaRao, S., & Venkataramana, S., (2008). Annual Report, SBI, Coimbatore.

Inman-Bamber, N. G., Lakshmanan, P., & Park, S., (2012). Sugarcane for water-limited environments: Theoretical assessment of suitable traits. *Field Crops Res., 134*, 95–104.

Jackson, M. B., (1990). Communication Between the Roots and Shoots of Flooded Plants. In: *Importance of Root to Shoot Communication in the Responses to Environmental Stress* (Davies, W. S., & Geffocat, B., Eds.), British Society for Plant Growth Regulation Bristol. U.K., pp. 115–133.

Jackson, M. B., & Richard, B., (2003). Physiology, biochemistry and molecular biology of plant root systems subjected to flooding of the soil. In: *Root Ecology.* H. de Kroon and E. J. W. Visser, 9th edn.), Springer Verlag, Berlin, Heidelberg.

Jones, H. G., & Corlett, J. E., (1992). Current topics in drought physiology. *Journal of Agricultural Science, 119*, 291–296.

Kennedy, R. A., Rumpho, M. E., & Fox, T. C., (1992). Anaerobic metabolism in plants. *Plant Physiology, 100*, pp. 1–6.

Liao, C. T., & Lin, C. H., (1995). Effect of flood stress on morphology and anaerobic metabolism of Momordicacharantia. *Environmental Experimental Botany, 35*, 105–113.

Lin, C. H., (1992). Physiological adaptation of wax apple to waterlogging. *Plant Cell Environment, 15*, 321–328.

Luo Jun, Zhang Mu Quing, Lu Jian Lin, & Lin Yan Quan, (1999). Chloroplast fluorescence parameters, MDA content and plasma membrane permeability in sugarcane and their relation to drought tolerance. *J. Fujian Agricultural University, 28*(3), 257–262.

Luo Jun, Zhang Mu Quing, Lu Jian Lin, & Lin Yan Quan, (2000). Effects of water stress on the chlorophyll a fluorescence induction kinetics of sugarcane genotypes. *J. Fujian Agricultural University, 29*(1), 18–22.

Mall, R. K., Lal, M., Bhatia, V. S., Rathore, L. S., & Singh, R., (2004). Mitigating climate change impact on soybean productivity in India: a simulation study. *Agric. For. Meteorol, 121*(1–2), 113–125.

Manoharan, M. L., Duraisamy, K., Krishnamurthy, S. V., Vijayaraghan, H., Muthkrishnan, K., (1990). Performance sugarcane varieties in waterlogged condition. *Maharastra Sugar, 15*(11), 39–45.

Mathieson, L., (2007). Climate change and the Australian Sugar Industry: Impacts, adaptation and R&D opportunities. *Sugar Research and Development Corporation*. Australia.

Mc Mahon, G. G., Chapple, P. A., Ham, G. J., Saunders, M., & Brandon, R., (1993). Planting sugarcane on heavy clay soils in the Burdekin. Proc.15th conference of the Australian Society of Sugarcane technologist, *Cairns, Queensland*, 27–30.

Molinari, H. B. C., Marur, C. J., Daros, E., Campos, M. K. F., Carvalho JFRP, Bespalhok Filho, J. C., Pereira, L. F. P., & Vieira, L. G. E., (2007). Evaluation of the stress-inducible production of proline in transgenic sugarcane (*Saccharum* spp.): osmotic adjustment, chlorophyll fluorescence and oxidative stress. *Physiol. Plant, 130*, 218–229.

Mongelrad, J. C., (1968a). Effects of planting methods on yield of sugarcane under subsurface irrigation. *Hawaiian Planter's Rec., 58*(22), 315–322.

Mongelrad, J. C., (1968b). Further notes on the use of Sin bar for the selection of drought resistant sugarcane varieties. *Annl. Rept. M S R I Mauritius*, 87–89.

Mujer, C. V., Rumpho, M. E., Lin, J. J., Kennedy, R. A., (1993). Constitute and inducible aerobic and anaerobic stress proteins in the *Echinochloa* complex and rice. *Plant Physiology, 101*, 217–226.

Naidu, K. M., (1987). Potential yield in sugarcane and its realization through varietal improvement: present status and future thrusts. Sreenivasan, T. V., & Premachandran, M. N. (eds.)., *Sugarcane Breeding Institute*, pp. 19–55.

Naidu, K. M., Ramana Rao, T. C., & Shunmugasundaram, S., (1983). Varietal stability in sugarcane under conditions of drought. *Proc. Int. Soc. Sugarcane Technol., 18*, 682–690.

Naidu, K. M., & Venkataramana, (1993). Sugarcane In *"Rooting pattern of tropical crops"* Ed. Salam, M. A., Tata McGraw Hill (India) New Delhi, pp. 169–187.

Pandey, D. M., Pant, R. C., & Roy, D. K., (2001). Changes in chlorophyll content of three sugarcane (*Saccharum officinarum* L.) cultivars in relation to waterlogging and planting methods. *Crop Research (Hisar), 21*(3), 360–363.

Prabhu, G., Kawar, P. G., Pagariya, M. C., & Prasad, D. T., (2011). Identification of water deficit stress up regulated genes in sugarcane. *Plant Mol. Biol. Rep., 29*.

Rahman, A. B. M. M., Martein, F. A., & Terry, M. E., (1989). Physiological response of sugarcane to flooding stress. *Proc. Int. Soc. Sugar. Tech., 20*(2), 668–676.

Ramanujam, T., & Venkataramana, S., (1999). Radiation interception and utilization at different growth stages of sugarcane and their influence on yield. *Indian J Plant Physiol., 4*, 85–89.

Ramesh, P., (2000). Effect of drought on nutrient utilization, yield and quality of sugarcane *(Saccharrum officinaram)* varieties. *Indian J. Agron., 45*(2), 401–406.

Rao, K. C., & Asokan, S., (1978). Studies on free proline association to drought resistance in sugarcane. *Sugar J., 40,* 23–24.

Rao, P. N., (2000). Cane management under drought conditions or water stress. P*roc. Annl. Conv. S. T. A. I., 61,* 3–8.

Riera, M., Valon, C., Fenzi, F., Giraudat, J., & Leung, J., (2005). The genetics of adaptive responses to drought stress: abscisic acid-dependent and abscisic acid-independent signaling components. *Physiol. Plant., 123*, 111–119.

Roach, B. T., & Mullins, R. T., (1985). Testing sugarcane for water logging tolerance. *Proc. Austr. Soc. Sugarcane Technol.,* 95–102.

Rocha, F. R., Papini-Terzi, F. S., Nishiyama, M. Y., Jr, Vencio, R. Z. N., Vicentim, R., Duarte, R. D. C., Rosa, V. E., Jr. Vinagre, F., Barsalobres, C., Medeiros, A. H., Rodrigues, F. A., Ulian, E. C., Zingaretti, S. M., Galbiatti, J. A., Almeida, R. S., Figueira, A. V. O., Hemerly, A. S., Silva-Filho, M. C., Menossi, M., & Souza, G. M., (2007). Signal transduction-related responses to phytohormones and environmental challenges in sugarcane. *BMC Genomics, 8*, 71.

Rodrigues, F. A., Laia, M. L., & Zingaretti, S. M., (2009). Analysis of gene expression profiles under water stress in tolerant and sensitive sugarcane plants. *Plant Sci., 176*, 286–302.

Russell, D. A., Wong, D. M. L., & Sachs, M. M., (1990). The anaerobic response of soybean. *Plant Physiology, 92,* 401–407.

Samuels, G., (1971). Influence of water deficiency and excess on growth and leaf nutrient element content of sugarcane. *Proc. International Soc. Sugar Cane Technol., 14,* 653–656.

Sarkar, R. K., Sahu, R. K., Suriya Rao, A. V., & De, R. N., (1999). Correlation and path analysis of certain morpho-physiological characters with submergence tolerance in Rain - fed lowland rice. Indian *Journal Plant Physiology, 4*(4), 346–348.

Sato, A. M., Catuchi, T. A., Ribeiro, R. V., & Souza, G. M., (2010). The use of network analysis to uncover homeostatic responses of a drought-tolerant sugarcane cultivar under severe water deficit and phosphorous supply. *Acta. Physiol. Plant., 32,* 1145–1151.

Selvi, A. R., Gomathi, K., Devi, P. T., Prathima & R. Manimekalai.(2014). Changes in Gene Expression in Sugarcane under Water Deficit Conditions. Proceed. in National Seminar on "Bio- energy for sustainable role of sugar crops" Organized by SSRD and *Sugarcane Breeding Institute*, Coimbatore during 23–25th June 2014. pp. 248.

Shih, S. F., & Gascho, G. J., (1980). Relationship among stalk elongation, leaf area and dry biomass of sugarcane. *Agron. J., 72,* 309–313.

Shinozaki, K., Yamaguchi-Shinozaki, K., Seki, M., (2003). Regulatory network of gene expression in the drought and cold stress responses. *Curr. Opin. Plant Biol., 6*, 410–417.

Sukchain, D. S., & Dhaliwal, L. S., (2005). Correlations and path coefficients analysis for aerial roots and various other traits in sugarcane under flooding. *Ann. of Bio., 21,* 43–46.

Sundara B, (1998). *Sugarcane Cultivation*, Vikash Pub. House, New Delhi, pp. 292.

Thomashow, M. F., (1999). Plant cold acclimation: freezing tolerance genes and regulatory mechanisms. Annu. *Rev. Plant Physiol. Plant Mol. Biol., 50,* 571–599.

Trujillo, L. E., Sotolongo, M., Menendez, C., Ochogavia, M. E., Coll, Y., Hernandez, I., Borras-Hidalgo, O., Thomma BPHJ, Vera, P., & Hernandez, L., (2008). SodERF3, a novel sugarcane ethylene responsive factor (ERF), enhances salt and drought tolerance when over expressed in tobacco plants. *Plant Cell Physiol., 49*, 512–525.

Van der Heyden, Ray, C. J., & Noble, R., (1998). Effects of waterlogging on young sugarcane plants. *Australian Sugarcane, 2*, 28–30.

Vasantha, Gururaja Rao, P. N., & Ramanujam, T., (2003). Effect of moisture stress on sucrose synthesizing enzymes in promising genotypes of sugarcane. *J. Plant Biol., 30*, 15–18.

Vasantha, S., & Gururaja Rao, (2003). Influence of moisture stress on the activity of oxidative enzymes in sugarcane *India Journal Plant Physiol., 8*, 405–407.

Venkataramana, S., Gururaja Rao, P. N., & Naidu, K. M., (1983). Evaluation of cellular membrane thermostability for screening drought resistant sugarcane varieties, Sugarcane (London) No. *4*, 13–15.

Venkataramana, S., Gururaja Rao, P. N., & Naidu, K. M., (1986). The effects of water stress during the formative phase on stomatal resistance and leaf water potential and its relationship with yield in ten sugarcane varieties. *Field Crops Res., 13*, 345–353.

Venkataramana, S., & Naidu, K. M., (1989). Root growth during formative phase in irrigated and water stressed sugarcane and its relationship with shoot development and yield. *Indian J. Plant Physiol., 32*, 43–50.

Venkataramana, S., & Naidu, K. M., (1993). Abscisic acid effect on water stress indicators in sugarcane *Plant Physiol., & Biochem., 20*, 1–4.

Venkataramana, S., Naidu, K. M., & Singh, (1987). Membrane thermostability and nitrate reductase activity in relation to water stress tolerance of young sugarcane plants. *New Phytol.,107*(2), 335–340.

Venkataramana, S., Shunmugasundaram, S., & Naidu, K. M., (1984). Growth behaviour of field grown sugarcane varieties in relation to environmental parameters and soil moisture stress. *Agric., & For. Meteorol., 31*, 251–260.

Webster, P. W. D., & Eavis, B. W., (1971). Effect of flooding on sugarcane growth 1. Stage of growth duration of flooding. *Proc. of Int. Soc. Sugarcane Tech., 14*, 708–714.

Weijun Zhou, & Xianqing Lin, (1995). Effects of waterlogging at different growth stages on physiological characteristics and seed yield of winter rape *(Brassica napus L.)*. *Field Crops Research., 44*, 103–110.

Wilkinson, S., & Davies, W. J., (2010). Drought, ozone, ABA and ethylene: new insights from cell to plant community. *Plant Cell Environ., 33*, 510–525.

Zende, G. K., (2002). Soil water management in waterlogged soils. *Bharatiya Sugar, 27*(5), 7–8.

Zende, N. A., & Hapase, D. G., (1986). Development of saline and alkaline soils in Maharashtra under sugarcane cultivation, their reclamation and management. *Bahratiya Sugar, 11*(11), 41–48.

Zhang, S. Z., Yang, B. P., Feng, C. L., Chen, R. K., Luo, J. P., Cai, W. W., & Liu, F. H., (2006). Expression of the *Grifola frondosa* trehalose synthetase gene and improvement of drought-tolerance in sugarcane (*Saccharum officinarum* L.). *J Integr. Plant Biol., 48*, 453–459.

CHAPTER 9

DISEASES AND PESTS AFFECTING SUGARCANE: METHODS OF CONTROL

YAQUELIN PUCHADES IZAGUIRRE,[1] MÉRIDA RODRÍGUEZ REGAL,[1] EIDA RODRÍGUEZ LEMA,[1] and MARÍA LA O. HECHAVARRÍA[1]

Institute of Sugarcane Research (INICA), Carretera ISPJAE, Km 1, Boyeros–19390, Havana, Cuba

CONTENTS

9.1   INTRODUCTION

Diseases and pests affecting sugarcane are considered as production constraints and their management is critical to increase profitability and competitiveness of this industry (Rott et al., 2013). The perennial nature of the crop and the fact that sugarcane is usually grown as a monoculture (i.e., it occupies most or all of the land within a given area for a prolonged period) also favors the build-up of diseases and pests.

Sugarcane is also a relatively low value crop in terms of income per hectare in comparison with many other crops. Costly intervention practices, such as the application of pesticides to the growing crop, are therefore usually not economically feasible (Bailey, 2011). Thus, identifying important diseases and pests, and knowing what to do about them is vital for the sustained production of sugarcane.

Sugarcane is susceptible to a myriad of bacterial, fungal, viral and phytoplasmal diseases (Table 9.1). Major fungal diseases like red rot, smut and rust pose a real challenge to successful cultivation of sugarcane. Ratoon stunting and leaf scald are two important bacterial diseases and cause considerable damage to the crop. The two viral diseases of sugarcane mosaic and yellow leaf occur throughout the crop cycle in almost all parts of the world and are accountable for the progressive deterioration of crop performance leading to reduced life-span of many promising sugarcane varieties (cultivars). Off late, sugarcane grassy shoot disease and leaf yellows caused by phytoplasma and leaf fleck have emerged as major constraints for sugarcane growers (Srivastava, 2016).

Numerous insects, nematodes, a few vertebrates and parasitic plants are pests of sugarcane (Table 9.2). With diverse feeding habits and nature of damage, these pests inflict varying levels of losses on cane and sugar yields in the tropics and subtropics. Despite the stability offered by the crop system and low pesticide usage, pests such as internode borer, white grubs and rats have shown a tendency to proliferate.

Most of the important sugarcane pathogens are widely distributed internationally. Other important and potentially damaging diseases and pests may have a more localized distribution. For all the above reasons the control measures are a critically important aspect of the crop management. Generally an integrated approach involving resistant varieties, disease and pests free seed material, appropriate control measures and strict quarantine procedures is used to manage pathogens in most sugarcane growing countries.

9.2   DESCRIPTION OF EARLIER TRENDS AND RESEARCH IN THE AREA

The use of classical plant breeding to control sugarcane pests and diseases has been productive. However, it is also generally recognized that a better knowledge of both pests and pathogens and of the host plants

TABLE 9.1 A Selected List of Important Sugarcane Disease and Their Causative Agents

Sugarcane disease	Classification	Causative pathogens
Leaf scald	Bacterial	*Xanthomonas albilineans*
Gumming	Bacterial	*Xanthomonas axonopodis* pv. *Vasculorum*
Ratoon stunting disease (RSD)	Bacterial	*Leifsonia xyli* subsp. *Xyli*
Brown rust	Fungal	*Puccinia melanocephala*
Orange rust	Fungal	*Puccinia kuehnii*
Tawny rust	Fungal	*Macruropyxis fulva* sp. *Nov*
Smut	Fungal	*Sporisorium scitamineum* = (*Ustilagos scitaminea*)
Fusarium sett or stem rot	Fungal	*Gibberella fujikuroi* (anamorph *Fusarium verticillioides*), *G. subglutinans* (anamorph *F. subglutinans*)
Wilt/Top rot/PokkahBoeng	Fungal	*Fusarium sacchari*
Red rot	Fungal	*Glomerella tucumanensis* (anamorph *Colletotrichum falcatum*)
Streak	Viral	*Sugarcane streak virus* (*SCSV*) (Geminiviridae)
Fiji leaf gall	Viral	*Fiji disease virus* (*FDV*) (Reoviridae)
Leaf fleck	Viral	*Sugarcane bacilliform virus* (*SCBV*) (Badna virus)
Mosaic	Viral	*Sugarcane mosaic virus* (*SCMV*), *Sorghum mosaic virus* (*SrMV*) (Potyviridae)
Mosaic streak	Viral	*Sugarcane streak mosaic virus* (*SCSMV*) (Potyviridae)
Yellow leaf	Viral	*Sugarcane yellow leaf virus* (*SCYLV*) (Luteoviridae)
Grassy shoot disease	Phytoplasmal	Sugarcane Phytoplasma
Leaf yellows	Phytoplasmal	*Sugarcane leaf yellows phytoplasma* (*ScLYP*)

defense mechanisms will lead to the development of novel and improved approaches to enhance the durability of resistance and more effective control measures.

TABLE 9.2   A Selected List of Important Sugarcane Insect and Nematode Pests

Common name	Nomenclature
Sugarcane borer	*Diatraea saccharalis* (Lepidoptera: Crambidae)
Sugarcane top borer	*Scirpophaga excerptalis* (Lepidoptera: Pyralidae)
Striped sugarcane borer	*Chilo sachariphagus* (Lepidoptera: Crambidae)
Mexican rice borer	*Eoreuma loftini* (Lepidoptera: Crambidae)
African sugarcane borer	*Eldana saccharina* (Lepidoptera: Pyralidae)
Early shoot borer	*Chilo infuscatellus* (Lepidoptera: Crambidae)
Sugarcane weevil borer	*Rhabdo scelus obscurus* (Coleoptera: Curculionidae)
Sugarcane grub	*Ligyrus subtropicus* (Coleoptera: Scarabaeidae)
Greyback cane grub	*Ancistrosoma argentinum* (Coleoptera: Melolonthidae)
	Dermolepida albohirtum (Coleoptera: Scarabaeidae)
Termites	*Odontotermes obesus*
Yellow sugarcane aphid	*Siphaflava* (Hemiptera: Aphididae)
Planthopper (vector of Fiji Leaf gall)	*Perkinsiella saccharicida* (Hemiptera: Delphacidae)
Planthopper (vector of Ramu stunt)	*Eumetopina flavipes* (Hemiptera: Delphacidae)
Grasshopper	*Hieroglyphus banyan*
Mealybug	*Saccharicoccus sacchari*
White flies	*Aleurolobus barodensis*
White woolly aphid	*Ceratavacuna lanigera* (Hemiptera: Aphididae)
Root knot nematode	*Meloidogyne spp.*
Dagger nematode	*Xiphinema spp.*
Lesion nematode	*Pratylenchus zeae*
Rodents	*Ratusrattus, R. sordidus*

The understanding of disease resistance has emerged with the pioneering observations and demonstrations on disease resistance since the beginning of 19th century (Sundar et al., 2015). Many serious outbreaks of diseases have occurred in sugarcane since the early days of commercial production. For example, Bourbon or Otaheite cane succumbed to a complex of diseases in Mauritius in the 1840s and to gumming disease in Brazil in 1869. The first outbreak of smut disease occurred in South Africa in 1877.

Since these times, the control of diseases in order to minimize their economic impact has remained essential to successful sugarcane production. In

2000, there was an outbreak of orange rust in the widely grown variety Q124 in Queensland which devastated large parts of that sugar industry (Magarey et al., 2001). Breeding and screening sugarcane for resistance is, therefore, a very important and a key process for the cultivation and production of sugarcane.

Although for most regions producing sugarcane the list of local pests and diseases that have been recorded is long, most are of little economic consequence. Invariably in most breeding program, only a few are regarded as current hazards that require attention by those involved in sugarcane breeding and selection. For these reasons and further considering local distribution, germplasm constitution, studies of resistance heritability, and the advances in the approaches of sugarcane-pathogen interaction, Cuban breeding program, leaded by Institute for Sugarcane Research (INICA), has established resistance against smut, brown rust, leaf scald, mosaic and borer as key criterion for selection of new sugarcane cultivars (INICA, 2011). This program has considered four strategies for resistance management:

(1) Disease of little economic consequence or of recent introduction, so it is not very distributed in the country and there is little information about it. Selection is not applied only monitoring the incidence of disease is carried out.

(2) Disease of major economic consequence, the inheritance of the disease is additive, there are a great number of parents with resistance in the germplasm collection and studies of host-pathogen information are well understood. In this case resistance-screening trials are performed in the last selection stage and susceptible cultivars are discarded or management strategies are applied if they are proven to be effective.

(3) Disease of major economic consequence, distributed all over the country and with a complex heritability, there are parents with resistance in the germplasm collection. Selection is achieved since early stages and resistance screening trials are performed in the last selection stage; susceptible cultivars are discarded in every selection step. Cultivars with intermediate resistance are selected but measures of management are adopted for their use in commercial practices.

(4) Disease with similar characteristics to those mentioned above (3) but there are a few parents with resistance in the germplasm collection, so it is needed a program for this objective. In early stages monitoring the incidence of disease is carried out and inoculated trials are

used to determine the resistance of clones at final stages. Susceptible cultivars are discarded in every selection step.

Currently a number of control options have been investigated in order to control diseases and pests in a sustainable and efficient way. These include:

- Genetic: using resistant and tolerant cultivars.
- Agronomic: planting diseases and pest free seed cane; management of planting or harvesting dates, effective destruction of the old crop before replanting; intercropping and rotation cropping; monitoring the incidence of diseases and pests to provide timely warning of any problem.
- Mechanic: roguing infected plants from nurseries and from fields (where economically feasible), use of traps for pests.
- Physical: water management and heat treatment of seed cane.
- Chemical: application of moderately and slightly hazardous fungicides and insecticides, also repellents for pests.
- Biological: use of predators, parasitoids and antagonist organisms.
- Legal: quarantine system and prevention policies.

Due to economic and ecological reasons, biological control has been the main practice implemented to pests control in Cuba (Rodríguez et al., 2016). Currently, four parasitoids and one entomopathogen are produced on the only alternative host, the great wax moth (*Galleria mellonella*). For *Diatraea saccharalis* there are a profitable control scheme covering all life cycle. It includes application of egg parasitoid *Trichogranma spp.*, use of larvae parasitoid *Lixophagadiatraeae* and *Tetrastichus howardi* for adult stage. Until present, this program has reduced the infestation index of *D. saccharalis* to 0.7, which is below the economic injury threshold.

Larvae control of lepidopterans, hemipterans and coleopterans through the use of the nematode entomophatogen, *Heterorhabditis bacteriophora* is another successful approach, in biological control. Although there are some parasitoids that in natural conditions keep those pest under relative control.

It is important highlighted the implementation of the Sugarcane Plant Health Service (SEFIT, in spanish) for all sugarcane growers in Cuba (Rodríguez et al., 2014). The service provides a permanent phytosanitary monitoring of the cultivars in sugarcane growing areas and preventive measures of diseases and pest controls. Since 2002, a varietal policy supported by a governmental resolution (616/2014) of the Agriculture Ministry, was applied to avoid the risk of outbreaks of the diseases. This policy establishes that any cultivar cannot occupy more than 20% of the production area for each sugar mill company (La et al., 2017). These measures had allowed maintaining an adequate balance of resistant cultivars since those that varied its behavior had been quickly replaced or decreased its planted area.

Further work is required to implementation of control strategies in the field. However, the multiple integrated control approach described here can be used as prevention techniques to improve the management of the sugarcane diseases and pest.

9.3 ONGOING WORK, RESEARCH, AND DEVELOPMENT

Major research topics regarding sugarcane diseases and pests are: diagnosis, resistance and screening, host/pathogen/insect interaction, integrated pest/disease management and biological control. Early detection, proper identification and accurate diagnosis of the causal organism and reasons for the prevalence of a disease/pest are prerequisites for appropriate and well-timed control and management of any plant disease. Conventional methods to identify pathogen have often relied on disease symptoms, isolation, culturing and morphology of pathogen and biochemical tests.

The development of diagnosis techniques has been conditioned by the evolution of novel technologies. Rapid advancements in biotechnology have strengthened efforts in recent years to develop innovative methods for detection and identification of disease causing agents. Diagnosis of pathogen has been very useful and significantly improved with techniques of Dot-blot immunoassay (DBIA) to detect *Leifsonia xyli* subsp. *xyli*, PCR, RT-PCR, DAS-ELISA and tissue-blot immunoassay (TBIA) for SCYP and SCYLV, SCMV and *Xanthomonas albilineans*, immunocapture-RT-PCR based assay to detect the presence of *Sugarcane streak mosaic virus* (SCSMV) (Srivastava, 2016).

Recently, high throughput next generation sequencing (NGS) technologies have been adopted for virus detection (Barba et al., 2014) and have been used to identify emerging sugarcane viruses (Candresse et al., 2014). Whereas traditional serological and molecular diagnostics assays are based on the identification of predefined targets, NGS requires no previous knowledge of the pathogens present in the samples.

Twelve samples of sugarcane (*Saccharum spp.*), one sample of St. Augustine grass (*Stenotaphrum secundatum*), and one sample of Columbus grass (*Sorghum almum*) of foreign and domestic origin maintained at the USDA-APHIS Plant Germplasm Quarantine Program (PGQP) were sequenced using Illumina technologies. BLASTn searches against a custom sugarcane virus database identified 11 of 14 samples to be positive for one or more viruses. NGS analysis detected all viruses that were identified by laboratory diagnostic tests used routinely by PGQP. Additionally, in two samples, NGS identified *Sugarcane bacilliform virus*, a pathogen that was not targeted by the lab-based tests. This study shows that NGS is a reliable diagnostic tool for rapid virus identification that can be applied in sugarcane quarantine, certification, and breeding programs (Mollov and Malapi-Wight, 2016).

Sugar Research Australia (SRA) recently resumed attempts to identify the causal agent of chlorotic streak using molecular techniques. By testing a wide range of 'universal' PCR primers, was identified a primer pair that generated a chlorotic streak-specific DNA fragment coding for actin, a conserved eukaryotic gene. Unfortunately, it was too conserved to identify the organism through the standard approach of sequencing followed by matching against GenBank using BlastN (Braithwaite and Croft, 2013).

However, the DNA fragment could form the basis of a diagnostic test, which was further developed into a qPCR format (quantitative PCR). NGS was used to identify the causal agent, with phylogenetic analyses suggesting that it is a unique *Cercozoa*. Now diagnostic PCR primers, developed from the NGS results, are available for the small and large subunits (SSU or 18S and LSU or 28S, respectively) of the ribosomal sequence. The ribosomal SSU primer set has become the standard diagnostic combination for routine screening for chlorotic streak (Braithwaite, et al., 2016).

In Cuba, the presence of *SCYLV* was first detected in 1999 (Arocha et al., 1999). Abu-Ahmad et al. (2006) further classified BRA-PER and CUB genotypes in C132-81 and Ja64-11 cultivars from Cuban germplasm. Recent sequence analyses of *SCYLV* isolates confirmed the presence of BRA-PER, CUB and REU genotypes in commercial sugarcane plantations. The REU genotype is reported first time in Cuba. Co-infections of *SCYLV* genotypes (BRA-PER + CUB, BRA-PER + REU and CUB + REU) were also reported. The most predominating *SCYLV* genotypes reported in Cuba were identified as CUB and BRA-PER (Aday et al., 2016).

Joomun and Dookun-Saumtally (2016) have developed three specific real-time Taqman® RT-PCR tests were designed and tested on 45 SCYLV infected clones in quarantine as well as on 392 samples collected from commercial fields. These tests proved highly specific for the intended clusters and were also more sensitive to conventional RT-PCR.

Sugarcane streak mosaic virus (*SCSMV*) is one of the most prevalent and rapidly spreading sugarcane diseases in Asia (Xu et al., 2010). Pest Risk Analysis suggests that *SCSMV* has a high likelihood of disease establishment and could cause significant economic impact in Australia. *SCSMV* was thought to be a strain of *Sugarcane mosaic virus* (*SCMV*), but molecular and host-range analyses confirmed that it is a distinct member of the Potyviridae and one of two members of the Poacevirus genus.

In a project funded by the Australian Centre for International Agriculture Research (ACIAR) ten sets of RT-qPCR primers were developed from the genome regions of high sequence conservation, and these primers tested for specificity on various specimens of *SCSMV*. Nine sugarcane leaf specimens from Indonesia and Australia were collected and processed onto FTA™ Plant Saver™ cards (Whatman, USA). One primer pair was identified for use in future studies of disease distribution and detection. Quantitative RT-PCR has the advantage of being very sensitive, specific and has the capacity for use as a high throughput assay on small samples. However, there is a need for the development of lower cost alternatives, for example, serological assays, for use in Indonesia (Thompson et al., 2016).

Pathogen genetics must be taken into consideration for breeding and screening new sugarcane varieties for disease resistance, and for efficient management of plant resistance. Thanks to high output and low cost sequencing, analysis of genome variation of sugarcane pathogens varying in virulence should result in the identification and comprehension of the molecular determinants involved in these genetic changes, and subsequently also facilitate breeding for resistance (Rott et al., 2013).

Amplified fragment length polymorphism (AFLP) display of complementary DNA (cDNA) was used to identify genes from sugarcane somaclones expressed during the interaction with the rust pathogen – *Puccinia melanocephala*. The isolated transcript-derived fragments (TDFs) correspond to genes involved in the resistance process. Genes related with recognition, signaling and general response were identified through BLAST search (Carmona et al., 2004).

Several studies employing various molecular techniques including cDNA-AFLP have also accumulated information on differentially expressed transcripts of sugarcane in response to *Sporisorium scitamineum* challenge.

La et al. (2008) reported differential expression of TDFs on the *Saccharum spp. – S. scitamineum* pathogenic interaction. A majority (67.2%) of the differential TDFs up-regulated was recorded in the smut resistant M31/45 cultivar, representing major genes involved in oxidative burst, defensive response, ethylene and auxins pathways during the first 72 h post-inoculation. Results obtained suggested a key role for genes involved in the oxidative burst and the lignin pathways in the initial sugarcane defense against the *S. scitamineum* infection.

Que et al. (2014) analyzed the transcriptome of smut resistant and susceptible sugarcane cultivars challenged with *S. scitamineum* using an Illumina-based platform HiSeqTM 2000. Transcriptome profiling at 24, 48 and 120 hours post inoculation and functional categorization of differentially expressed transcripts indicated that up-regulation of defense related genes occurred earlier in the resistant variety. Pathway enrichment analysis indicated that majority of differentially expressed genes were related to plant hormone signal transduction, flavonoid biosynthesis, cell wall fortification and other defense-associated metabolic pathways. Transcriptome profiling during distinct stages of infection (5 and 200 dpi) has resulted in identification of several effectors and other genes with putative roles in pathogenicity and virulence (Taniguti et al., 2015).

Medeiros et al. (2014) studied changes in the transcription profile obtained by cDNA-AFLP analysis involving two sugarcane varieties contrasting to *SCMV* resistance, when challenged with a severe virus strain. A total of 392 TDFs were verified in the resistant variety against 380 in the susceptible one. Ten out of 23 sequenced TDFs (unique from the resistance variety), showed identity with known plant sequences, mostly related to plant defense mechanisms against pathogens.

La et al. (2016) screened 11 sugarcane cultivars with different level of resistance to *SCMV* with eight pair of primer for resistance genes. Results showed basal levels of endoglucanase, Zn-binding protein and nucleic acid binding protein in all cultivars tested. These genes seem to be regulated by viral silencing. However, fragments similar to the polyubiquitin and cyclin genes avoid such mechanism, but expressed auxin-independent growth promoter endopeptidase and phophoglucomutase are induced by viral infection. This result has potential for application in germplasm characterization and use as resistance indicators in sugarcane-*SCMV* screening.

Studies using semi-quantitative RT-PCR after pathogen inoculation from sugarcane cultivars varying in red rot resistance detected transcripts of resistant gene analogues (RGAs), transcription factors (TFs), defense-related genes and few signaling-related genes up-regulated specifically during sugarcane – *Colletotrichum falcatum* interaction (Viswanathan et al. 2009). Further works revealed induction of pathogenesis related (PR) proteins and 3-deoxyanthocyanidin phytoalexins as a defense and induced response (Ganesh Kumar et al., 2015).

Recently, a chitinase gene from sugarcane has been characterized as a class IV glycosyl hydrolase and its differential expression in sugarcane cultivars varying in red rot resistance was monitored through qRT-PCR (Rahul et al., 2015). Further, molecular techniques such as differential display (DD)-RT-PCR and suppression subtractive hybridization (SSH) were applied to study differential expression of the transcripts during the host-pathogen interaction.

Up-regulation of transcripts associated with jasmonic acid (JA), ethylene (ET), phosphoinositide (PI) and calcium (Ca^{2+}) signals, defense, reactive oxygen species (ROS) and the secretory pathway in red rot resistant cultivar were identified (Sathyabhama et al., 2016). In the experimental trials, some cultivars of sugarcane exhibited differential responses to various *C. falcatum* pathotypes over time. The cultivars varying in red rot resistance may have variable genetic background and that may have a profound influence on identifying the genes/proteins of sugarcane involved in host defense.

A high throughput analysis of interaction between sugarcane and *C. falcatum*, was made by comparing transcriptomes of compatible and incompatible interactions in a tropical sugarcane cultivar Co7805 exhibiting differential reaction to the pathotypes Cf94012 and Cf87012, respectively. SSH combined with Illumina 2000 high throughput sequencing was used to identify the differential transcripts in resistance response library (RRL) and susceptible response library (SRL). Based on the transcripts mapping to KEGG-KASS database, the presence of MAPK (mitogen activated protein kinase), ET (ethylene), PI (phosphoinositide) signals and JA (jasmonic acid) amino conjugation were found only in the incompatible interaction and expression of 10 transcripts involved in these pathways was validated using qRT-PCR (Viswanathan et al., 2016).

The study concludes that perception of PAMPs (pathogen-associated molecular patterns) occurs in both systems, but downstream signaling through MAPK, ET, PI and JA amino conjugation and activation of R genes occurs only in the incompatible interaction. This is the first detailed transcriptomic analysis of compatible and incompatible interactions in sugarcane with two different *C. falcatum* pathotypes through SSH and the next generation sequencing (NGS) platform.

Two resistance genes for brown rust, Bru1 and Bru2, have been reported. Molecular markers were developed (Costet et al., 2012) that then allowed breeders to assess the frequency of *Bru1* in parent and selection populations. Breeding programs apparently have unknowingly selected for *Bru1*, and resistance in some industries may rely heavily on this gene that has shown to be durable and effective against geographically diverse *P. melanocephala* populations. In contrast, the frequency of *Bru1* was found to be low in germplasm in Louisiana (Parco et al., 2014) and Argentina (Racedo et al., 2013).

The identification of transcripts differentially expressed in response to infection by *P. melanocephala* and expression analysis of genes in cultivar L99-233, a host genotype without *Bru1* but expressing quantitative resistance could provide information about other possible mechanisms involved in durable brown rust resistance. Therefore, a differential gene expression study of the resistance response to brown rust in L99-233 was performed using SSH technology to elucidate genes associated and infer mechanisms of resistance (Hoy et al., 2016).

The study has identified genes involved in primary metabolism, signal transduction, nucleic acid binding and protease activities to be differentially expressed in response to fungal infection. Analysis of expression kinetics of a selected set of genes showed transient upregulation of their mRNA accumulation in susceptible cultivars, but their transcripts were also upregulated up to 1 week post-inoculation in the resistant cultivars. The maintenance of high amounts of mRNAs of the genes for a prolonged time period appeared to be the contributing factor for resistance to brown rust. Breeding and selection for *Bru1* and other genes for quantitative resistance could provide effective and durable resistance to brown rust in future cultivars.

Screening for disease resistance has been an essential component of the plant improvement program. Ratings applied to test clones and varieties depend on the reliability of the reaction of standard varieties of known resistance status. Varieties are often classed as resistant (R), intermediate (I) or susceptible (S) based on relative severity of disease symptoms or yield loss. These ordinal categories can be represented by the numbers 1–3 (R), 4–6 (I) and 7–9 (S) as recommended by the ISSCT Pathology Committee for sugarcane diseases. Hutchinson et al. (1971) suggested that the ratings be calculated from a regression between the disease score of the standards in the trial and their historical rating.

Pachymetra root rot (caused by *Pachymetra chaunorhiza*) is the most important of the root diseases affecting commercial sugarcane production in Australia and is found in almost all districts. The disease was first recognized in the late 1970s when the susceptible variety Q90 was widely grown over northern Queensland production areas. Low yields, failed ratoons and increased extraneous matter in the cane supply required research and development to identify management options. Varietal resistance is strongly related to soil inoculum levels in the field (Magarey and Bull, 2011).

Resistance screening method was developed that incorporated the growing of sugarcane test varieties in *Pachymetra*-infested potting mix in pots for six weeks under controlled glasshouse conditions. Disease assessments were based on quantifying the number (percentage) of rotted primary shoot roots. To increase the capacity to screen larger numbers of clones, modifications were made that involved growing test plants in potting mix for 12 weeks and visually estimating the disease severity (0–5 basis) for each plant based on the extent of root rotting in unwashed root systems.

All data obtained using both objective and subjective resistance screens were carefully analyzed and the resistance of the standard varieties refined. It was found that the method based on disease severity resulted in some variation in the ratings for the standard varieties. These differences are reflected in improved coefficients of determination for trials where the new standard ratings have been substituted. Significant changes in the coefficient occurred with the revised standard ratings. Changes in the ratings for individual test clones also resulted and therefore affected the discard of some clones from the selection program (Magarey et al., 2011).

'Standard varieties' are important for resistance screening trials as they provide a guide to the likely commercial crop disease reaction of clones. A divergence of the data from the 'norm' for the standard varieties suggests a problem with trial methods or conditions and enables a decision to be made to discard unreliable trial data (Magarey and Bull, 2011).

In smut screening trials, stalks of clones from various stages of breeding programs are dip-inoculated with a smut spore suspension (10^6 spores/mL water) and planted in the field. Two methods of rating smut resistance screening trials were compared in a series of eight trials in 2008 and seven trials in 2009 in Bundaberg, Australia (Bhuiyan and Croft, 2011). Disease incidence (percent of infected stools) and disease severity (visual assessment of the severity of stunting and smut whip production) were measured in the plant and ratoon crops.

Both incidence and severity were strongly correlated with the long term rating of the standard varieties in both years. Rating clones based on disease severity would have retained more clones than rating clones on disease incidence. In both years, 5–6% of clones were discarded using the incidence rating scale but would have been kept if the severity ratings scale had been used. Most of these clones would fall into intermediate or intermediate-susceptible categories with the severity rating scale. There is the potential to discard some agronomically superior clones unnecessarily if the rating scale is too severe. Observations on the smut reaction of existing commercial varieties in the field may help in determining the level of resistance required and the best method of estimating this resistance.

For sugarcane diseases, various statistical methods are used to help compare clones to the standard varieties and to assign ratings. Application of the Additive Main effects and Multiplicative Interaction (AMMI) model for common rust evaluation in the Cuban breeding program, allowed appreciating likeness and differences among evaluation moments, leaf number and cultivars. Also it was possible to identify genotypes with large contribution to genotype by environment interaction, greater susceptibility and the most stable in the response to brown rust (Tamayo et al., 2012).

In a multi-environment trial for evaluating the response of sugarcane genotypes to *SCMV* and to determine stability of standard varieties tested across environments AMMI models were used (Puchades et al., 2016). The results showed significant differences among locations and genotypes. The most important source of variation was location. The sources of inoculum showed a similar pattern of performance at the three test sites.

A biplot of the AMMI model showed the instability of the standard genotypes used for SCMV evaluation. Differential sensitivities of the standard varieties across all tested environments were rated as 39MQ832 (INT) < B42231(R) < C236-51(S).The ranking for stability was 39MQ832 > C236-51> B42231. The standard B42231(R) showed the greatest instability in their response to *SCMV*. This shows that there is potential for unreliable classification of genotypes. These results highlighted the importance of improving the current system for *SCMV* evaluation based on statistical methods that provide more accurate ratings for genotypes

Up until recently, most of breeding program used the Hutchinson method for rating varieties in disease resistance trials. Stringer et al. (2012) described a new technique based on application of linear mixed models to rate varieties in disease screening trials. In this method, the standard varieties are separated into groups based on the least significant difference test and the test varieties are placed in the group to which they best fit. A correlation is performed on the score of the standards in the new trial against the average score from all previous trials to determine if the current trial can be considered to provide a reliable estimate for the standards.

The new method of analysis of disease resistance trials will provide more accurate ratings for varieties and is more conservative than the Hutchinson method. The method also recognizes and accounts for the variability within and between the rating groups. One of the advantages of using the linear mixed model to analyze data is that ratings can be refined for varieties that proceed to the release stage by conducting an analysis of data from different years and multiple trials. This estimate of the accuracy of the ratings shows that assigning ratings to test clones to nine separate groups is unrealistic in most trials (Stringer et al., 2013).

Regarding insect pests, studies conducted by Sugar Research Australia (SRA) showed that *Erianthus arundinaceus* and *Saccharum spontaneum* are sources of resistance to root knot (*Meloidogyne javanica*) and lesion nematodes (*Pratylenchus zeae*). The average levels of resistance tended to decrease with successive backcrosses between the wild species and commercial hybrids. Thirteen accessions of *S. spontaneum* were screened twice against root-knot nematode and once against root-lesion nematode in 2012 and 2013 in a glasshouse (Bhuiyan et al., 2016).

Nematode reproduction (both *M. javanica* and *P. zeae*) were significantly lower on basic *S. spontaneum* clones compared to backcross progenies and commercial standards. Some progenies from *S. spontaneum* and commercial varieties also had low nematode reproduction. This study provided valuable information that can be used to develop nematode-resistant varieties for the Australian sugar industry.

Plant resistance to stem borer is often negatively associated with yield potential. There exists then, a need to identify sources of resistance that have no adverse effect on sugar yield. Clones of the genus *Erianthus*, a taxa of sugarcane that possess leaf insertion (throat) hairs, may provide that source of resistance. White et al. (2013) evaluated six *Erianthus* accessions in the glasshouse and laboratory to determine if these throat hairs represent a barrier to young *Diatraea saccharalis* (Lepidoptera: Crambidae) larvae and thereby confer resistance.

Results showed that the throat hairs represent a physical barrier to young larvae, because test larvae were unable to successfully negotiate those accessions with dense throat hairs. However, larvae were able to easily negotiate hairs on those accessions with less dense hair patterns. Fewer larvae were recovered after 21 days of feeding and less stalk tunneling occurred on plants with dense hair patterns than on plants with less dense hairs. Before this trait can be successfully exploited, additional research is needed to determine if the leaf throat hairs can be transmitted to high yielding cultivars in both a form and density that confers resistance.

A new pest mite was found on sugarcane leaf sheath on the farm of the Sugarcane Research Institute, Guangxi Academy of Agricultural Sciences in 2011(Lin et al., 2013). The pest mite on sugarcane was identified as *Aceria sacchari* (Acari: Eriophyidae). The mite was mainly present on the tender leaf sheath during the middle to later growth period. Dark red sponginess appeared on the inner leaf sheath at the peak of infestation with the exterior markedly distorted. This species was first recorded on sugarcane in India, and was found to be widely distributed in that country. It has also been recorded in Australia, Philippines and Taiwan province of China. It is believed that the mite might have been introduced into mainland China from Taiwan province in the ROC22 germplasm and plantlets.

In 2008, *Ancistro soma argentinum* (Moser) (Coleoptera: Scarabaeidae) was detected for the first time in sugarcane in Jujuy Province, Argentina. This insect has a high potential to cause losses on the sugarcane. Fortunately, the growth potential has been low. However, yield losses ranging from 70% to 100% occurred. Chemical control was performed using imidacloprid, chlorpirifos, thiametoxam, and teflutrin while biological control with the fungus *Metarhizium anisopliae*, and cultural practices were also attempted. The best control in terms of grub mortality was obtained by ploughing out infested fields followed by a long fallow period (more than 4 months), crop rotation with soybean, and replanting after a minimum of 1 year after tillage (Easdale et al., 2015).

The Mexican rice borer (*Eoreuma loftini* (Dyar), Lepidoptera: Crambidae) was observed in Texas in 1979 and in Louisiana in 2008. It is a pest of sugarcane and rice. In 2012, it was discovered in Florida and is spreading naturally and by movement of infested plant material. There are concerns of its spread into the Poaceae family (Nuessly et al., 2015).

Some cultural strategies have been implemented for Mexican Rice Borer Control. These includes: reduce spread by eliminating movement of seed pieces from infested areas to areas not infested; maintain irrigation where available to reduce plant stress, mow (<18 cm) and disk down infested rice stubble following harvest to reduce pest reservoirs; use pheromone traps to monitor spread and to alert growers for the presence of moths in their area (20 months per trap correlates well with 5% treatment threshold).

Sugarcane borers are widely distributed in Guangxi Province, China, and control is very difficult. In recent years, continuous application of pesticides has not only affected the environment and natural enemies, but also resulted in pesticide resistance in the cane borers (Pan et al., 2016). There is growing attention to the ecological effects of these chemicals and to biocontrol, which is regarded as an interesting alternative. Better knowledge of the distribution and natural parasitism rate of egg parasitoids of sugarcane borers would supply more diverse germplasm resources for biological control programs.

The distribution and natural parasitism rate of egg parasitoids in 11 sugarcane-planting areas of seven cities in Guangxi were investigated by collecting sugarcane borer eggs from 2011 to 2014 (Pan et al., 2016). Results showed that there were mainly three kinds of egg parasitoids in Guangxi: *Trichogramma chilonis* Ishii, *Trichogramma ostriniae* Pang and Chen, and *Telenomussp.* *T. chilonis* was widely distributed in Guangxi, while *T. ostriniae* and *Telenomus sp.* were only found in some of the areas surveyed. From 2011 to 2014, the parasitism rate of *Trichogramma spp.* was significantly lower at the early growth stage than mid-late growth stage ofsugarcane. In Chongzuo, the parasitism rate of *Telenomus sp.* was higher than that of *Trichogramma* and there was little difference in the parasitism rate between the early growth stage and the middle-late growth stage.

Sugarcane white leaf (SCWL) disease is an important disease of sugarcane caused by unculturable plant- pathogenic phytoplasma that lives in the phloem of host plants and in the insect vectors. The disease is a threat to the economic stability sugarcane industry in Thailand. The disease can spread through the use of infected cane stalks for planting material and by the leafhopper vectors *Matsumuratettix hiroglyphicus* (Matsumura), *Yamatotettix flavovittatus* (Matsumura), *Maiestas portica* (Melichar), *Exitianus indicus* (Ross) and *Cofana unimaculata* (Signoret). A long-term research has generated basic knowledge for an integrated disease management practice that is aimed at both insect vectors and host plants (Hanboonsong, et al., 2016; Rao et al., 2014; Tiwari et al., 2016, 2017).

The studies indicated that no alternative host plants are known as a reservoir of SCWL phytoplasma. Instead, insect vectors can transmit the pathogen from generation to generation through eggs. The natural dispersal of these insect vectors (*M. hiroglyphicus* and *Y. flavovittatus*) was found to be minimal. On the other hand, the disease is quickly spread via the use of infected planting material (Tiwari et al., 2012). Therefore, using clean cane stalks for planting is the most effective prevention strategy. Knowledge of vector population dynamics, ecology and dispersal is required to complement disease management strategies (Rao et al., 2014).

The advantages offered by geostatistical techniques for spatial variability analysis of the incidence of *D. saccharalis* in sugarcane was explored in Cuba (Loddo, et al., 2016). Data from the damage recorded in commercial fields during 2011 and 2012 were used. Additionally, records of field releases of both parasitoids *Lixophagadiatraeae* (Diptera: Tachinidae) and *Tetrastichus howardi* (Hymenoptera: Eulophidae) to control the stem borer were considered. Data analyzed using Mapinfo Vertical Mapper and Variowin geostatistical tools showed that the most adequate distances for pest intensity evaluation determined by resulting correlations in the variogram were 1968.75 m and 4804.84 m. Estimation maps displaying the incidence of *D. saccharalis,* allowed understanding and improve the strategies for pest control in real time. One of the most important results is the optimization and redirection of the application of biological control to the most infested areas.

Satellite imagery was investigated as a method for detecting infestations over large areas of white grubs (Coleoptera: Scarabaeidae) in Australia (Samson et al., 2015). High spatial resolution multispectral and panchromatic satellite images were acquired in May-June, corresponding with the months when symptoms of feeding by greyback canegrubs are most visible. Images taken over three years were processed using geographic object-based image analysis (GEOBIA). Results indicated that disturbances within cane fields could be detected by very high resolution imagery. However, specifying the type of disturbance such as cane grub damage was more difficult because other problems such as water-logging, pig damage, or weed infestation appeared similar.

9.4 FUTURISTIC APPROACH

To increase sugarcane productivity researchers and development workers should be actively involved in enhance innovative pest and disease management strategies. Regular monitoring and detection of all major diseases and pests in the nursery and commercial plots, will keep them free from all major diseases. Pathogen population genetics must therefore be taken into consideration for breeding and screening new sugarcane varieties for resistance to diseases.

Knowing the genetic diversity of a pathogen is very important, especially when plants are artificially inoculated with pathogens for screening. Furthermore, sugarcane genetics is very complex and few specific resistance genes have been identified so far, and selection for these traits is also complex. Sequencing and analysis of genome variation of sugarcane pathogens varying in virulence should result, in the future, in the identification and comprehension of the molecular determinants involved in these genetic changes, and subsequently also facilitate breeding for resistance.

In addition to detecting sugarcane pathogens in seed canes, the recent approaches in the disease diagnosis using serological and molecular approaches have applications in the field of developing virus-free seedlings, germplasm exchange and quarantine, diseases and pests surveillance and integrated pest management in sugarcane. However, more robust detection assays, focusing on strains/biotypes in the region allow early and accurately pathogen identification.

The application of novel studies based on software design, satellite imagery, global position system and image-processing mobile application can be used to support strategic decisions Addition of knowledge on pathology and disease control is envisaged to identify insect pests and diseases rapidly and accurately, leading to reliable recommendations to manage these problems.

Furthermore, increased international collaboration, a policy of prevention, incursion risk analysis and biosecurity planning are needed to be adopted for ensuring rapid response to disease/pest outbreak.

9.5 CONCLUDING REMARK

The present scenario has made it very clear that continuous research efforts are required to understand the evolution of pathogenicity of sugarcane pathogens and the mechanism of diseases and pests resistance in sugarcane, so as to sustain the productivity of the released varieties for commercial cultivation. The knowledge thus gained would help in unlocking the secrets underlying any host-pathogen interaction.

Regarding insect pests, studies for prospections, biological control and Integrated Pest Management (IPM) are key strategies to adopt a preventive solution before implementing expensive treatments. The IPM strategies need to become a decision support system that would enable the identification of the multiple and interdependent factors that trigger pest outbreaks.

KEYWORDS

- **biological control**
- **control measures**
- **diagnosis**
- **disease resistance**
- **integrated pest management**
- **sugarcane**

REFERENCES

Abu-Ahmad, Y., Royer, M., Daugrois, J. H., Costet, L., Lett, J. M., Victoria, J. I., Girard, J. C., & Rott P., (2006). Geographical distribution of four *Sugarcane yellow leaf virus* genotypes. *Plant Disease*, *90*, 1156–1160.

Aday, O. C., La, O. M., Puchades, Y., Zardón, M. A., Mesa, J. M., Lissbrant, S., & Arencibia, A. D., (2016). Genotyping of *Sugarcane Yellow Leaf Virus* in Commercial Cultivars and the Cuban Germplasm Collection. *Publication online Sugar Tech*, doi: 10.1007/s12355-016-0447-9.

Arocha, Y., Gonzalez, L., Peralta, E. L., & Jones P., (1999). First report of virus and phytoplasma pathogens associated with Yellow Leaf Syndrome of sugarcane in Cuba. *Plant Disease, 83,* 1171.

Bailey R., (2011). Disease Control. In: Meyer Y. (Ed). Good management practices manual for the cane sugar industry. International Finance Corporation (IFC), *PGBI Sugar and Bioenergy*, Johannesburg, South Africa, 696 p.

Barba, M., Czosnek, H., & Hadidi A., (2014). Historical perspective, development and application of next-generation sequencing in plant virology. *Viruses, 6,* 106–136.

Bhuiyan, S. A., & Croft, B. J., (2011). Incidence versus severity—what difference would it make to smut screening. *Proc Aust Soc Sugar Cane Technol., 33,* 8 p.

Bhuiyan, S. A., Croft, B. J., Stirling, G., Jackson, P. A., Piperidis, G., & Aitken KS.(2016). Resistance of *Saccharum spontaneum* and its backcross progenies to root-knot and root-lesion nematodes. *Proc. Aust. Soc. Sugar Cane Technol., 29,* 1792–1804.

Braithwaite, K. S., & Croft, B. J., (2013). A diagnostic test for chlorotic streak disease. Proc. Australian *Soc. Sugar Cane Technol., 35,* 8 pp.

Braithwaite, K. S., Ngo, C., Croft, B., Magarey, R., & Young A., (2016). Progress in understanding and managing chlorotic streak of sugarcane. *Proc. Int. Soc. Sugar Cane Technol., 29,* 11768–1775.

Candresse, T., Filoux, D., & Muhire B. Appearance can be deceptive: revealing a hidden viral infection with deep sequencing in a plant quarantine context. *PlosOne, 9, e102945.*

Carmona, E., Vargas, D., Borroto, C. J., Lopez, J., Fernandez, A. I., Arencibia, A., & Borras O., (2004). cDNA-AFLP analysis of differential gene expression during sugarcane–*Puccinia melanocephala* interaction. *Plant Breeding, 123,* 499–501.

Costet, L., Le Cunff, L., & Royaert, S. (2012). Haplotype structure around Bru1 reveals a narrow genetic basis for brown rust resistance in modern sugarcane cultivars. *Theoretical and Applied Genetics, 125,* 825–836.

Easdale, C., Bravo, J. L., & Iriarte A., (2015). A new pest in Argentina sugarcane fields: *Ancistro soma argentinum* Moser. ISSCT XI Pathol.and, I. X., Entomol. *Workshops, Guayaquil, Ecuador,* 10p.

Ganesh Kumar, V., Viswanathan, R., Malathi, P., Nanda Kumar, M., & Ramesh Sundar A., (2015). Differential Induction of 3-deoxyanthocyanidin phytoalexins in relation to *Colletotrichum falcatum* resistance in sugarcane. *Sugar Tech., 17*(3), 314–321.

Hanboonsong, Y., Wangkeeree, J., & Kobori Y., (2016). Integrated management of the vectors of sugarcane white leaf disease in Thailand: an update. *Proc. Int. Soc. Sugar Cane Technol., 29,* 1258–1263.

Hoy, J. W., Baisakh, N., Avellaneda, M. C., Kimbeng, C. A., & Hale, A. L., (2016). Detection, breeding and selection of durable resistance to brown rust in sugarcane. *Proc. Int. Soc. Sugar Cane Technol., 29,* 1034–1039.

Hutchinson, P. B., Daniels, J., & Husain, A. A., (1971). The interpretation of results from Fiji disease resistance trials. *Sugarcane Pathologists' Newsletter, 6,* 19.

INICA, (2011). Manual de Normas y Procedimientos del Programa de Mejoramiento. *Boletín Especial Cuba & Caña,* 200 p.

Joomun, N., & Dookun-Saumtally A., (2016). Importance of using highly specific and sensitive diagnostics for *Sugarcane yellow leaf virus* in quarantine. *Proc. Int. Soc. Sugar Cane Technol., 29,* 1740–1746.

La, O. M., Arencibia, A. D., Carmona, E. R., Acevedo, R., Rodriguez, E., Leon, O., & Santana I., (2008). Differential expression analysis by cDNA-AFLP of *Saccharum spp.* after inoculation with the host pathogen *Sporisorium scitamineum. Plant Cell Reports, 27,* 1103–1111.

La, O. M., Perera, M. F., Bertani, R. P., Acevedo, R., Arias, M. E., Casas, M. A., Pérez, J., Puchades, Y., Rodríguez, E., Alfonso, I., & Castagnaro, A. P., (2017). An overview of sugarcane brown rust in cuba. *Scientia Agricola.*

La, O. M., Puchades, Y., Zardón, M. A., Pérez, J., Casas, M., & Mesa J., (2016). Detection of resistance related gene in sugarcane cultivars inoculated with *Sugarcane mosaic virus. Proc. Int. Soc. Sugar Cane Technol., 29,* 728–731.

Lin S. H., Huang C. H., Shang X. K., Pan X. H., & Wang G. Q., (2013). *Aceria sacchari,* anew pest mite recorded on sugarcane in Guangxi Province, China. *Proc. Int. Soc. Sugar Cane Technol., 28,* 4p.

Loddo, Z., Granado, C., & Rodriguez M., (2016). Geostatistical tools for the assessment of the incidence of *Diatraea saccharalis* in sugarcane in Cuba. *Proc. Int. Soc. Sugar Cane Technol., 29,* 1362–1368.

Magarey, R. C., & Bull, J. I., (2011). The relationship between glasshouse and field-derived resistance ratings for Pachymetra root rot. *Proc. of the Aust. Soc. Sugar Cane Technol.,* 8 p.

Magarey, R. C., Bull, J. I., Lonie, K. J., & Croft, B. J., (2011). The influence of two Pachyme-
    tra root rot resistance screening methods on ratings of standard varieties. *Proc. Aust.
    Soc. Sugar Cane Technol., 33,* 6 p.

Magarey, R. C., Croft, B. J., & Willcox, T. G., (2001). An epidemic of orange rust in Austra-
    lia. *Proc. Int. Soc. Sugar Cane Technol., 24,* 410–416.

Medeiros, C. N. F., Gonçalves, M. C., Harakava, R., Creste, S., Nóbile, P. M., Pinto, L.
    R., Perecin, D., & Landel MGA., (2014). Sugarcane transcript profiling assessed by
    cDNA-AFLP analysis during the interaction with *Sugarcane mosaic virus. Advances in
    Microbiology, 4,* 511–520.

Mollov D., Malapi-Wight M., (2016). Next Generation Sequencing: a useful tool for detec-
    tion of sugarcane viruses in quarantine programs. *Proc. Int. Soc. Sugar Cane Technol.,
    29,* 1631–1635.

Nuessly G., Beuzelin, J., Wilson, B., Reagan, T., & Way, M. O., (2015). Status of the Mexican
    Rice Borer (*Eoremaloftini*) in the USA. *ISSCT XI Pathol. and IX Entomol. Workshops,*
    Guayaquil, Ecuador, 8p.

Pan, X. H., Huang, C. H., Wei, J. L., Shang, X. K., Lin, S. H., & Xin D. Y., (2016). Distribu-
    tion and natural parasitism rate of egg parasitoids of sugarcane borers in China. *Proc.
    Int. Soc. Sugar Cane Technol., 29,* 481–484.

Parco, A. S., Avellaneda, M. C., & Hale, A. H., (2014). Frequency and distribution of the
    brown rust resistance gene Bru1 and implications for the Louisiana sugarcane breeding
    programme. *Plant Breeding, 133,* 654–659.

Puchades, Y., La, O. M., Carvajal, O., Montalván, J., Rodríguez, J., Rodríguez, E., & Rodrí-
    guez M., (2016). Multi-environment trials for evaluating resistance to *Sugarcane
    mosaic virus. Proc. Int. Soc. Sugar Cane Technol (ISSCT), 29,* 915–920.

Que, Y., Xu, L., Wu, Q., Liu, Y., & Ling, H., (2014). Genome sequencing of *Sporisorium sci-
    tamineum* provides insights into the pathogenic mechanisms of sugarcane smut. *BMC
    Genomics.* Vol. *15,* n. 1, p. 996, 2014.

Racedo, J., Perera, M. F., & Bertani, R., (2013). Bru1 gene and potential alternative sources
    of resistance to sugarcane brown rust disease. *Euphytica, 191,* 429–436.

Rahul, P. R., Ganesh Kumar, V., Sathyabhama, M., Viswanathan, R., Ramesh Sundar, A., &
    Malathi P., (2015). Characterization and 3D structure prediction of chitinase induced
    in sugarcane during pathogenesis of *Colletotrichum falcatum. J. Plant Biochem. and
    Biotechnol., 10,* 154–157.

Rao, G.P., Madhu Priya., Tiwari, A.K., Kumar, S., & Baranwal, V.K., (2014). Identification of
    sugarcane grassy shoot-associated phytoplasma and one of its putative vectors in India.
    Phytoparasitica 42: 349–354.

Rodríguez, M., Fuentes, A., Loddo, Z., Rodríguez, E., Álvarez, J. F., & Jiménez, A. L., (2016).
    Impact of the biological control program in sugarcane pest management in Cuba. *Proc.
    Int. Soc. Sugar Cane Technol., 29,* 910–914.

Rodríguez, M., Rodríguez, E., Alfonso, I., & Fuentes A., (2014). Enfermedades y plagas. En:
    *Santana, I., González, M., Guillén, S., Crespo R. Instructivo Técnico para el Manejo
    de la Caña de Azúcar. Grupo Azucarero Azcuba.* INICA. ISBN: 978-959-300-036-9,
    pp. 209–256.

Rott, P. C., Girard, J. C., & Comstock, J. C., (2013). Impact of pathogen genetics on breed-
    ing for resistance to sugarcane diseases. *Proc. Int. Soc. Sugar Cane Technol., 28,* 11p.

Samson, P., Johansen, K., Sallam, N., & Robson A., (2015). Detection of white grub infesta-
    tions by satellite imagery in the central cane- growing region of Queensland, Australia.
    ISSCT XI Pathol. and IX, Entomol. Workshops, Guayaquil, Ecuador, 10p.

Sathyabhama, M., Viswanathan, R., Malathi, P., & Ramesh Sundar A., (2016). Identification of differentially expressed genes in sugarcane during pathogenesis of *Colletotrichum falcatum* by suppression subtractive hybridization. *Sugar Tech., 18,* 176–183.

Srivastava S., (2016). Molecular diagnostics and application of DNA markers in the management of major diseases of sugarcane. In: Kumar P., et al. (eds.), *Current Trends in Plant Disease Diagnostics and Management Practices, Fungal Biology.* Springer International Publishing, Switzerland, pp. 299–315.

Stringer, J., Croft, B., Bhuiyan, S., Deomano, E., Magarey, R., Cox, M., & Xu X., (2012). A new method of statistical analysis for disease screening trials. *Proc. Aust. Soc. Sugar Cane Technol., 34,* 6 p.

Stringer, J., Croft, B., Deomano, E., & Bhuiyan S., (2013). Analysis of sugarcane disease screening trials over years with a mixed model to improve ratings of varieties. *Proc. Int. Soc. Sugar Cane Technol., 28,* 1–6.

Sundar, A. R., Ashwin, N. M. R., Leonard-Barnabas, E., Malathi, P., & Viswanathan R., (2015). Disease resistance in sugarcane: an overview. *Scientia Agraria Paranaensis, 14,* 200–212.

Tamayo, M., Puchades, Y., Rodríguez, R., González, R., Suárez, H. J., Alfonso, I., Rodríguez, E., & La O M., (2012). Modelo de efectos principales aditivos e interacción multiplicativa aplicado a laevaluación de la roya común de la caña de azúcar. *Fitosanidad, 16,* 129–135.

Taniguti, L. M., Schaker, P. D. C., Benevenuto, J., Peters, L. P., Carvalho, G., & Palhares A., (2015). Complete genome sequence of *Sporisorium scitamineum* and biotrophic interaction transcriptome with Sugarcane. *PLoS ONE, 10*(6), 1–31.

Thompson, N., Wilson, E. J., Magarey, R. C., Putra, L. K., & Hidayat S. H., (2016). Development of diagnostic tests for *Sugarcane streakmosaic virus. Proc. Int. Soc. Sugar Cane Technol., 29,* 1606–1613.

Tiwari, A.K., Vishwakarma, S.K., & Rao, G.P. (2012). Increasing incidence of sugarcane grassy shoot disease in Uttar Pradesh, India and its impact on yield and quality of sugarcane. *Phytopathogenic Mollicutes 2,* 63–67.

Tiwari, A.K., Madhupriya., Srivastava, V.K., Pandey, K.P., Sharma, B.L., & Rao, G.P. (2016). Detection of sugarcane grassy shoot phytoplasma (16SrXI-B subgroup) in *Pyrilla perpusilla* Walker in Uttar Pradesh, India. *Phytopathogenic Mollicutes, 6,* 56–59.

Tiwari, A.K., Kumar, S., Mall, S., Jadon, V., & Rao, G.P., (2017). New Efficient natural leafhopper vectors of sugarcane grassy shoot phytoplasma in India. *Sugar Tech 19,* 191–197.

Viswanathan, R., Ramesh Sundar, A., Malathi, P., Rahul, P. R., Ganesh Kumar, V., Banumathy, R., Prathima, P. T., Raveendran, M., Kumar, K. K., & Balasubramanian P., (2009). Interaction between sugarcane and *Colletotrichum falcatum* causing red rot: understanding disease resistance at transcriptional level. *Sugar Tech., 11,* 44–50.

Viswanathan, R., Sathyabhama, M., Malathi, P., & Ramesh Sundar A., (2016). Transcriptome analysis of host-pathogen interaction between sugarcane and *Colletotrichumfalcatum* by Suppression Subtractive Hybridization and Illumina sequencing. *Proc. Int. Soc. Sugar Cane Technol., 29,* 1639–1644.

White, W. H., Richard, R. T., & Hale, A. L., (2013). *Erianthus*: a sugarcane relative with potential as a source of resistance to the stem borer *Diatraea saccharalis* (F.) (Lepidoptera: Crambidae). *Proc. Int. Soc. Sugar Cane Technol., 28,* 10p.

CHAPTER 10

BIOLOGICAL CONTROL OF TWO SUGARCANE STALK BORERS IN THE UNITED STATES

ALLAN T. SHOWLER

Knipling-Bushland, U.S. Livestock Insects Research Laboratory, USDA-ARS, 2700 Fredericksburg Road, Kerrville, TX78028, USA, E-mail: allan.showler@ars.usda.gov

CONTENTS

ABSTRACT

In the United States, two exotic stalk boring crambid moths, the sugarcane borer, *Diatraea saccharalis* (F.); and the Mexican rice borer, *Eoreuma loftini* (Dyar), are the two most economically important arthropod pests of sugarcane, *Saccharum* spp. Although the two pest species are closely related, their

---

*Mention of trade names or commercial products in this publication is solely for the purpose of providing specific information and does not imply recommendation or endorsement by the U.S. Department of Agriculture. USDA is an equal opportunity provider and employer.

behaviors are sufficiently dissimilar to present different challenges to sugar-cane production. *D. saccharalis* is the key pest in relatively wet sugarcane growing regions such as Florida and Louisiana, but *E. loftini* displaced *D. saccharalis* in the drier South Texas climate after it arrived there in the early 1980s. Because *D. saccharalis* undergoes its life stages in less protected places than *E. loftini*, it is more vulnerable to insecticide applications and to biological control agents. Although a few introduced parasitoid wasp species appear to have become established in Florida and Louisiana (releases have occurred since 1915), control of *D. saccharalis* through their action has been modest. Exclusion of predators using insecticides has resulted in increased *D. saccharalis*-induced injury to sugarcane in Louisiana, and enhancement of the red imported fire ant, *Solenopsis invicta* Buren, using vegetational diversification reduced injury to sugarcane. While the weed species that diversify sugarcane fields themselves can decrease sugarcane production, they are associated with heightened predator diversity and abundances. *Eoreuma loftini* is more cryptic than *D. saccharalis*, protecting it from nat-ural enemies and many insecticides. The pest is spreading from Texas to Louisiana, and it has been found in Florida as well. Because *E. loftini* prefers to lay its eggs on water deficit-stressed sugarcane plants, and because red imported fire ants are relatively sparse in South Texas, it has thrived there. It is not yet clear whether it will displace *D. saccharalis* in wetter environ-ments. The effects of efficient predators like the red imported fire ant on *E. loftini* populations is also not yet well understood, nor the effect of enhanc-ing predator numbers using vegetational diversification. Many parasites and parasitoids have been released for *E. loftini* control in South Texas but to negligible effect. Despite consistent attempts to control the two stalk borers with biological control agents, including entomopathogens, use of resistant cultivars is of more value for protecting sugarcane from them. Insecticides are generally effective against *D. saccharalis* and less so against *E. loftini*. Biological control, however, has value in some conditions as a tactic inte-grated into stalk borer control strategies.

10.1   THE STALK BORERS

The sugarcane borer, *Diatraea saccharalis* (F.), and the Mexican rice borer, *Eoreuma loftini* (Dyar), both crambid moths, are economically important and invasive stalk boring pests of sugarcane, *Saccharum* spp., in the United

States (Showler and Reagan, 2012). *D. saccharalis* was introduced to the United States in about 1855 from tropical regions in the Western Hemisphere (Capinera, 2001). The pest's range extends from the United States Gulf Coast across the Caribbean and Central America to subtropical parts of northern Argentina (Capinera, 2001). In the United States *D. saccharalis* is an incidental pest of other graminaceous crops including maize, *Zea mays* L.; rice, *Oryza sativa* L.; sorghum, *Sorghum bicolor* (L.) Moench; and sudan grass, *Sorghum bicolor* (L.) Moench spp. *drummondi* (Nees ex Steud.) de Wet and Harlan (Fuller and Reagan, 1989; Lv et al., 2008; Moré et al., 2003; Sosa, 1990). Sugarcane and large-diameter stems of some grassy weeds, including Johnson grass, *Paspalum* sp., *Panicum* spp., *Holcus* sp., and *Andropogon* sp., are used by *D. saccharalis* as overwintering sites (Capinera, 2001). In Texas and Louisiana overwintering larvae pupate and emerge as adults by April and May (Showler and Reagan 2012). Four to five overlapping (non-synchronous) generations occur through autumn (Reagan and Martin, 1989; Rodriguez-del-Bosque et al., 1995). Completion of a generation requires 25–40 days in the summer and >200 days in the winter (Capinera, 2001; Reagan and Martin, 1989).

Each female can produce ≈700 eggs in a lifetime and the egg clusters, comprised of 2 to >80 eggs (Capinera, 2001), are deposited unconcealed on the upper and lower sides of green leaf blades near the apex of the plant (Browning and Melton, 1987). In 4–6 days the eggs hatch and first instars feed on leaf tissue or tunnel into the midrib; second and third instars bore into the stalk (Reagan and Martin, 1989). Larvae develop in 25–30 days in warm weather, up to 5 days longer during cool weather, and development stops during cold winter weather (Capinera, 2001). Late (4th or 5th) instars enlarge a place inside the stalk with a thin exterior layer of plant tissue remaining to protect pupae from predators, parasites, and toxic chemicals (Showler and Reagan, 2012). After 8–9 days (up to 22 days in cool weather), adults emerge and exit the stalk. The moth is nocturnal and females deposit their eggs for up to 4 nights and die in 3–8 days (Capinera, 2001).

Serious injury to sugarcane occurs when *D. saccharalis* larvae tunnel into the stalks, which can cause lodging and death, and the inner whorl of young plants can be killed resulting in a condition known as "dead heart" (Reagan and Martin, 1989). Sucrose yield can decline by 10–20% (Capinera, 2001). *Diatraea saccharalis* is responsible for >90% of all insect-related sugarcane losses in Louisiana (Hensley, 1971a; Ogunwolu et al., 1987) although *E. loftini* has recently arrived there and it might supplant *D. saccharalis* as the state's key pest of sugarcane (Reay-Jones et al., 2008). Mechanical damage inflicted by *D. saccharalis* tunneling also facilitates infection of sugarcane by red rot, *Colletrotrichum falcatum* Went, and other deleterious fungal pathogens (Ogunwolu et al., 1991). Ulloa et al. (1982) calculated that one bored internode per stalk results in a loss of ≈2.5 kg of sugar per ton of harvested sugarcane and they also found that losses intensify as infestation levels increase.

Eoreuma loftini is indigenous to western Mexico (Morrill, 1925; Van Zwaluwenberg, 1926) and the pest was detected in the United States in the Lower Rio Grande Valley of Texas in 1980 (Johnson, 1981, 1984; Johnson and Van Leerdam, 1981), and by 2012 it had invaded most of South Texas, rice growing parts of east Texas (Browning et al., 1989; Reay-Jones et al., 2008), and into Louisiana rice and sugarcane (Hummel et al., 2008, 2010; Wilson et al., 2011). *E. loftini* is also found in Arizona and in southern California (Johnson, 1984), but it is not economically important in those two places. The pest was first detected in March 2012 at a light trap in Levy County, Florida, and in 2013 larvae were found in wild grasses there (Hayden, 2012; University of Florida, 2014). *E. loftini* attacks [3]23 species of common gramineous crop and non-crop host plants (Table 10.1), many of which, especially maize, can serve as reservoirs for the pest (Showler et al., 2011, 2012). Because *E. loftini* prefers maize over other crop plants (Showler et al., 2011), its assumed range might be underestimated, like that of the eldana borer, *Eldana saccharina* Walker (Pyralidae), in South Africa (also reported to prefer maize) (Assefa et al., 2008).

Mean fecundity of *E. loftini* can be as high as 400 eggs (Legaspi et al., 1997; Showler and Reagan, 2012). On sugarcane, the female oviposits within folds of dry leaves and dry portions of otherwise green leaves, although the pest also uses folded green living tissue if available (Showler and Castro, 2010b). Unlike *D. saccharalis*, *E. loftini* prefers to lay eggs on drought-stressed sugarcane plants (Showler and Castro, 2010a; Showler and Reagan, 2012) and on sugarcane plants grown from nitrogen-rich soil (Showler, 2015). Early instars feed on leaf tissue, under fresh leaf sheaths, and some bore into leaf midribs; later instars tunnel into the main stalk (Wilson, 2011). Injury includes dead heart, diminished sugar yield, and stalk stunting or lodging to the extent that sometimes harvest is uneconomical (Hummel et al., 2008; Johnson, 1985; Legaspi et al., 1997). *E. loftini* tunnels are horizontal and vertical and packed with frass, blocking predators and parasites

TABLE 10.1   *Eoreuma loftini* Host Plants

Scientific name	Common name
Arundo donax L.	Giant river cane
Bromus sp.	---
Canna sp.	Canna
Cortaderia selloana (Schult. and Schult. F.)	Pampas grass
Cymbopogon citratus (DC, ex Nees) Stapf	Lemon grass
Cynodon dactylon (L.) Pers.	Bermuda grass
Echinochloa crus-galli (L.)	Barnyard grass
Hordeum vulgare L.	Barley
Leptochloa panicoides P. Beauv.	Amazon sprangletop
Lolium sp.	---
Oryza sativa L.	Rice
Panicum miliaceum L.	Broom corn-millet
Panicum sp.	Panicum grass
Paspalum urvillei Steud.	Vasey's grass
Saccharum spp.and hybrids	Sugarcane
Scripus sp.	Bulrush
Setaria lutescens (Weigel) Hubb.	Bristle grass
Sorghum bicolor (L.) Moench	Sorghum
Sorghum bicolor (L.) Moench ssp. *drummondi* (Nees ex Steud.) de Wet and Harlan	Sudan grass
Sorghum halepense (L.) Pers.	Johnson grass
Triticum aestivum L.	Wheat
Urochloa platyphylla (Munro ex C. Wright)	Broadleaf signal grass
Zea mays L.	Maize

(Reprinted from Showler, A. T.; Beuzelin, J. M.; and Reagan, T. E. Alternate crop and weed host plant oviposition preferences by the Mexican rice borer (Lepidoptera: Crambidae). *Crop Prot.* **2011**, *30*, 895-901. © 2011 Elsevier.)

(Showler and Reagan, 2012). Like *D. saccharalis*, the last instar makes an emergence hole with a thin outer plant tissue "window" before pupating (Showler and Reagan, 2012). In the Lower Rio Grande Valley of Texas *E. loftini*'s life cycle takes 30–45 days, and 4–6 overlapping generations occur in a year (Johnson, 1985; Legaspi et al., 1997).

Tunneling and the insect's prevalence made it the most destructive sugarcane pest in South Texas, displacing *D. saccharalis* (Legaspi et al., 1997;

Showler and Reagan, 2012; Van Leerdam et al., 1984). *E. loftini* injures ≈20% of South Texas sugarcane internodes and larval tunneling also facilitates red rot infection (Johnson, 1985; Osborn and Phillips, 1946; Van Zwaluwenberg, 1926), but up to 50% bored internodes have been reported on some varieties of sugarcane (Johnson, 1981). Theborer causes losses to the United States sugarcane industry of US$575 per hectare and US$10–20 million annually (Legaspi et al., 1997, 1999b; Meagher et al., 1994). An economic projection of *E. loftini*-induced loss to Louisiana sugarcane is US$220 million per year (Reay-Jones et al., 2008).

10.2   CHEMICAL, VARIETAL, AND CULTURAL STALK BORER CONTROL TACTICS

Diatraea saccharalis is vulnerable to insecticide sprays because its life stages tend to be more exposed than those of *E. loftini* (Showler and Reagan, 2012). Fall insecticide applications are discontinued against *D. saccharalis* as early as possible to protect predators that might survive winter to the following year's growing season (Pollet et al., 1978) and because late season infestations are mostly in the upper portions of the stalks that contain relatively little sugar, often discarded at harvest (Hensley, 1971b). Bessin et al. (1990) reported that insecticidal (fenvalerate) control provided 42% control of *D. saccharalis* while varietal resistance provided 47% control and combining the two tactics gave >60% control. As often occurs in pests after repeated exposure to an insecticide, resistance can develop for which pesticide rotation has been recommended to avert the buildup of resistance (Reay-Jones et al., 2005b). In Louisiana, it is also important to select pesticides that conserve natural enemy populations (Showler and Reagan, 1991; Showler et al., 1991) because integration of best-fit tactics against *D. saccharalis* is the most effective approach to its control (Bessin and Reagan, 1990; Showler and Reagan, 1991).

In contrast, insecticide use against *E. loftini* was discontinued in the Lower Rio Grande Valley because yield was not increasing in response to applications (Legaspi et al., 1997, 1999; Meagher et al., 1994). Foliar applications of tebufenozide, flubendiamide, beta cyfluthrin, and chlorantraniliprole were more recently reported to reduce percentages of *E. loftini*-bored internodes by 3- to 8.5-fold (Akbar et al., 2009; Van Weelden et al., 2013). Novaluron, an insect growth regulator, was also shown, although not consistently, to maintain *E. loftini* injury at economically acceptable levels and to elevate sugar yield by 14% (Wilson et al., 2012). Tebufenozide, novaluron, flubendiamide, and several pyrethroids are registered for applications against *E. loftini* but their use for *E. loftini* control is limited (Showler and Reagan, 2012).

Although cultural practices are not specifically aimed at controlling *D. saccharalis*, some cultural practices have been demonstrated to protect against large infestations of *E. loftini*. Adequate irrigation of sugarcane, for example, reduces *E. loftini* infestations (Reay-Jones et al., 2005a; Showler and Castro, 2010a). Avoiding excessive application of nitrogen to sugarcane soil also exacerbates *E. loftini* attack (Showler, 2015). Integration of plant resistance, irrigation, and applications of the insect growth regulator tebufenozide was shown to reduce *E. loftini* damage to sugarcane from 70% bored internodes to <10% (Reay-Jones et al., 2005a).

10.3   BIOLOGICAL CONTROL OF *DIATRAEA SACCHARALIS*

10.3.1   *PREDATORS*

Louisiana sugarcane harbors large predator populations, most of which are indigenous (Showler and Reagan, 1991). Twenty predatory arthropod families have been reported from Louisiana sugarcane fields, including 10 spider families, on the soil surface, weed foliage, sugarcane leaves, and sugarcane stalks (Showler and Reagan, 1991; Table 10.2). Hall (1988) found 11 predatory insect families in five orders (spiders were not included in the survey) that were associated with Florida sugarcane (Table 10.3). The effects of most individual predatory arthropod taxa on *D. saccharalis* infestations have not been assessed in the United States, but in Argentina, populations of a dermapteran (Forficulidae), *Doru luteipes* Scudder, in maize are positively correlated with *D. saccharalis* mortality (Fenoglio and Trumper, 2007). In Louisiana sugarcane, soil-applied aldicarb greatly reduced some predatory arthropod populations which was associated with 19.4% to 32.9% more *D. saccharalis* damage than where aldicarb was not applied (Showler and Reagan, 1991). Chemical suppression of predatory arthropods using nonselective and persistent insecticides (e.g., mirex, heptachlor, and chlordane) has also been associated with as much as 6% more bored internodes than where predators were not suppressed (Charpentier et al., 1967; Hensley et al., 1961; Long et al., 1958; Negm and Hensley, 1967, 1969; Reagan et al., 1972).

TABLE 10.2　Abundances of Predatory and Parasitic/Parasitoid Arthropods Detected on the Soil Surface, Weeds, and Sugarcane Foliage and Stalks at an Assumption Parish, Louisiana, Sugarcane Plantation in 1985 and 1986

Predators	Common name	Soil surface[a]	Weeds[b]	Sugarcane foliage[b]	Sugarcane stalk[c]
Aranaea	Spider	+	+	+	+
Aranaeaidae	Orb-weaver spider	–	+	+	–
Clubionidae	Sac spider	+	+	–	+
Linyphiidae	Sheetweb spider	+	+	+	+
Lycosidae	Wolf spider	+	+	–	–
Nesticidae	Scaffold spider	+	+	+	–
Oxyopidae	Lynx spider	+	+	+	–
Pisuridae	Fishing spider	+	+	+	+
Salticidae	Jumping spider	+	+	+	+
Theridiidae	Comb-footed spider	+	+	+	+
Thomisidae	Crab spider	–	+	+	–
Carabidae	Ground beetle	+	–	–	+
Cicindelidae	Tiger beetle	+	–	–	–
Coccinelidae	Ladybird beetle	+	+	+	–
Dermaptera	Earwig	+	–	–	+
Elateridae	Click beetle	+	–	–	–
Enicocephalidae	Unique-headed bug				
Formicidae	Ant	+	+	+	+
Paratrechina vividus	--	+	+	+	+
Pheidole moerens	--	+	+	+	+
Solenopsis invicta	Red imported fire ant	+	+	+	+
Geocoris sp.	Big-eyed bug	+	+	+	–
Reduviidae	Assassin bug	+	+	+	–
Staphylinidae	Rove beetle	+	–	–	+

+, present; –, absent.

[a] Based on pitfall trap captures, July–October.

[b] Based on sweep netting, July–August.

[c] Based on visual leaf sheath and stalk inspection, July–October.

(Reprinted from Showler, A. T.; and Reagan, T. E. Effects of sugarcane borer, weed, and nematode control strategies in Louisiana sugarcane. *Environ. Entomol.* **1991**, *20*, 358-370. By permission of Oxford University Press.)

TABLE 10.3 Predatory Insect Families Associated with Sugarcane in Florida Based on Review of Literature and Observation*

Orders and families	No. of species
Coleoptera	
Alleculidae	1
Carabidae	2
Coccinelidae	7
Elateridae	12
Staphylinidae	12
Dermaptera	
Labiduridae	1
Diptera	
Asilidae	3
Syrphidae	1
Hymenoptera	
Formicidae	>30[c]
Neuroptera	
Chrysopidae	1
Hemerobiidae	1

[a] Includes *Solenopsis invicta*.
*Sampling methods not described.
Source: Hall, (1988).

Many species of spiders feed on *D. saccharalis* eggs and other life stages (Negm and Hensley, 1969). At least 84 spider species in 18 families have been reported from Louisiana sugarcane fields (Ali and Reagan, 1985b), and vegetational diversity involving weeds is associated with increasing spider diversity (Ali and Reagan, 1986). Collecting specimens by hand, and with aspirators and sweep nets, Breene et al. (1993) found 18 spider families, 43 genera, and 37 species on sugarcane in southern Texas (Table 10.4).

Many ant species are predatory and they are particularly important for reducing *D. saccharalis* incidence in Florida sugarcane (Adams et al., 1981), and combined predation by different ant species can reduce damage, including dead hearts, from *D. saccharalis* by >90% (Bessin and Reagan, 1993). Eighteen species of ants were reported in Louisiana sugarcane (Showler et al., 1990) and eight on Texas sugarcane (Breene et al., 1993) (Table 10.5).

TABLE 10.4   Spider Families Collected From Sugarcane Fields in Cameron, Hidalgo, and Willacy Counties, Lower Rio Grande Valley, Texas, 1990-1991[a]

Spider family	No. of species[b]
Anyphaenidae	2
Araneidae	≥8
Clubionidae	≥5
Dictynidae	4
Gnaphosidae	≥2
Linyphiidae	≥4
Lycosidae	≥2
Mysmenidae	1
Nesticidae	1
Oxyopidae	1
Philodromidae	1
Pholcidae	≥1
Pisauridae	1
Salticidae	≥12
Tetragnathidae	2
Theridiidae	≥11
Thomisidae	3
Uloboridae	2

[a] Collected manually and using aspirators and sweep nets.
[b] ≥, some species in the family were not identified to genus and species.
Source: Breene et al. (1993).

The red imported fire ant, *Solenopsis invicta* Buren, can decisively suppress *D. saccharalis* populations particularly where the ant's colonies are dense, such as in Louisiana (Showler and Reagan, 1991; Showler et al., 1989). In sugarcane fields treated with insecticides that exclude red imported fire ants, *D. saccharalis*-induced damage increases significantly (Adams et al., 1981; Hensley et al., 1961; Long et al., 1958; Negm and Hensley, 1967, 1969; Reagan et al., 1972; Rossi and Fowler, 2000; Showler et al., 1991). Based on ant populations in sugarcane in Brazil, the extent of ant predation against *D. saccharalis* is also influenced by the time of day and year (Oliveira et al., 2012), and by vegetational diversity in the form of weed growth (Showler and Reagan, 1991).The red imported fire ant, however, has been implicated in the displacement of many native North American formicidae species (Showler and Reagan, 1987).

TABLE 10.5   Ant Species Found in Louisiana and Texas Sugarcane Fields

Louisiana[a]

Cardiocondyla minutoir Forel

Crematogaster clara Mayr

Iridomyrmex humilis Mayr

Iridomyrmex pruinosus (Roger)

Monomorium minimum (Buckley)

Monomorium pharaonis (L.)

Paratrechina melanderi Wheeler

Paratrechina vividula Nylander

Pheidole sp. *flavens* group

Pheidole dentata Mayr

Pheidole moerens Wheeler

Ponera opaciceps Mayr

Ponera trigona opacior Forel

Solenopsis invicta Buren

Solenopsis saevissima richteri Forel

Solenopsis xyloni McCook

Strumigenys louisianae Roger

Tetramorium guineense (L.)

Texas[b]

Crematogaster clara

Forelius sp.

Hypoponera opaciceps (Mayr)

Paratrehina vividula

Paracondlyla harpax (F.)

Pheidole sp.

Pogonomyrmex barbatus (F. Smith)

Solenopsis geminata (F.)

[a] Ants were collected in Louisiana by pitfall trapping, sweep netting weed growth in the furrows and sugarcane foliage, and visually inspecting stalks and leaves.
[b] Ants were collected in Texas manually and by using aspirators and sweep nets.
Source: Breene et al. (1993); Showler et al. (1990).

10.3.1.1   Vegetational Diversification and Natural Enemies

Weeds in sugarcane fields can be diverse, with 53 species reported from southern Louisiana (Ali and Reagan, 1985a) and they support substantial populations of herbivorous prey arthropods that maintain heightened predatory arthropod populations (Ali and Reagan 1985a; Showler, 2013; Showler and Reagan, 1991; Showler et al., 1989) (Table 10.6). Numbers of red imported fire ant mounds in Louisiana sugarcane were 77.8% to 85.7% lower where weeds were eliminated; pitfall collected red imported fire ants and spiders were as much as 43.6% and 36.3% fewer, respectively; spiders were as much as 48.1% less abundant on sugarcane foliage, and upto 66% fewer red imported fire ants were collected as they foraged on sugarcane stalks (Showler and Reagan, 1991). Red imported fire ant colonies "labeled" with an ingested rare earth element (samarium) were used in an experiment to reveal that foraging territories were 40% larger in weedless sugarcane plots than in weedy plots, hence, weedy habitats supported 4.9-fold more colony mounds per hectare (Showler et al., 1989). Further, in the weedy plots, 2.25- and 1.6-fold more red imported fire ants were collected in pitfall traps and on sugarcane stalks, respectively, than in the weedless plots (Showler et al., 1989). In Louisiana, percentages of *D. saccharalis*-bored internodes were ≈50% to 57% less abundant in weedy sugarcane plots than in weedless plots (Showler and Reagan, 1991).

Weeds, on the other hand, are problematic in sugarcane because, in contrast with short season row crop systems, weeds compete more effectively with sugarcane largely as a result of the crop's wide row spacing (1.8 m) and relatively slow seedling growth (Peng, 1984). Showler and Reagan (1991) found that typical stands of annual summer weeds, mostly Bermuda grass, *Digitaria sanguinalis* Scop., were associated with up to 24%, 19%, and 15% reductions in sugarcane density, biomass, and commercial sugar yield, respectively. Despite the detrimental aspects of unchecked weed growth in crops, strategic preservation of weeds in noncompetitive densities and in furrows or "islands" might provide reservoirs of prey items that sustain enhanced predator populations, including redimported fire ants, to augment other control measures as part of an integrated strategy (Ali et al., 1984; Showler and Reagan, 1991; Showler et al., 1989).

10.3.2 PARASITES AND PARASITOIDS

Parasites and parasitoids have been released for control of *D. saccharalis* in many countries with varying degrees of success (Table 10.7) (Alam et

TABLE 10.6 Weed Species Found in Southwestern Louisiana Sugarcane Fields, 1982

Family and species
Alzoaceae
Mollugo verticillata L.
Amaranthaceae
Amaranthus hybridus L.
A. retroflexus L.
A. spinosus L.
Asteraceae
Aster subulatus Michaux
Conyza bonariensis (L.) Cronquist
Eclipta alba (L.) Hasskarl
Eupatorium capillifolium (Lamarck) Small
E. serotinum Michaux
Gnaphalium purpurium L.
Iva annua L.
Senecio glabellus Poiret
Solidago canadensis L.
Sonchus sp.
Boraginaceae
Heliotropium indicum L.
Brassicaceae
Corononpus didymus (L.) Smith
Lepidium virginicum L.
Convolvulaceae
Ipomoea coccinea L.
I. hederacea (L.) Jacquin
I. quomoclit L.
Jacquemontia tamnifolia (L.) Grisebach
Cucurbitaceae
Cucumis melo L.
Cyperaceae
Cyperus esculentus L.
C. iria L.
C. rotundus L.
Fimbristylis miliacea (L.) Vahl
Euphorbiaceae

TABLE 10.6   (Continued)

Family and species
Acalypha ostryaefolia Riddell
Euphorbia nutans Legasca
Phyllanthus sp.
Poinsettia sp.
Fabaceae
Sesbania exaltata (Raf.) Rydlberg ex. A. W. Hill
Hypericaceae
Hypericum mutilum L.
Malvaceae
Modiola caroliniana (L.) G. Don
Onagraceae
Ludwigia docurrens Walter
Oenothera laciniata Hill
Oxalidaceae
Oxalis stricta L.
Poaceae
Brachiaria platyphylla (Grisebach) Nash
Cynodon dactylon (L.) Persoon
Digitaria ciliaris (Retzius) Koeler
Echinochloa colonum (L.) Link
Eleusine indica (L.) Gaerner
Leptochloa filiformis (Lamarck) Beauvois
L. panicoides A. S. Hitchcock
Panicum dichotomiflorum Michaux
Setaria viridis (L.) Beauvois
Sorghum halepense (L.) Persoon
Portulaceae
Portulaca oleracea L.
Scrophulariaceae
Lindernia crustacean (L.) F. Mueller
Solanaceae
Physalis angulata L.
Tiliaceae
Corchorus aestuans L.
C. hirtus L.

TABLE 10.6   (Continued)

Family and species
Urtacaeae
Bohemia cylindrical (L.) Wartz
Verbenaceae
Verbena brasiliensis Vellozo

Source: Ali and Reagan (1985a).

al., 1971; Bennett, 1971; Showler and Reagan, 2012). *D. saccharalis* is not known to have effective parasitoids that are indigenous to Florida, hence, attempts to control the pest with parasitoids have involved releases of species exotic to Florida's sugarcane growing areas (Showler and Reagan, 2012). The earliest introduction of a parasitoid to Florida for use against *D. saccharalis* involved the Cuban fly, *Lixophaga diatraea* (Townsend) (Diptera: Tachinidae), a parasite of late instar larvae (King et al., 1981; Roth et al., 1982) from the Caribbean and Central America, in 1926 (Gifford and Mann, 1967). *Bassus stigmaterus* Cresson (Hymenoptera: Braconidae) was released in Florida in 1932 and it was reported to parasitize 3.4% to 41.4% of *D. saccharalis* larvae, depending on the area sampled (Wilson, 1941). In 1938 the Amazon fly, *Metafonistylum minense* Townsend (Diptera: Tachinidae) from Trinidad, was released in Louisiana but it did not appear to have become established there (Wilson, 1941). *Trichogramma minutum* (Riley) (Hymenoptera: Trichogrammatidae) was released in Florida in 1940 but it did not reduce damage to sugarcane stalks (Wilson, 1941). Wasps of the genus *Trichogramma* are naturally-occurring egg parasitoids that provide weak early-season suppression, but late-season egg parasitism can sometimes be high (Browning and Melton, 1987). While in other countries trichogrammatids have occasionally been reported as being successful, three indigenous and five exotic species that develop on *D. saccharalis* in the laboratory were ineffective under field conditions in the United States (Browning and Melton, 1987; Long and Hensley, 1972). *Apanteles flavipes* (Cameron) (Hymenoptera: Braconidae), from India, was introduced to Florida in 1963, but by 1964 the parasitoid could not be found (Gifford and Mann, 1967). Gifford and Mann (1967) reported that by 1967 only two parasitoids were recovered in Florida, *Alabagrus stigma* (Brullé) (=*Agathis stigmaterus* [Cresson]) (Hymenoptera: Braconidae) and *Trichgramma fasciatum* (Perkins) (Hymenoptera: Trichogrammatidae), and suggested that all

TABLE 10.7 Parasite and Parasitoid Species Released in Barbados, 1930–1969, and Those That Became Established

Species	Family	Origin	Established Barbados
Apanteles chilonis (Munakata)	Braconidae	India	No
Apanteles diatraea (Muesebeck)	Braconidae	Trinidad	No
Apanteles flavipes (Cameron)	Braconidae	India	Yes
Apanteles sesamiae (Cameron)	Braconidae	East Africa	No
Bracon chinensis (Walker)	Braconidae	India	No
Campyloneurus mutator (F.)	Braconidae	India	No
Leskiopalpus diadema (Wied.)	Tachinidae	Trinidad	No
Lixophaga diatraea (Townsend)	Tachinidae	Cuba, Dominican Rep.	Yes
		Jamaica, Antigua	
Metagonistylum minense T. T.	Tachinidae	Brazil	No
Palpozenillia sp.	Tachinidae	Bolivia	No
Paratheresia claripalpus (Wulp)	Tachinidae	Bolivia, Peru, Mexico	No
		Columbia, Trinidad	
Pediobius furvus (Gahan)	Eulophidae	East Africa	No
Sturmiopsis inferens T. T.	Tachinidae	India	No
Trichogramma australicum Gir.	Trichogrammatidae	India	No
Trichogramma japonicum Ashm.	Trichogrammatidae	India	No

Source: Alam et al. (1971).

of the other parasitoids were ineffective. *Digonogastra kimballi* Kirkland, *D. solitaria* Wharton and Quicke (Hymenoptera: Braconidae), and *Conura acuta* (F.) (Hymenoptera: Chalcididae), indigenous to Mexico and South Texas, attack *D. saccharalis* but they failed to suppress it (Wharton et al., 1989). By 2007, only *Cotesia flavipes* (Cameron) (Hymenoptera: Braconidae) and *A. stigma* were considered to be important exotic parasitoids of *D. saccharalis* (Hall et al., 2007). Of the two, the most important parasitoid of *D. saccharalis* in Louisiana and Texas is *A. stigma*, which can reduce, on average, <12% of *D. saccharalis* populations (King et al., 1981), and other parasitoids such as *Allorhogas pyralophagus* Marsh (Hymenoptera: Braconidae) and *C. flavipes* attack *D. saccharalis* in the field, although to a lesser extent (Harbison et al., 2001; Meagher et al., 1998). *Lixophaga diatraea*, released in South Texas, tended to disappear after multiple releases (Legaspi et al., 2000b, c). In southern Louisiana, the average rate of parasitism after *L. diatraea* releases at several different locations ranged between 4.1% and 35.7% (King et al., 1981), and in Florida, the fly parasitized 20% to 78% of larval *D. saccharalis* populations within three weeks of release (Summers et al., 1976). Parasitism by *A. stigma* at the same locations ranged between 2.6% and 11.9% (King et al., 1981). In the Lower Rio Grande Valley, despite the presence of indigenous parasitoids, average parasitism was found to be only 8.9%, mostly by the exotic *C. flavipes* (Meagher et al., 1998). Stalkborer parasitism in South Texas has been described as being low and ineffectual (Youm et al., 1990), but *C. flavipes* parasitism of *D. saccharalis* in maize of the adjacent Mexican state of Tamaulipas was somewhat greater at 13% (Rodriguez-del-Bosque et al., 1990). Despite the attempts to establish exotic biological control agents in Louisiana since 1915, relatively recent surveys there have found only *L. diatraea*, *A. stigma*, and *Trichogramma* spp. parasitizing *D. saccharalis*. Because 15 releases of *Cotesia flavipes* did not result in establishment, "release refuges" were created to provide abundant *D. saccharalis* larvae, minimized red imported fire ant predation, absence of insecticide exposure, regular parasitoid augmentation, and overwintering sites in nonharvested sugarcane stands, but the braconid failed to become well established, in part, because of red imported fire ants (White et al., 2004). In contrast, suppression of *D. saccharalis* with releases of *Cotesia flavipes* augmented by indigenous parasitic tachinid flies *Lydella minense* (Townsend) and *Paratheresia claripalpis* Wulp is claimed to be one of the best instances of classical biological control in Brazil (Rossi and Fowler, 2004). *Cotesia chilonis* (Matsumura) (Hymenoptera: Braconidae), *Rhaconotus roslinensis* Lal (Hymenoptera: Braconidae), and other parasitoids have also been shown to parasitize *D. saccharalis*, but principally in the laboratory (Hawkins and Smith, 1986; Wiedenmann and Smith, 1995).

10.3.3   ENTOMOPATHOGENS

Entomopathogen applications against *D. saccharalis*, including *Beauveria bassiana* (Balsamo) Vuillemin, have been ineffective (Legaspi et al., 2000c), and while *Metarhizium anisopliae* (Metsch.) Sorok. infects *D. saccharalis* (Alves et al., 1984), it has not been successful under field conditions. A spray formulation of *Bacillus thuringiensis* Berliner, however, reduced numbers of *D. saccharlis*-injured stalks by 75% (a granular formulation was ineffective) (Rosas-Garcia, 2006).

10.4   BIOLOGICAL CONTROL OF EOREUMA LOFTINI

10.4.1   PREDATORS

Predators have not been identified as effective natural enemies of *E. loftini*. The red imported fire ant is sparse under the dry conditions of South Texas (Showler and Greenberg, 2003) relative to the large populations under southern Louisiana's wet conditions (Showler and Reagan, 1991; Showler et al., 1989) and it is not yet definitively known whether or not red imported fire ants can suppress *E. loftini* in wet habitats. As an example using a different pest, red imported fire ant predation in the wet Texas Coastal Bendcotton-growing region accounts for 58% of boll weevil, *Anthonomus grandis grandis* Boheman, mortality (Sturm and Sterling, 1990), and Fillman and Sterling (1983) reported that red imported fire ant predation on immature boll weevils averaged 84% compared to only 0.14% and 6.9% mortality caused by parasitism and desiccation, respectively. In the drier subtropics of South Texas, however, cotton fields with rank weed growth harbored few red imported fire ants and boll weevil infestations were not reduced (Showler and Greenberg, 2003). A preliminary study conducted in the relatively wet environment of East Texas showed that suppression of red imported fire ants using pesticides did not significantly reduce numbers of *E. loftini*-bored sugarcane internodes, but exit holes were 44% fewer than where red imported fire ants were not suppressed (Van Weelden et al., 2012). Enhancing arthropod predator populations by diversifying the flora in or around sugarcane fields in dry regions, such as the Lower Rio Grande Valley, is less likely to exert control of *E. loftini* than in wetter regions.

10.4.2 PARASITES AND PARASITOIDS

A relatively large number of parasitic and parasitoid species have been reared on *E. loftini* in the laboratory, including eight *Trichogramma* spp. (three local and five introduced) (Browning and Melton, 1987; Legaspi et al., 1997), and newer species have been investigated (Bernal et al., 2002). Exploration for biological control agents to use against *E. loftini*, aside from nine parasitoids indigenous to the Lower Rio Grande Valley and 18 nonindigenous species that were released there from 1982 to 1997 (Table 10.8). Some parasitoids were not recovered after releases (e.g., *Pediobius furvus* Gahan [Hymenoptera: Eulophidae]) (Pfannenstiel et al., 1992), and the average seasonal level of parasitized larvae and pupae was ≈6.2%, mostly from the indigenous braconids *Chelonus sonorensis* Cameron (Hymenoptera: Braconidae) and *D. solitaria*, and two exotic braconids, *A. stigma* and *A. pyralophagus* (Meagher et al., 1998).

Although *E. loftini* injury to sugarcane in Jalisco, Mexico, is relatively low, ranging between 3% and 4.4% bored internodes, only 6.9% of larvae (out of 3,000) were parasitized by the Jalisco fly, *Lydella jalisco* Woodley, a tachinid native to that area (Legaspi et al., 2000c). Introduced to the Lower Rio Grande Valley (Browning and Smith, 1988), most parasitism by the Jalisco fly occurs between mid-August to mid-October, sometimes reaching ≈30%; after that populations decline and parasitism by *C. sonorensis* predominates (Legaspi et al., 2000c). Because the two species only afflict *E. loftini* late in the growing season (sugarcane is mostly harvested from November to December in South Texas), benefits conferred are likely to be relatively slight. Parasitism of stalkborers in the Lower Rio Grande Valley, initially described as being low and ineffectual (Youm et al., 1990), has remained weak despite many attempts (Meagher et al., 1998). Further, *E. loftini* populations in the Lower Rio Grande Valley are still not controlled by any means other than use of resistant cultivars (Showler and Reagan, 2012).

10.4.3 ENTOMOPATHOGENS

Nosema pyrausta Paillot was isolated from *E. loftini* (Solter et al., 2005) but it has not been useful against the pest, nor has *B. bassiana* (Legaspi et al., 1997). Endemic *Steinernema riobravis* Cabanillas, Pinar and Raulston

TABLE 10.8 Indigenous and Exotic Parasites and Parasitoids of *Eoreouma loftini* in the Lower Rio Grande Valley of Texas from 1982 to 1997

Origin	Species	Family
Indigenous		
	Bracon melitor (Say)	Braconidae
	Bracon rhyssaliformis (Quicke and Wharton)	Braconidae
	Chelonus sonorensis (Cameron)	Braconidae
	Digonogastra kimballi (Kirkland)	Braconidae
	Digonogastra solitaria (Wharton and Quicke)*	Braconidae
	Eucelatoria sp.	Tachinidae
	Glyptapanteles sp.	Braconidae
	Orgilus gelechiaevorus (Cushman)	Braconidae
	Spilochalcis sp.	Chalcididae
Exotic		
Bolivia	*Alabagrus stigma* (Brullé)	Braconidae
Mexico	*Allorhogas pyralophagus* (Marsh)	Braconidae
Bolivia	*Apanteles minator* (Muesebeck)	Braconidae
Indo-Australia	*Cotesia flavipes* (Cameron)	Braconidae
Kenya	*Dentichasmias busseolae* (Heinrich)	Ichneumonidae
South Africa	*Goniozus natalensis* (Gordh)	Bethylidae
Mexico	*Lydella jalisco* (Woodley)	Tachinidae
Mexico	*Macrocentris prolificus* (Wharton)	Braconidae
Mexico	*Mallochia pyralidis* (Wharton)	Ichneumonidae
Mexico	*Palpozenillia diatraeae* (Townsend)	Tachinidae
Kenya	*Pediobius furvus* Gahan	Eulophidae
Pakistan	*Rhaconotus roslinensis* Lal (Wilkinson)	Braconidae

TABLE 10.8   (Continued)

Origin	Species	Family
Mexico	*Trichogramma atopovirilia* (Oatman and Platner)	Trichogrammatidae
Canada	*Trichogramma minutum* (Riley)	Trichogrammatidae
Ivory Coast	*Trichogrammatoidea eldanae* (Viggiani)	Trichogrammatidae
Pakistan	*Trichogrammatoidea chilonis* (Ishii)	Trichogrammatidae
Florida	*Trichospilus diatraeae* (Cherian and Margabandhu)	Eulophidae
Mauritius	*Xanthopimpla stemmator* Thunberg	Ichneumonidae

Source: Showler and Reagan (2012).

nematodes caused 100% *E. loftini* larval mortality at all concentrations from 20 to 240 individuals per larva in the laboratory, but it was ineffective under field conditions mainly because of desiccation and poor delivery to the pest (Legaspi et al., 2000a). While under laboratory conditions, *Steinernema carpocapsas* (Weiser) and *Steinernema feltae* Filipjev also appeared to be good candidates for biological control of *E. loftini* larvae (Ring and Browning, 1990), their utility has not yet been demonstrated under field conditions.

10.5   CONCLUSIONS

Although *D. saccharalis* and *E. loftini* are both sugarcane stalk borers in the family Crambidae, they differ in many respects, including their vulnerability to predators and to parasites and parasitoids. It is not yet clear regarding the extent to which, during its range expansion from Texas to Louisiana and Florida, *E. loftini* will become established in relatively wet regions. Because *E. loftini* prefers to oviposit on dry sugarcane leaf tissue and on water deficit-stressed sugarcane plants, the pest might prove better suited to the more arid conditions of South Texas than the wetter conditions of Louisiana and Florida. While *E. loftini* displaced *D. saccharalis* in South Texas sugarcane that might not occur in Louisiana and Florida. *Eoreuma loftini*'s cryptic habits contribute toward its success in South Texas, protecting it from predators, parasites, parasitoids, and from many insecticides. The effects of predators, in particular the red imported fire ant, on *E. loftini* have not been well studied in wet regions, and the red imported fire ant is not sufficiently abundant in South Texas to suppress *E. loftini* injury to sugarcane below economically acceptable levels. *Diatraea saccharalis*, however, is well established in wet sugarcane growing regions and it is more vulnerable than *E. loftini* to the red imported fire ant than to any other biological control agent, including indigenous and exotic parasites and parasitoids, and entomopathogens, than *E. loftini*.

While biological control of *D. saccharalis* and *E. loftini* in the United States has not been resoundingly successful, the tactic has value under some conditions and it can serve as a component of integrated sugarcane stalk borer control strategies. Planting sugarcane cultivars that show resistance to *D. saccharalis* and encouraging predation, for example, have additive protective effects against the incidence of deadheart (Bessin and Reagan, 1993). It is possible, especially if the red imported fire ant is found to have a strong suppressive effect against *E. loftini*, that the predator can be incorporated into an approach that involves resistant cultivars, adequate irrigation to avoid water deficit stress, and rational application of tebufenozide (or another similarly effective insecticide). Integrated use of cultivar resistance, irrigation, and tebufenozide was more effective at protecting sugarcane from *E. loftini* than tebufenozide alone (Reay-Jones et al., 2005a). Keeping sugarcane fields adequately irrigated, thereby providing a wet environment, might also enhance predatory arthropod populations, particularly the red imported fire ant.

From a somewhat broader perspective, the use of [3]27 parasites and parasitoids against *E. loftini*, and the efforts to use biological control agents against *D. saccharalis* for the last 90 years resulted in marginal effects and failure. It is not without some irony that the most efficacious biological control agent of *D. saccharalis* in sugarcane, the red imported fire ant, was not intentionally introduced to the United States where it is universally regarded as a pest for a variety of legitimate reasons.

While it remains important to continue searching for and assessing new parasites and parasitoids as candidates for controlling the two stalks of boring species, enhancing biological control with sugarcane cultivars should receive greater attention than it has before. As an example, sugarcane leaves that do not curl tightly at the edges upon drying would deny *E. loftini* its preferred oviposition sites and expose eggs to predators, parasites, parasitoids, and entomopathogens (Showler and Castro, 2010b).

At present, biological control against *D. saccharalis* and *E. loftini* in the United States has yielded limited promise (it has been more effective against *D. saccharalis* than *E. loftini*), and, as for most biological control efforts against agricultural pests, the tactic is not a panacea. Further research is needed on new biological control agents and on methods for enhancing the impacts of predators, parasites, parasitoids, and entomopathogens against the two stalk borer species. The research should not only focus on discovering untried biological control agents, it should also concentrate on developing cultural practices and varieties that favor the establishment and increase the impact of known biological control agents.

KEYWORDS

- *diatraea saccharalis*
- entomophathogens
- *eoreuma loftini*
- mexican rice borer
- natural enemies
- parasites
- predators
- red imported fire ant
- *saccharum*
- *solenopsis invicta*
- spiders
- sugarcane borer
- vegetational diversity
- weeds

REFERENCES

Adams, C. T., Summers, T. E., Lofgren, C. S., Focks, D. A., & Prewitt, J. C., (1981). Interrelationship of ants and the sugarcane borer in Florida sugarcane fields. *Environ. Entomol, 10,* 415–418.

Akbar, W., Ottea, J. A., Beuzelin, J. M., Reagan, T. E., & Huang, F., (2008). Selection and life history traits of tebufenozide-resistant sugarcane borer (Lepidoptera: Crambidae). *J. Econ. Entomol, 101*, 1903–1910.

Alam, M. M., Bennett, F. D., & Carl. K. P., (1971). Biological control of *Diatraea saccharalis* (F.) in Barbados by *Apanteles flavipes* Cam. and *Lixophaga diatraea* T. T. *Entomophaga, 16*, 151–158.

Ali, A. D., & Reagan, T. E. (1985a). Vegetation manipulation impact on predator and prey populations in Louisiana sugarcane ecosystems. *J. Econ. Entomol, 78*, 1409–1414.

Ali, A. D., & Reagan, T. E. (1985b). Spider inhabitants of sugarcane ecosystems in Louisiana: an update. *Proc. La. Acad. Sci., 48*, 18–22.

Ali, A. D., & Reagan, T. E., (1986). Influence of selected weed control practices on araneid faunal composition and abundance in sugarcane. *Environ. Entomol, 15*, 527–531.

Ali, A. D., Reagan, T. E., & Flynn, J. L., (1984). Influence of selected weedy and weed-free sugarcane habitats on diet composition and foraging activity of the imported fire ant (Hymenopera: Formicidae). *Environ. Entomol, 13*, 1037–1041.

Alves, S. B., Risco, S. H., Neto, S. S., & Neto, R. M., (1984). Pathogenicity of nine isolates of *Metarhizium anisopliae* (Metsch.) Sorok. to *Diatraea saccharalis* (Fabr.). *J. Appl. Entomol., 97*, 403–406.

Assefa, Y., Conlong, D. E., Van den Berg, J., & Le Rü, B. P., (2008). The wider distribution of *Eldana saccharina* (Lepidoptera: Pyralidae) in South Africa and its potential risk to maize production. *Proc. South Afr. Sugarcane Technol. Assoc., 81*, 290–297.

Bennett, F. D., (1971). Current status of biological control of the small moth borers of sugar cane *Diatraea* spp. (Lep. Pyralididae) *Entomophaga, 16*, 111–124.

Bernal, J. S., Griset, J. G., & Gillogly, P. O., (2002). A new species of Braconidae (Hymenoptera) from Mexico introduced into Texas to control a sugar cane borer, *Eoreuma loftini* (Lepidoptera: Pyralidae). *J. Entomol. Sci., 37*, 27–40.

Bessin, R. T., & Reagan T. E., (1993). Cultivar resistance and arthropod predation of sugarcane borer (Lepidoptera: Pyralidae) affects incidence of deadhearts in Louisiana sugarcane. *J. Econ. Entomol, 86*, 929–932.

Bessin, R. T., & Reagan, T. E., (1990). Fecundity of sugarcane borer (Lepidoptera: Pyralidae), as affected by larval development on graminous host plants. *Environ. Entomol, 19*, 635–639.

Bessin, R. T., Reagan, T. E., & Martin, F. A., (1990). A moth production index for evaluating sugarcane cultivars for resistance to the sugarcane borer (Lepidoptera: Pyralidae). *J. Econ. Entomol, 83*, 221–225.

Breene, R. G., Meagher, R. L., & Dean, D. A., (1993). Spiders (Araneae) and ants (Hymenoptera: Formicidae) in Texas sugarcane fields. *Fla. Entomol, 76*, 645–650.

Browning, H. W., & Melton, C. W., (1987). Indigenous and exotic trichogrammatids (Hymenoptera: Trichogrammatidae) evaluated for biological control of *Eoreuma loftini* and *Diatraea saccharalis* (Lepidoptera: Pyralidae) borers on sugarcane. *Environ. Entomol, 16*, 360–364.

Browning, H. W., Smith, J. W., Jr., (1988). Progress report of the Mexican rice borer research program, 1980–1989. *Annual Mexican Rice Borer Research Review,* 22 March 1988. College Station, Texas, USA.

Browning, H. W., Way, M. O., & Drees, B. M., (1989). Managing the Mexican rice borer in Texas. *Texas Agric. Ext. Serv. Bull.* No. B-1620.

Capinera, J. L., (2001). Sugarcane borer, *Diatraea saccharalis* (Fabricius) (Insecta: Lepidoptera: Pyralidae). Entomology and Nematology Department, *Florida Cooperative Extension Service*, University of Florida, Gainesville, Florida, USA.

Charpentier, L. J., McCormick, W. J., & Mathes, R., (1967). Beneficial arthropods inhabiting sugarcane fields and their effects on borer infestations. *Sugar Bull.*, *45,*276–277.

Fenoglio, M. S., & Trumper, E. V. (2007). Influence of weather conditions and density of *Doru luteipes* (Dermaptera: Forficulidae) on *Diatraea saccharalis* (Lepidoptera: Crambidae) egg mortality. *Environ. Entomol, 36*, 1159–1165.

Fillman, D. A., & Sterling W. L., (1983). Killing power of the red imported fire ant (Hym.: Formicidae): a key predator of the boll weevil (Coleoptera: Curculionidae). *Entomophaga, 28*, 339–344.

Fuller, B. W., & Reagan, T. E., (1989). The relationship of sweet sorghum plant fiber and survival of the sugarcane borer, *Diatraea saccharalis* (F.) (Lepidoptera: Pyralidae). *J. Agric. Entomol, 6*, 113–118.

Gifford, J. R., & Mann, G. A., (1967). Biology, rearing, and a trial release of *Apanteles flavipes* in the Florida Everglades to control the sugarcane borer. *J. Econ. Entomol, 60*, 44–47.

Hall, D. G., (1988). Insects and mites associated with sugarcane in Florida. *Fla. Entomol, 71*, 138–150.

Hall, D. G., Nuessly, G. S., & Gilbert, R. A., (2007). *The Sugarcane Borer in Florida.* University of Florida, Gainesville, Florida, USA.

Harbison, J. L., Legaspi, J. C., Fabritius, S. L., Saldana, R. R., Legaspi, B. C., Jr., & Enkegaard, A., (2001). Effects of age and host number on reproductive biology of *Allorhogas pyralophagus* (Hymenoptera: Braconidae) attacking the Mexican rice borer (Lepidoptera: Pryalidae). *Environ. Entomol, 30*, 129–135.

Hawkins, B. A., & Smith, J. W., (1986). Jr. *Rhaconotus roslinensis* (Hymenoptera: Braconidae), a candidate for biological control of stalkboring sugarcane pests (Lepidoptera: Pyralidae): development, life tables, and intraspecific competition. *Ann. Entomol. Soc. Am.*, *79*, 905–911.

Hayden, J. E., (2012). Mexican rice borer, *Eoreuma loftini* (Dyar) (Lepidoptera: Crambidae) in Florida. Pest Alert, Florida Department of Agric and Consumer Serv., DACS-P-01827, *http://www.freshfromflorida.com/content/download/23854/eoreuma-loftini. pdf.*

Hensley, S. D. (1971a). Management of sugarcane borer populations in Louisiana – a decade of change. *Entomophaga, 16*, 133–146.

Hensley, S. D. (1971b). Control of the sugarcane borer, *Diatraea saccharalis* (F.), in Louisiana. *Proc. Internat. Soc. Sugar Cane Technol.*, *14*, 454–461.

Hensley, S. D., Long, W. H., Roddy, L. R., McCormick, W. J., & Cancienne, E. J., (1961). Effects of insecticides on the predaceous arthropod fauna of Louisiana sugarcane fields. *J. Econ. Entomol, 54*, 146–149.

Hummel, N. A., Hardy, T., Reagan, T. E., Pollet, D., Carlton, C., Stout, M. J., Beuzelin, J. M., Akbar, W., & White, W. H., (2010). Monitoring and first discovery of the Mexican rice borer, *Eoreuma loftini* (Lepidoptera: Crambidae) in Louisiana. *Fla. Entomol, 93*, 123–124.

Hummel, N., Reagan, T. E., Pollet, D., Akbar, W., Beuzelin, J. M., Carlton, C., Saichuk, J., Hardy, T., & Way, M. O., (2008). Mexican rice borer, *Eoreuma loftini* (Dyar). Louisiana State University Ag Center Pub. 3098, *Baton Rouge*, LA.

Johnson, K. J. R., & Van Leerdam, M. B. (1981). Range extension of *Acigona loftini* into the Lower Rio Grande Valley of Texas. *Sugar Azucar, 76,* 119.

Johnson, K. J. R., (1981). *Acigona loftini* (Lepidoptera: Pyralidae) in the Lower Rio Grande Valley of Texas, 1980–1981. In: *Proc. 2nd Inter-American Sugar Cane Seminar (Insect and Rodent Pests)*, Miami, Florida, USA, pp. 166–171.

Johnson, K. J. R., (1984). Identification of *Eoreuma loftini* (Dyar) (Lepidoptera: Pyralidae) in Texas, 1980: forerunner for other sugarcane boring pest immigrants from Mexico? *Bull. Entomol. Soc. Am.*, *30*, 47–52.

Johnson, K. J. R., (1985). Seasonal occurrence and insecticidal suppression of *Eoreuma loftini* (Lepidoptera: Pyralidae) in sugarcane. *J. Econ. Entomol*, *78*, 960–966.

King, E. G., Sanford, J., Smith, J. W., Jr., & Martin, D. F., (1981). Augmentative release of *Lixophaga diatraea* (Dip.: Tachinidae) for suppression of early-season sugarcane borer populations in Louisiana. *Entomophaga*, *26*, 59–69.

Legaspi, J. C., Legaspi, B. C., Jr., Lauziere, I., Smith, J. W. Jr., Rodriguez-del-Bosque, L. A., Jones, W. A., & Saldana, R. R. (2000b). Incidence of Mexican rice borer (Lepidoptera: Pyralidae) and Jalisco fly parasite (Diptera: Tachinidae) in Mexico. *Southwest Entomol*, *25*, 21–30.

Legaspi, J. C., Legaspi, B. C., Jr., & Saldana, R. R. (2000a). Evaluation of *Steinernema riobravis* (Nematoda: Steinernematodae) against the Mexican rice borer (Lepidoptera: Pyralidae). *J. Entomol. Sci.*, *35*, 141–149.

Legaspi, J. C., Legaspi, B. C., Jr.; Irvine, J. E., & Saldana, R. R., (1997). Mexican rice borer, *Eoreuma loftini* (Lepidoptera: Pyralidae) in the Lower Rio Grande Valley of Texas: its history and control. *Subtrop. Plant Sci.*, *49*, 53–64.

Legaspi, J. C., Legaspi, B. C., Jr.; Irvine, J. E., Meagher, R. L., Jr., & Rozeff, N., (1999). Stalkborer damage on yield and quality of sugarcane in the Lower Rio Grande Valley of Texas. *J. Econ. Entomol*, *92*, 228–234.

Legaspi, J. C., Poprawski, T. J., & Legaspi, B. C., (2000c). Jr. Laboratory and field evaluation of *Beauveria bassiana* against sugarcane stalkborers (Lepidoptera: Pyralidae) in the Lower Rio Grande Valley of Texas. *J. Econ. Entomol*, *93*, 54–59.

Long, C. S., Cancienne, E. A., Cancienne, E. J., Dobson, R. N., & Newsom, L. D., (1958). Fire ant eradication program increases damage by the sugarcane borer. *Sugar Bull.*, *37*, 62–63.

Long, W. H., & Hensley, S. D., (1972). Insect pests of sugar cane. *Annu. Rev. Entomol*, *17*, 149–176.

Lv, J., Wilson, L. T., & Longnecker, M. T., (2008). Tolerance and compensatory response of rice to sugarcane borer (Lepidoptera: Crambidae) injury. *Environ. Entomol*, *37*, 796–807.

Meagher, R. L., & Jr., Smith, J. W., (1994). Jr.; and Johnson, K. J. R. Insecticidal management of *Eoreuma loftini* (Lepidoptera: Pyralidae) on Texas sugarcane: a critical review. *J. Econ. Entomol*, *87*, 1332–1344.

Meagher, R. L., & Jr. Smith, J. W., (1998). Jr., Browning, H. W., & Saldana, R. R. 1998. Sugarcane stemborers and their parasites in southern Texas. *Environ. Entomol*, *27*, 759–766.

Moré, M., Trumper, E. V., & Prola, M. J., (2003). Influence of corn, *Zea mays*, phenological stages in *Diatraea saccharalis* F. (Lep. Crambidae) oviposition. *J. Appl. Entomol.*, *127*, 512–515.

Morrill, A. W., (1925). Commercial entomology on the west coast of Mexico. *J. Econ. Entomol*, *18*, 707–716.

Negm, A. A., & Hensley, S. D., (1969). Evaluation of certain biological control agents of the sugarcane borer in Louisiana. *J. Econ. Entomol*, *62*, 1008–1013.

Negm, A. A., & Hensley, S. D., (1967). The relationship of arthropod predators to crop damage inflicted by the sugarcane borer. *J. Econ. Entomol*, *60*, 1503–1506.

Ogunwolu, E. O., Reagan, T. E., & Damann, K. E., (1987). Jr. Efficacy of pest control strategies in Louisiana sugar cane: a preliminary survey. *J. Agric. Sci. Cambr.*, *108*, 661–665.

Ogunwolu, E. O., Reagan, T. E., Flynn, J. L., & Hensley, S. D., (1991). Effects of *Diatraeasaccharalis* (F.) (Lepidoptera: Pyralidae) damage and stalk rot fungi on sugarcane yield in Louisiana. *Crop Prot.*, *10*, 57–61.

Oliveira, R. F., Almeida, L. C., Souza, D. R., Munhae, C. B., Bueno, O. C., & Morini, M. S. C., (2012), Ant diversity (Hymenoptera: Formicidae) and predation by ants on the different statges of the sugarcane borer life cycle *Diatraea saccharalis* (Lepidoptera: Crambidae). *Eur. J. Entomol*, *109*, 381–387.

Osborn, H. T., & Phillips, G. R., (1946). *Chilo loftini* in California, Arizona, and Mexico. *J. Econ. Entomol*, *39*, 755–759.

Peng, S. Y., (1984). *The Biology and Control of Weeds in Sugarcane*. Elsevier, Amsterdam, the Netherlands.

Pfannenstiel, R. S., Browning, H. W., & Smith, J. W., (1992). Jr. Searching behavior of *Pediobius furvus* (Hymenoptera: Eulophidae) for *Eoreuma loftini* (Lepidoptera Pyralidae) in sugarcane. *J. Econ. Entomol*, *85*, 384–388.

Pollet, D. K., Reagan, T. E., White, W. H., & Rester, D. C., (1978). Pest management of sugarcane insects. LA *Coop. Ext. Publ.* no. 1982.

Reagan, T. E., & Martin, F. A., (1989). Breeding for resistance to *Diatraea saccharalis* (F.). In Naidu, K. M., Sreenivasan, T. V., Premachandran, M. N. (Eds.) *Sugarcane Varietal Improvement*, Sugarcane Breeding Institute, Coimbatore, India, pp. 313–331.

Reagan, T. E., Coburn, G., & Hensley, S. D., (1972). Effects of mirex on the arthropod fauna of a Louisiana sugarcane field. *Environ. Entomol*, *1*, 588–591.

Reay-Jones, F. P. F., Akbar, W., McAllister, C. D., Reagan, T. E., & Ottea, J. A. (2005b). Reduced susceptibility to tebufenozide in populations of the sugarcane borer (Lepidoptera: Crambidae) in Louisiana. *J. Econ. Entomol*, *98*, 955–960.

Reay-Jones, F. P. F., Showler, A. T., Reagan, T. E., Legendre, B. L., Way, M. O., & Moser, E. B., (2005a) Integrated tactics for managing the Mexican rice borer (Lepidoptera: Crambidae) in sugarcane. *Environ. Entomol*, *34*, 1558–1565.

Reay-Jones, F. P. F., Wilson, L. T., Reagan, T. E., Legendre, B. L., & Way, M. O., (2008). Predicting economic losses from the continued spread of the Mexican rice borer (Lepidoptera: Crambidae). *J. Econ. Entomol*, *101*, 237–250.

Ring, D. R., & Browning, H. W., (1990). Evaluation of entomopathogenic nematodes against the Mexican rice borer (Lepidoptera: Pyralidae). *J. Nematol.*, *22*, 420–422.

Rodriguez-del-Bosque, L. A., & Smith, J. W., (1995). Jr.; Martinez, A. J. Winter mortality and spring emergence of corn stalkborers (Lepidoptera: Pyralidae) in subtropical Mexico. *J. Econ. Entomol*, *88*, 628–634.

Rodriquez-del-Bosque, L. A., Browning, H. W., & Smith, J. W., (1990). Jr. Seasonal parasitism of cornstalk borers (Lepidoptera: Pyralidae) by indigenous and introduced parasites in northeastern Mexico. *Environ. Entomol*, *19*, 393–402.

Rosas-Garcia, N. M., (2006). Laboratory and field tests of spray-dried and granular formulations of a *Bacillus thuringiensis* strain with insecticidal activity against the sugarcane borer. *Pest Manag. Sci.*, *62*, 855–861.

Rossi, M. N., & Fowler H. G., (2000). Ant predation of larval *Diatraea saccharalis* Fab. (Lep., Crambidae) in new sugarcane in Brazil. *J. Appl. Entomol*, *124*, 245–247.

Rossi, M. N., & Fowler, H. G., (2004). Spatial and temporal population interaction between the parasitoids *Cotesia flavipes* and Tachinidae flies: considerations on the adverse effects of biological control practice. *J. Appl. Entomol*, *128*, 112–119.

Roth, J. P., King, E. G., & Hensley, S. D., (1982). Plant, host and parasite interactions in the host selection sequence of the tachinid *Lixophaga diatraea*. *Environ. Entomol*, *11*, 273–277.

Showler, A. T., & Castro, B. A. (2010a). Influence of drought stress on Mexican rice borer (Lepidoptera: Crambidae) oviposition preference and development to adulthood in sugarcane. *Crop Prot.*, *29*, 722–727.

Showler, A. T., & Castro, B. A. (2010b). Mexican rice borer (Lepidoptera: Crambidae) oviposition site selection stimuli on sugarcane, and potential field applications. *J. Econ. Entomol*, *103*, 1180–1186.

Showler, A. T., & Greenberg, S. M., (2003). Effects of weeds on selected arthropod herbivore and natural enemy populations, and on cotton growth and yield. *Environ. Entomol*, *32*, 39–50.

Showler, A. T., & Reagan, T. E., (1987). Ecological interaction of the red imported fire ant in the southeastern United States. *J. Entomol. Sci. Suppl.*, *1*, 52–64.

Showler, A. T., & Reagan, T. E., (1991). Effects of sugarcane borer, weed, and nematode control strategies in Louisiana sugarcane. *Environ. Entomol.*, *20*, 358–370.

Showler, A. T., & Reagan, T. E., (2012). Ecology and tactics for control of three sugarcane stalk-boring species in the western Hemisphere and Africa. In: J. F. Goncalves and K. D. Correia (eds.), *Sugarcane: Production and Uses*, Nova, New York, USA, pp. 1–15.

Showler, A. T., (2013). Beneficial and detrimental interactions between weeds and other pests of sugarcane. In: A. Taab (ed.), *Weeds: Cultivation, Ecological Benefits and Role in Biological Diversity*, Nova, New York, USA, pp. 153–188.

Showler, A. T., (2015). Effects of compost and chicken litter on soil nutrition, and sugarcane physiochemistry, yield, and injury caused by Mexican rice borer, *Eoreuma loftini* (Dyar) (Lepidoptera: Crambidae). *Crop Prot.*, *71*, 1–11.

Showler, A. T., Beuzelin, J. M., & Reagan, T. E., (2011). Alternate crop and weed host plant oviposition preferences by the Mexican rice borer (Lepidoptera: Crambidae). *Crop Prot.*, *30*, 895–901.

Showler, A. T., Knaus, R. M., & Reagan, T. E., (1989). Foraging territoriality of the imported fire ant, *Solenopsis invicta* Buren, in sugarcane as determined by neutron activation analysis. *Insectes Sociaux*, *36*, 235–239.

Showler, A. T., Reagan, T. E., & Flynn, J. L., (1991). Augmentation and aldicarb treatment of nematodes in selected sugarcane weed habitats. *Ann. Appl. Nematol.*, *23*, 717–723.

Showler, A. T., Reagan, T. E., & Knaus, R. M., (1990). Sugarcane weed community interactions with arthropods and pathogens. *Insect Sci. Applic.*, *11*, 1–11.

Showler, A. T., Wilson, B. E., & Reagan, T. E., (2012). Mexican rice borer, *Eoreuma loftini* (Dyar) (Lepidoptera: Crambidae), injury to corn in relation to sorghum and sugarcane in the field. *J. Econ. Entomol*, *105*, 1597–1602.

Solter, L. F., Maddox, J. V., & Vossbrinck, C. R., (2005). Physiological host specificity: a model using the European corn borer, *Ostrinia nubilalis* (Hübner) (Lepidoptera: Crambidae) and microsporidia of row crop and other stalkboring hosts. *J. Invert. Pathol.*, *90*, 127–130.

Sosa, O. Jr. , (1990). Oviposition preference by the sugarcane borer (Lepidoptera: Pyralidae). *J. Econ. Entomol*, *83*, 866–868.

Sturm, M. M., & Sterling, W. L., (1990). Geographical patterns of boll weevil mortality: observations and hypotheses. *Environ. Entomol*, *19*, 59–65.

Summers, T. E., King, E. G., Martin, D. F., & Jackson, R. D., (1976). Biological control of *Diatraea saccharalis* (Lep.: Pyralidae) in Floridae by periodic releases of *Lixophaga diatraea* (Dipt.: Tachinidae). *Entomophaga, 21,* 359–366.

Ulloa, M., Bell, M. G., & Miller, J. D., (1982). Losses caused by *Diatraea saccharalis* in Florida. *Am. Soc. Sugar Cane Technol., 1,* 8–10.

University of Florida. Mexican rice borer: an introduction. *http://erec.ifas.ufl.edu/mexican_rice_borer_introduction.pdf.(2014).*

Van Leerdam, M. B., Johnson, K. J. R., & Smith, J. W., (1984). Jr. Effects of substrate physical characteristics and orientation on oviposition by *Eoreuma loftini* (Lepidoptera: Pyralidae). *Environ. Entomol, 13,* 800–802.

Van Zwaluenberg, R. H., (1926). Insect enemies of sugarcane in western Mexico. *J. Econ. Entomol, 19,* 664–669.

Van Weelden, M. T., Beuzelin, J. M., Ragan, B. E., Reagan, T. E., & Way, M. O., (2012). Impact of red imported fire ant on Mexican rice borer in sugarcane and noncrop hosts. *Imported Fire Ant Conf. Proc.,* pp. 17–19.

Van Weelden, M. T., Wilson, B. E., Reagan, T. E., & Beuzelin, J. M., (2013). Aerial insecticide control of the Mexican rice borer in sugarcane. *Arthr. Manage. Tests, 38,* 771.

Wharton, R. A., Smith, J. W. Jr., Quicke, D. L. J., & Browning, H. W., (1989). Two new species of *Digonogastra* Viereck (Hymenoptera: Braconidae) parasitic on Neotropical pyralid borers (Lepidoptera: Pyralidae) in maize, sorghum, and sugarcane. *Bull. Entomol. Res., 79,* 401–410.

White, W. H., Reagan, T. E., Smith, J. W. Jr., & Salazar, J. A., (2004). Refuge releases of *Cotesia flavipes* (Hymenoptera: Braconidae) into the Louisiana sugarcane ecosystem. *Environ. Entomol, 33,* 627–632.

Wiedenmann, R. N., & Smith, J. W., (1995). Jr. Parasitization of *Diatraea saccharalis* (Lepidoptera: Pyralidae) by *Cotesia chilonis* and *C. flavipes* (Hymenoptera: Braconidae). *Environ. Entomol, 24,* 950–961.

Wilson, B. E., (2011). Advanced management of the Mexican rice borer (*Eoreuma loftini*) in sugarcane. M.S. thesis, Louisiana State University, Baton Rouge, Louisiana, USA.

Wilson, B. E., Showler, A. T., Reagan, T. E., & Beuzelin, J. M., (2011). Improved chemical control for the Mexican rice borer (Lepidoptera: Crambidae) in sugarcane: larval exposure, a novel scouting method, and efficacy of a single aerial insecticide application. *J. Econ. Entomol, 105,* 1998–2006.

Wilson, J. W., (1941). Biological control of *Diatraea saccharalis* in the Florida Everglades during 1940 and 1941. *Fla. Entomol, 24,* 52–57.

Youm, O., Browning, H. W., & Gilstrap, F. E., (1988). Population dynamics of stalk borers attacking corn and sorghum in the Texas Rio Grande Valley. *Southwest. Entomol, 13,* 199–204.

CHAPTER 11

MANAGEMENT OF PLANT-PARASITIC NEMATODES ON SUGARCANE UNDER TROPICAL CONDITIONS

ANDRÉA CHAVES FIUZA PORTO,[1] E. M. R. PEDROSA,[2]
L. M. P. GUIMARÃES,[3] S. R. V. L. MARANHÃO,[3] and
M. O. CARDOSO[2]

[1]*Universidade Federal Rural de Pernambuco, Estação Experimental de Cana-de-açúcar de Carpina, Rua Angela Cristina C. P. de Luna, S/N. Bairro de Santa Terezinha, Carpina, Pernambuco, 55810-000, Brazil, E-mail: achavesfiuza@yahoo.com.br*

[2]*Universidade Federal Rural de Pernambuco, Departamento de Engenharia Agrícola, Rua Dom Manoel de Medeiros, S/N, Dois Irmãos, Recife, Pernambuco, 52.171-900, Brazil*

[3] *Universidade Federal Rural de Pernambuco, Departamento de Engenharia Agronomia, Rua Dom Manoel de Medeiros, S/N, Dois Irmãos, Recife, Pernambuco, 52.171-900, Brazil*

CONTENTS

11.1 INTRODUCTION

Brazil is the third largest country in area and the largest in South America, occupying 66% of the South American territorial area. The northeast of Brazil (NE) occupies 18% of the country territorial area, embracing 1,554,291.744 km², extending between the parallel of 07°12'35" of north latitude and 48°20'07" of south. Located between the meridians 34°47'30" and 48°45'24" to west of Greenwich, it embraces a vast semi-arid steppe area comprising thorn scrub and dry deciduous forest, as well as rocky outcrops and isolated rain forest patches, bounded on the east coast by the Atlantic Ocean.

In terms of rainfall, there is considerable climatic variation in the NE, ranging from a semi-arid climate interior, with an accumulated precipitation lower than 500 mm/year, to a rainy climate mainly observed on the east coast, with an accumulated annual precipitation above 1,500 mm. The main rainy season in the NE, including the east of the region, which accounts for 60% of the annual rainfall, occurs from April to July and the dry season, for the greatest part of the region, takes place from September to December.

The sugarcane (*Saccharum* spp.) plantations extend in a coastal strip of up to 200 km wide. The uneven relief of the coastal zone adds still more variability to this ecosystem, which includes coastal forests and scrub on sandy soils. The climate of this area is tropical humid with usually high temperature and precipitation during the autumn and winter. The native vegetation was Atlantic Forest, nowadays almost extinct and replaced by sugarcane plantations introduced since the beginning of the colonization in the 16th century.

Although sugarcane agro industry is the main socioeconomic activity of northeast of Brazil, the NE productivity is low when compared to other producing country regions. The high temperatures, precipitation irregularity along with prolonged drought, in addition to the high incidence of plant-parasitic nematodes, have caused significant losses in agricultural production, especially in the sandy soils of the coastal zone.

Control measures for plant-parasitic nematodes adopted in the region are mainly based on the use of systemic nematicides, crop rotation, soil inversion and fallow. The use of resistant varieties, although it represents the main component for efficient nematode management in integrated systems, has not been used because there is no record of commercial sugarcane varieties effectively resistant to one or more species and races of the major plant-parasitic nematodes (i.e., *Meloidogyne* and *Pratylenchus*), which occur frequently in mixed populations.

Although the use of nematicides has been increasing in the region, their efficiency has not been always confirmed, and, thus high plant-parasitic nematode populations have been frequently reported at harvest, compromising the development in successive ratoon crops. On the other hand, jointed *Meloidogyne* and *Pratylenchus* infestations have make the crop rotation a non-viable alternative, and despite the weather conditions are favorable, the soil inversion is underutilized. Moreover, the fallow, generally used when the sugarcane plantation becomes unproductive, is a low efficiency management technic because of the high weeds diversity, many of which are hosts of the plant-parasitic nematodes.

In this chapter, we will focus on the events contributing to the development of the plant-parasitic nematode integrated management in sugarcane plantations under the tropical conditions of the northeast of Brazil, highlighting facts, main studies, and management alternatives.

11.2 EVOLUTION OF THE SUGARCANE PLANT-PARASITIC NEMATODE INTEGRATED MANAGEMENT IN NORTHEAST OF BRAZIL

The first record of a plant-parasitic nematode feeding on sugarcane (*Saccharum officinarum* L.) occurred in 1885, when Treub reported *Heterodera javanica*, later classified by Chitwood (1949) as *Meloidogyne javanica* (Moura; Guimarães, 2004), in Java, Indonesia (Winchester, 1969). In Brazil, the first studies onplant-parasitic nematode injuring sugarcane were carried out by Lordello (Lordello and Mendonça, 1970; Monteiro and Lordello, 1977).

In northeast of Brazil, Liu H. Pin, a plant pathologist from the National Program for Sugarcane Breeding (Planalsucar), was the first researcher to associate the importance of plant-parasitic nematodes for sugarcane production. First surveys demonstrated that nematode communities in sugarcane fields generally were comprised of numerous endoparasitic and ectoparasitic species with *Meloidogyne* and *Pratylenchus* being the most important. The initial surveys also evidenced that nematodes played a crucial role in the crop losses stimulating studies to their control (Moura and Guimarães, 2004). Moura and Almeida (1981) associated the low yield of sugarcane plantations in Pernambuco State, Brazil, with the presence of genera such as *Meloidogyne, Pratylenchus* and *Helicotylenchus*. In the following years new nematode occurrences and reductions on crop yield were reported in other states of NE (Moura, Regis, and Moura, 1990).

Pin (1986) also performed the first studies to control plant-parasitic nematodes on sugarcane plantations. The work carried out by this outstanding researcher showed that although the systemic nematicides increased plant cane yield, the plant-parasitic nematodes population densities were equivalent or superior to the initial density at harvest and, therefore, there was no protection to the successive ratoons bringing on reductions in the ratooning ability of the crop (Moura and Guimarães, 2004; Pin, 1986).

Later, Moura (1991) carried out researches aiming nematode integrated control in the sugarcane plantations. One of the first experiments using crop rotation as alternative for decrease in plant-parasitic nematode population in sugarcane plantations was designed by Moura (1991) in areas highly infested with *M. incongita* (Kofoid and White) Chitwood race 1 and *M. javanica* (Treub) Chitwood, using peanut (*Arachis hypogaea* L.), corn (*Zea mays* L.), vetiver (*Chrysopogon zizanioides* L.) and *Crotalaria juncea* L. as options within the crop rotation management for nematode control. The author reports that use of a row for two years with peanut + peanut, corn + corn, corn + peanut, and *C. juncea* reduced plant-parasitic nematodes population. Studies on nematode population dynamics have characterized the nematode populations in sugarcane soil and demonstrated that monoculturing of sugarcane can foster the accumulation of diverse nematode communities.

In pioneer studies, using antagonistic plants as control alternative to reduce plant-parasitic nematodes population density in sugarcane plantations, Moura (1991, 1995) found promising results of *C. junceae*. Rosa, Pedrosa and Moura (2004) observed that the use of *C. junceae* for a year followed by soil incorporation reduced dense mixed populations of *Meloidogyne* in sugarcane plantations, demonstrating to be an efficient alternative in the management.

The adequate fertilization is another component that plays important role in the integrated management system. Nutrients availability in an equilibrium soil decreased the plant vulnerability to attack by pathogens (Silva, 2009). Asano and Moura (1995) observed that the use of macronutrients in sugarcane plants significantly decreased *M. incognita* race 1 population, but micronutrients had no effect on nematode reproduction.

The application of vinasse as fertirrigation is another technique often used in sugarcane plantations of NE. Some researches involving this technique have already been performed. As a starting point for other studies conducted in the field, Albuquerque, Pedrosa and Moura (2001) observed that vinasse decrease-hatching rate of *Meloidogyne* species.

Several soil organisms are potential harmful agents to plant-parasitic nematodes, however, there are no practical reports of biological control applicability in sugarcane in the region, and few studies have been developed. Oliveira et al., (2011) tested the efficiency of the natural inoculant based on *Paecilomyces lilacinus* (Nemout®), isolated or interacting with insecticides commonly applied in sugarcane plantations, and a nematicide (carbofuran) in the control of *Pratylenchus* spp. and reported that Nemout® had a potential to control the plant-parasitic nematodes in sugarcane plants under greenhouse conditions.

Based on studies carried out between 1980s and 1990s, Moura (2000) proposed an integrated control system for the main plant-parasitic nematodes of sugarcane, coded by the acronym SIC/CANA, composed of easy to apply, environmentally clean and economically viable. In this system, the author used his own criterion, selecting plots during the off-season for renewal, which should have yields below 40 tonnes per hectare (determined by estimate near harvest) and high population densities in the soil of one or both plant-parasitic nematodes (*Meloidogyne* or *Pratylenchus*). Thus, commercial stands for renovation with these two negative points should not be renewed.

11.3 INTEGRATED MANAGEMENT OF PLANT-PARASITIC NEMATODES IN SUGARCANE PLANTATIONS UNDER TROPICAL CONDITIONS: TECHNIQUES TO BE ADOPTED

Several genera and species of plant-parasitic nematodes parasitize sugarcane plants in the nine states from northeast of Brazil (Brazilian Institute of Geography and Statistics, 2016). However, the greatest severity of the damage occurs in the costal tables, characterized by sandy soils, irregular precipitation, and high temperatures, conferring differentiated cultivation condition of other Brazilian regions.

The genera *Meloidogyne* and *Pratylenchus* stand out for the high incidence, potentiating the yield losses, usually low (Barros et al., 2002; Chaves et al., 2009; Chaves et al., 2016; Guimarães et al., 2008; Maranhão, 2008) (Figure 11.1).

About 2,500 soil and sugarcane root samples were analyzed by Laboratory of Nematology from Estação Experimental de Cana-de-Açúcar deCarpina/UFRPE and showed a higher frequency (67%) of *Pratylenchus* in Pernambuco State, while the frequency of *Meloidogyne* is higher (73 and 66%) in Paraíba and Rio Grande do Norte State, respectively.

Meloidogyne incognita, M. javanica and *P. zeae* Graham are ordinarily found in the sugarcane plantations (Mattos et al., 2011; Moura, 2000; Rodrigues et al., 2011). However, reports of *M. Hispanica* Hirschmann (Chaves et al., 2007), *P. brachyurus* (Godfrey) Filipjev & S. Stekhoven and *P. penetrans* (Cobb, 1917) Chitwood & Oteifa (Moura, 2012) are less frequent. The incidence of *M. hispanica* is observed mainly in areas with problems related to fungi-nematode interactions (Figure 11.2), resulting in large yield losses leading to the death of sugarcane plants (Chaves et al., 2007).

Considering that in Brazil, there are two sugarcane growing seasons, summer planting (dry season) and winter planting (rainy season), the farmers

FIGURES 11.1   Symptoms of nematode attack in northeastern sugarcane plantations.

FIGURE 11.2   Characteristics of sugarcane plantations with fungi-nematode interaction in São José Farm, Igarassu Municipality, Pernambuco State, Brazil (Chaves et al, 2007).

needs to better understanding which techniques can be used more efficiently in these periods and which can be used throughout the year.

In general, the use of resistant varieties would be the best option both economically and environmentally for the farmers. However, in practice, the factors that determine high agricultural and industrial yields to sugarcane genotypes appear to be inversely proportional to the resistance to pests and diseases (Ferraz et al., 2010), in addition to the potential presence of different species of *Meloidogyne* and *Pratylenchus* in the fields restricts the use of resistant varieties (Chaves et al., 2007).

Usually 2,500 *P. zeae* individuals (Dinardo-Miranda et al., 1996) and 400 *M. incognita* juveniles in 50 g of roots (Novaretti, 1997) justify the introduction of integrated management in sugarcane plantations. Studies on nematode damage levels on sugarcane in the northeast of Brazil have been performed.

A system of integrated management of plant-parasitic nematodes in sugarcane plantations in northeast of Brazil was proposed by Moura (2000), however, the need for adjustments in the cultivation of sugarcane related to intensification of agricultural mechanization, prohibition of burning, problems with labor acquisition and requirements for efficient use of water has brought many changes to plant pathological problems, including nematodes.

For integrated nematode management for sugarcane in the northeast of Brazil conditions, it is necessary to use some techniques, according to the growing season. These techniques are proposed here based on field experiences, directly linked to the research and extension for local farmers.

11.3.1 MANAGEMENT OF PLANT-PARASITIC NEMATODES IN SUMMER (DRY SEASON) AND WINTER (RAINY SEASON) PLANTINGS

Directly using organic matter influences composition and balance of nematodes in soil (Eisenhauer et al., 2011; Ferris, Sanchéz-Moreno and Brennan, 2012). In this context, fertirrigation is a usual practice during sugarcane planting for crop renewal in the dry season of the costal table. Through this technique the vinasse produced during sugarcane milling for ethanol production is applied along with the irrigation water (Vasconcelos et al., 2010). Thus, in summer planting, the fertirrigation should be considered mainly in areas highly infested with plant-parasitic nematodes.

Many are the benefits of using this technique improving crop nutritional aspect and reducing the plant-parasitic nematodes population density (Cordeiro Neto, 2013). Pedrosa et al. (2005) observed that the exposure of nematode eggs to vinasse in soils from costal tables exerts negative effect on the hatch, reducing the population density of *Meloidogyne* spp. juveniles in sugarcane plantations.

Vinasse acidifies the soil when directly applied, favoring fast colonization by decomposing organisms such as bacteria and fungi due to pH increases (Almeida, 1955), leading to the appearance of several nematode antagonists. The lower dominance of *Meloidogyne* and *Pratylenchus* in irrigated areas with vinasse is a strong indication of the suppressive effect of vinasse on these important plant pathogens (Matos et al., 2011). However, this effect did not extend to other plant-parasitic nematodes, such as *Criconemella,* which showed tolerance to the residue. On the other hand, most of the taxa found by Matos et al. (2011) were sensitive to variations in levels of Fe, Cu, Zn and Mn in the soil.

Filter cake, also a by-product of sugarcane processing, improves the physical, chemical and biological conditions of the soil, retaining moisture and incorporating nutrients and antagonists to the plant-parasitic nematodes, and also has corrective properties of soil acidity due to the chelating effects of organic matter on aluminum. In areas with high population densities of *P. zeae* in northern coast of Pernambuco, using 50 tonnes per hectare of filter cake in summer planting resulted in gains of 17% in tonnes of sugarcane per hectare (TSH) (Chaves et al., 2012).

For sugarcane planting during winter, a period characterized by a longer time between the harvesting and planting of renewed areas, the use of *Crotalaria* can be considered after the destruction of the ratoon as suggested by Moura (2000) and Rosa, Moura and Pedrosa (2004). This leguminous is efficient controlling plant-parasitic nematodes mainly *Meloidogyne* (Moura, 1995) and also increases sugarcane yield due to the action against these organisms and the incorporation of organic matter in the soil (Dinardo-Miranda, 2010).

Despite the benefits, the use of *Crotalaria* for plant-parasitic nematode control has been little explored in NE. In Alagoas State, only one farm has used *C. spectabillis* Roth. at a dose of 20 kg per hectare. At the end of the cycle, the leguminous is incorporated into the soil with minimal cultivation. Increases in sugarcane yield are between 4 and 6 TSH in these locations (Figure 11.3).

FIGURE 11.3 Use of *Crotalaria spectabilis* and minimum sugarcane cultivation in Coruripe Farm – Alagoas/Brazil.

The sugarcane induced resistance to plant-parasitic nematodes has been explored (Chaves et al., 2009, 2016; Guimarães et al., 2008). Inducers with satisfactory results in other pathosystems, such as methyl jasmonate and potassium silicate, were not efficient in the variety RB863129 against *P. zeae* (Guimarães et al., 2008). However, Chaves et al. (2016) in a greenhouse experiment observed a positive effect of pyraclostrobinon *M. Incognita* populations in sugarcane plants which showed yield increments even in high population density of this nematode (Chaves and Pedrosa, 2015). These results show there was an increase in peroxidase activation at 5 days after treatment of the nematode inoculated plants, as well as the catalase and ascorbate peroxidase enzymes at 20 days after treatment and inoculation with the nematodes (Chaves et al., 2016). Although there is a need for further studies, this is a good indication of the resistance of sugarcane to plant-parasitic nematodes.

Considering the intensification of agricultural mechanization in northeast of Brazil among other factors, such as the prohibition of the burning and the difficulty in acquiring labor for cultivation and cutting of sugarcane, it should be pointed out that the use of pyraclostrobinmay reduce consumption of pesticides. In addition, pyraclostrobin-based products are potent fungicides, influencing the problems of sprouting caused by soil pathogens, such as *Ceratocystis paradoxa* (Dade) C. Moreau [teleomorph] (Dade) C. Moreau, the causal agent of the rot on pineapple. Thus, a possible combination of protection of more than one potential pathogen in the field would be an interesting alternative in plant pathological and environmental aspects.

Chemical control (i.e., nematicides) is a usual practice among farmers because they achieve favorable results in a short term, however, several studies have been conducted with the objective of directing farmers to the safety standards for the use of these products in the sugarcane plantations due to environmental problems arising from its overuse (Chaves, Pedrosa and Moura, 2002; Chaves et al., 2012).

Use of nematicides increases sugarcane yield (Barros, Moura and Pedrosa, 2005; Chaves et al., 2012; Dinardo-Miranda, 2010). Making their application technically and economically feasible, it is necessary to take certain precautions, such as use only in infested areas and reducing of applications and doses of the products. Thus, the solubility of the nematicides is one of the most important factors for the success of the use according to the dry or rainy season in which it will be used. These products can be applied to both plant cane and ratoon cane (Dinardo-Miranda, 2010), bringing good results to the farmer.

In the northeast of Brazil, some nematicides registered for sugarcane plantations have shown good performance, such as Pottente® (Benfuracarbe), Furadan® (Carbofuran), Marshall® (Carbosulfan) and Rugby® (Cadusafós). The first three nematicides belong to the carbamates chemical group, and the latter are organophosphates and are directly linked to the inhibition of acetyl cholinesterase in nematodes (Oppermann and Chang, 1990). It should be noted that both Rugby® and Pottente® are products characterized by lower aggressive toxicological classes (III and II, respectively).

The interruption of this system will affect the penetration of the juvenile into the roots in order to delay or even reduce the initial penetration of the nematode, but the effect of the carbamates in plants are reversible (Sikora and Hartwuig, 1991). As previously discussed, Pin (1986) was one of the first researchers to work in Tabuleiros Costeiros in northeast of Brazil, and performed studies in areas with high nematode population densities, observing that the application of carbofuran protected the plant in the first months of the cycle when the nematodes re-parasitized the roots and multiplied freely with high population densities at the end of the cycle. Therefore, nematicides, although contributing to productivity gains in plant cane, did not protect subsequent cycles.

Barros et al. (2005), Ferreira Lima (1998), and Moura et al. (1998) summarizes earlier works in this area and discuss nematicide effects on sugarcane plantations in northeast of Brazil.

11.3.2 FUTURE PERSPECTIVES FOR INTEGRATED NEMATODE MANAGEMENT IN COSTAL TABLES OF NORTHEAST OF BRAZIL

Soil inversion for eggs and juveniles exposures has been successful adopted in many crops (Campos et al., 2005), but in sugarcane it is not a usual practice in NE, although the place is favorable to this practice. Due to the short time between cutting and planting of sugarcane in the dry season, soil inversion has been adopted in very low extension. However, in the rainy season this practice could be suitable because there is greater time between renewal and planting of plant cane.

In costal tables, soil inversion can be used mainly in the winter planting, but can occur in the summer as well. However, these soils are easily weathered with a risk of erosion. Thus, it is recommended a light harrowing. In rainy season, a fallow for 20 to 30 days is recommended. If dry season, irrigation after mechanization is essential, providing favorable environment to eggs hatch.

Some nematicides have been successful tested in sugarcane plantations. Recently, a product based on fluensulfone showed a reduction in nematode population densities from sugarcane plantations (Chaves, 2016; Chaves, Pedrosa and Silva, 2015). This product has a mechanism of action different from carbamates, being known as a true nematicide, because it acts selectively, killing only nematodes, affecting their reproduction, development, feeding and motility (Kearn et al., 2014).

Recent studies showed fluensulfone effects on *M. javanica* (Oka et al., 2009), *M. incognita* (Oka et al., 2011) and *M. arenaria* (Neal) Chitwood (Csinos et al., 2010). Therefore, it is expected that fluensulfone together with integrated management in sugarcane decrease plant-parasitic nematodes population at the end of the plant cycle and thereby increasing sugarcane yield due to greater roots protection.

In Brazil, Nemix® is registered as a soil inoculant based on *Bacillus subtilis* (Ehrenberg) Cohn + *Bacillus licheniformis* (Weigmann) Chester, but has shown good commercial results in sugarcane plantations infested by plant-parasitic nematodes, although tests still are in begin. This product forms a biofilm around the roots, hampering the nematode penetration into the plant root (Nunes et al., 2010). Thus, it has a high potential to be used in the future as an important tool in the integrated nematode management with biological products.

Azamax® is a natural insecticide of the Tetranortriterpenoid group. The active substance azadirachtin is being tested in sugarcane areas with high nematode population densities in NE. Initial results have shown decrease on *P. zeae* population density and increase on sugarcane yield (data not yet published).

A major problem discussed by local farmers is the use of chemical products in isolation to recover sugarcane plantations infected by high plant-parasitic nematode population densities. It is essential that technicians and researchers clarify the nematode problem, emphasizing that the adoption of integrated measures are necessary to shift high population densities of plant-parasitic nematodes to biological balance in the soil.

Considering questions above, it is proposed the use of different techniques as options for integrated nematode management according to the season planting and the intensification of agricultural mechanization as stated in the Figures 11.4 and 11.5.

11.4   CONCLUDING REMARKS

There are two major problems with sugarcane cultivation in northeast Brazil: the first is that the soil and climate conditions are very favorable to the development of nematodes. The second problem is that the producers believe that the use of nematicides alone is enough to control these organisms.

Many studies are being developed to change this idea, contributing to an awareness that the integrated management of nematodes is the outlet to maintain the balance of populations in the field. Thus, we envisage that only

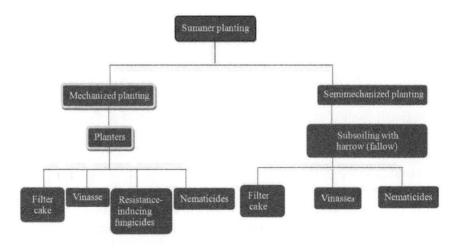

FIGURE 11.4   Integrated nematode management for summer planting (dry season) in costal table of northeast of Brazil.

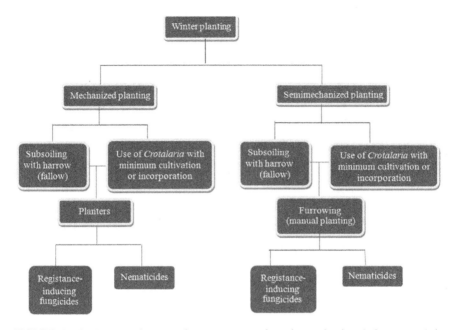

FIGURE 11.5 Integrated nematode management for winter planting (rainy season) in costal table of northeast of Brazil.

through the union of producers, technicians and researchers can this battle be overcome.

KEYWORDS

- **costal tables**
- **integrated management**
- **nematodes**
- **sugarcane**
- **tropical conditions**

REFERENCES

Albuquerque, P. H. S., Pedrosa, E. M. R., & Moura, R. M., (2001). Effect of vinasse and filter cake on *Meloidogyne incognita* race 1 and *M. javanica* hatching. *Brazilian Nematology.*, *25*(2), 175–183.

Almeida, J. R., (1955). The vinasse problem. *Sugar Brazil.*, *46*(2), 72–77.

Asano, S., & Moura, R. M., (1995). Effects of macro and micronutrients on severity of *Meloidogyne* in sugarcane. *Brazilian Nematology*, *19*(1), 15–20.

Barros, A. C. B., Moura, R. M., & Pedrosa, E. M. R., (2002). Effect of terbufos on three ecto-parasite nematode populations in sugarcane. *Brazilian Phytopathology.*, *27*(3), 309–311.

Barros, A. C. B., Moura, R. M., & Pedrosa, E. M. R., (2005). Study of sugarcane variety and nematicide interaction in soil naturally infested with *Meloidogyne incognita, M. javanica* and *Pratylenchus zeae*. *Brazilian Nematology.*, *29*(1), 39–46.

Campos, V. P., Dutra, M. R., Silva, J. R. C., & Valério, C. R., (2005). Soil inversion and irrigation on plant parasitic nematodes control. *Agricultural Bulletin.*, 6. UFLA press. 36.

Chaves, A., Maranhão, S. R. V. L., Pedrosa, E. M., & Guimarães, L. M. P., (2009a). Incidence of *Meloidogyne* spp. and *Pratylenchus* sp. in sugarcane in Pernambuco State. *Brazilian Nematology.*, *33*(4), 345–349.

Chaves, A., Melo, L. J. O. T., Simões Neto, D. E., Costa, I. G., & Pedrosa, E. M. R., (2007). Severe development decrease of sugarcane in costal tables of Pernambuco State. *Brazilian Nematology.*, *31*(1), 10–12,

Chaves, A., Pedrosa, E. M. R., & Moura, R. M., (2002). Effects of terbufos on population density of endoparasite nematodes in five sugarcane varieties in Northeast. *Brazilian Nematology.*, *26*(2), 167–176.

Chaves, A., Pedrosa, E. M. R., & Silva, F. M. L., (2015). Efficiency of fluensulfone on *Meloidogyne incognita* control in Northeastern costal tables.. *In: XXXII Brazilian Congress of Nematology,* 152–152.

Chaves, A., Pedrosa, E. M. R., & SimoesNeto, D. E., (2014). Pyraclostrobin as resistance inducer of *Meloidogyne incognita* in sugarcane In: VI International Congress of Nematology, 2014. Cape Town. Annals of *VI International Congress of Nematology*, pp. 235–235.

Chaves, A., Pedrosa, E. M. R., Coelho, R. S. B., Guimarães, L. M. P., Maranhão, S. R. V. L., & Gama, M. A. S., (2012). Alternatives for integrated management of plant parasitic nematodes in sugarcane. *Agrarian.*, *7*(1), 73–80.

Chaves, A., Pedrosa, E. M. R., Pimentel, R. M. M., Coelho, R. S. B., Guimarães, L. M. P., & Maranhão, S. R. V. L., (2009b). Resistance Induction to *Meloidogyne incognita* in sugarcane through mineral organic fertilizers. *Brazilian Archives of Biology and Technology*, *52*(6), 1393–1400.

Chaves, A., Pedrosa, E. M. R., Wiladino, L. G., & Cardoso, M. S. O., (2016). Activation of resistance to *Meloidogyne incognita* in sugarcane treated with pyraclostrobin. *Nematoda.*, *3*(1), 1–7.

Cordeiro Neto, A. T., (2012). Effect of soil compaction and vinasse on initial development and nutrition of sugarcane infected by plant parasitic nematodes. Dissertation (*Master in Agricultural Engineering*) - *Federal Rural University of Pernambuco*, Recife, *67p*.

Csinos, A., Whitehead, J., Hickman, L. L., & LaHue, S. S., (2010). Evaluation of fluensulfone for root knot nematode on tobacco. *Phytopathology.*, *100*, 28–28.

Dinardo-Miranda, L. L., (2010). Nematodes: Sugarcane. 1 ed., *Agronomic Press*. p. 405–422.

Dinardo-Miranda, L. L., Morelli, J. L., Landell M. G. A., & Silva, M. A., (1996). Comportment of sugarcane genotypes in relation to *Pratylenchus zeae*. *Brazilian Nematology.*, *20*(2), 52–58.

Dutra, M. R., & Campos, V. P. (2003). Soil management and irrigation as a new tactic for *Meloidogyne incognita* control in common beans. *Brazilian Phytopathology.*, *28*(6), 608–614.

Ferraz, S., Freitas, L. G., Lopes, E. A., & Dias-Arieira, C. R., (2010). Sustainable management of plant parasitic nematodes. Vol. *1*, UFV Press. 304p.

Ferreira Lima, R., (2000). Influence of the nematicide Terbufos on fluctuation of nematode population and productive parameters of two sugarcane (*Saccharum* spp.) varieties. *Society of Sugar and Alcohol Technicians of Brazil.*, *19*(2), 36–39.

Ferris, H., Sanchez-Moreno, S., & Brennan, E. B., (2012). Structure, functions and interguild relationships of the soil nematode assemblage in organic vegetable production. *Applied Soil Ecology*, *61*, 16–25.

Guimarães, L. M. P., Pedrosa, E. M., Coelho, R. S. B., Chaves, A., Maranhão, S. R. V. L., & Miranda, T. L., (2008). Effect of methyl jasmonate and silicate of potassium on *Meloidogyneincognita* and *Pratylenchus zeae* parasitism in sugarcane. *Brazilian Nematology.*, *32*(1), 50–55.

Brazilian Institute, (2016). Of Geography and Statistics – IBGE. Systematic survey of agricultural production LSPA. Available in: http://ftp.ibge.gov.br/Producao_Agricola/Levantamento_Sistematico_da_Producao_Agricola_[mensal]/Fasciculo/lspa_201611. pdf Access in December 27.

Kearn, J., Ludlow, E., Dillon, J., O'Connor, V., & Holden-Dye, L., (2014). Fluensulfone is a nematicide with a mode of action distinct from anticholinesterases and macrocyclic lactones. *Pesticide Biochemistry and Physiology*, *109*, 44–57.

Lordello, L. G. E; & Mendonça, M. M., (1970). Nematodes associated with roots of sugarcane in Piauí, Brazil. *Brazilian Journal of Biology*, *30*(4), 617–618.

Maranhão, S. R. V. L., (2008). Community, population dynamic and spatial variability of nematodes in sugarcane commercial fields under different soil and climatic conditions (*Doctorial Thesis*). *Federal Rural University of Pernambuco,* Recife, *126p.*

Matos, D. S. S., Pedrosa, E. M. R., Guimarães, L. M. P., Rodrigues, C. V. M. A., & Barbosa, N. M. R., (2011). Relationship between nematode community and soil chemical attributes under vinasse application. *Nematropica*, *41*(1), 28–38.

Monteiro, A. R., & Lordello, L. G. E., (1977). Two new nematodes associated with sugarcane. *Brazilian Journal of Agriculture.*, Piracicaba, *52*, 05–11.

Moura, R. M., & Almeida, A. V., (1981). Preliminary study on plant parasitic nematode occurrence associated with low production sugarcane areas in Pernambuco State. *In: Brazilian Nematology Meeting*, *5*, 213–220,

Moura, R. M., & Guimarães, L. M. P., (2004). Historic data and evolution of plant parasitic nematology of sugarcane. *Annals of the Pernambuco Academy of Agronomic Science*, *1*, 69–78.

Moura, R. M., (1991). Two years of crop rotation in sugarcane fields for control of plant parasitic nematodes: 1. Effect of the treatments on the nematode population. *Brazilian Nematology.*, *15*(1), 1–7.

Moura, R. M., (1995). Two years of crop rotation in sugarcane fields for control of plant parasitic nematodes: 2. Considerations on methods and reflexess on agro industrial productivity in sugarcane plant. *Brazilian Phytopathology.*, *20*(4), 597–600.

Moura, R. M., (2000). Integrated control of sugarcane nematodes in Northeast Brazil. In: XXII *Brazilian Congress of Nematology*, 88–94.

Moura, R. M., Macedo, M. E. A., Silva, E. G., & Silva, L. P., (1998). Effect of carbofuran application in sugarcane variety CB45–3. *Brazilian Phytopathology.*, *23*, 503.

Moura, R. M., Regis, E. M. O., & Moura, A. M., (1990). Species and races of signaled in sugarcane in the State of Rio Grande do Norte. *Brazilian Nematology*, *14*(1), 33–38.

Moura, R. M., Silva, J. C., & Silva, J. V. C. L., (2013). Population densities of lesion nematode and root-knot nematode in sugarcane research areas in the State of Pernambuco. *Annals of the Pernambuco Academy of Agronomic Science*, *10*, 227–239.

Novaretti, W. R. T., (1997). Control of *Meloidogyne incognita* and *Pratylenchus zeae* (Nemata: Tylenchoidea) in sugarcane with nematicides, associated or no to organic matter. Thesis (*Doctorial*) - Superior College of Agriculture Luiz de Queiroz – *University of São Paulo*, Piracicaba, *51*p.

Nunes, H. T., Monteiro, A. C., & Pomela, A. W. V., (2010). Use of microbial and chemical agents for *Meloidogyne incognita* control in soybean. *Acta Scientiarum Agronomy*, *32*(3), 403–409.

Oka Y., Shuker, S., & Tkachi, N., (2011). Systemic nematicidal activity of fluensulfone against the root-knot nematode *Meloidogyne incognita* on pepper. Pest. *Management Science*, *68*(2), pp. 268–275.

Oka, Y., Shuker, S., & Tkachi, N., (2009). Nematicidal efficacy of MCW-2, a new nematicide of the fluoroalkenyl group, against the root-knot nematode *Meloidogyne javanica*. *Pest Management Science*, *65*(10), pp. 1082–1089.

Oliveira, M. K. R. S., Chaves, A., Vieira, D. A. N., Silva, E. J., & Rodrigues, W. D. L., (2011). Biological control of plant-parasitic nematodes of genus *Pratylenchus* through natural inoculants in sugarcane. *Brazilian Journal of Agriculture Sciences*, *6*(2), 203–207.

Oppermann, C. H., & Chang, S., (1991). Effects of Aldicarb and Fenamiphos on Acetylcholinesterase and Motility of *Caenorhabditis elegans*. *Journal of Nematology*, *23*(1), 20–27.

Pedrosa, E. M. R., Rolim, M. M., Albuquerque, P. H. S., & Cunha, A. C., (2005). Suppression of nematodes in sugarcane through vinasse addition in soil. *Brazilian Journal of Agricultural and Environmental Engineering.*, v.9 (Supplement), 197–201.

Pin. H. P., (1986). Effect of Furadan 5G in sugarcane planting in relation with productivity of plant and ratoon. *In: X Brazilian Meeting of Nematology,* 110–111.

Rodrigues, C. V. M. A., Pedrosa, E. M. R., Oliveira, A. K. S., Leitão, D. A. H. S., Barbosa, N. M. R., & Oliveira, N. J. V., (2011). Vertical distribution of the nematode community associated with sugarcane. *Nematropica.*, *41*(1), 5–11.

Rosa, R. C. T., Moura, R. M., & Pedrosa, E. M. R., (2004). Effects of the use of *Crotalaria juncea* and carbofuran in plant-parasitic ectonematodes of sugarcane. *Brazilian Phytopathology*, *29*(2), 447–449.

Sikora, R. A., & Hartw, J., (1991). Mode-of-action of the carbamate nematicides cloethocarb, aldicarb and carbofuran on *Heterodera schachtii*. 2. Systemic activity. *Revue Nematologie.*, *14*(4), 531–536.

Silva, G. S., (2009). Alternative methods for control of plant-parasitic nematodes. *Annual Revision of Plant Pathology.*, *19*, 81–152.

Vasconcelos, R. F. B., De Cantalice, J. R. B., Oliveira, V. S., Costa, Y. D. J., & Cavalcante, D. M., (2010). Aggregate stability of a Latosoil Yellow Distrocoeso of costal table under different input of organic residues from sugarcane. *Brazilian Journal of Soil Science.*, *34*(2), 309–316.

Winchester, J. A., (1969). Sugarcane Nematode Control. In: Peachery, J. E. (ed.). *Nematodes of Tropical Crops*, 1 ed. C.A.B.: St. Albans, Hearts, U.K.204–209.

CHAPTER 12

INSECT VECTORS AND SUGARCANE WHITE LEAF DISEASE MANAGEMENT

YUPA HANBOONSONG

Entomology Division, Faculty of Agriculture, Khon Kaen University, Thailand, E-mail: yupa_han@kku.ac.thyupa_han@yahoo.com

CONTENTS

12.1   INTRODUCTION

Sugarcane white leaf (SCWL) disease is one of the most harmful disease threats to sugarcane production in the Asia region. The disease was first reported in India in 1947 (Masumoto et al., 1968). The disease has since spread widely to other countries through importing infected cane. The first report of SCWL disease in Lampang province in the northern part of Thailand in 1954 was from imported sugarcane variety Co 421 from India (Hanboonsong et al., 2016). The disease was also discovered in imported sugarcane variety NCo 310 in Taiwan in 1958 (Ling, 1962). SCWL disease was also recorded in 1986 in Japan, but later disappeared (Nakashima and Murata 1993; Nakashima et al., 2001). It was also reported in Sri Lanka in 2001 (Kumarasinghe and Jones, 2001), and recently it was reported that SCWL disease has spread into Lao PDR, Vietnam and Myanmar (Hoat et al., 2012; Thein et al., 2012).

12.2 DISEASE SYMPTOMS

The typical symptom of SCWL is white colored leaves. The infected plants show variable leave chlorosis symptoms from light green to yellow, light yellow to cream strips parallel to the midribs. As the disease progresses, the entire leaf become white (Figure 12.1). It was reported that the white leaf symptom is directly related to the development of chloroplast and chlorophyll biosynthesis (Wu et al., 1969). Moreover, there seems to be a correlation between the phytoplasma load and the chlorophyll loss typical of white leaf disease. In general, the highest amount of phytoplasmaare found in sugarcane disease plants showing the complete white leaf symptoms. (Nakashima et al., 1994). The infected

FIGURE 12.1 Sugarcane white leaf disease plant and various disease symptoms development.

leaves are narrower and smaller than those of healthy plants. Other disease symptoms are stunted stalks and rapidly producing tillers (Ling, 1962). The infected sugarcane plants show disease symptoms at an early growth stage (1–6 months old) and in the case of severe infection, the plants die at a young age. Infected plants with no white leaf symptoms show shortened internodes and cannot prolong growth until the second ratoon crop as the disease symptoms will appear and the plant dies, with high yield loss. However, the disease severity is influenced by many factors such as the cane-sett quality, sugarcane cultivars as some varieties maybe resistant to insect vectors and caused disease resistant, cultivation practice, soil type and fertilization and the containing of SCWL phytoplasma quantity (Wongkaew, 2012).

12.3 CAUSAL ORGANISM

The disease is caused by a plant-pathogenic phytoplasma, a bacterium without a cell wall. It is uncultural pathogen localized in the phloem tissue in sieve tube elements of infected sugarcane. The phytoplasma is pleomorphic in shapes and vary in size from 80 to 800 nm (Chen, 1974; Maramorosch et al., 1975). The SCWL phytoplasma belongs to the class Mollicutes and was classified as 'Candidatus Phytoplasma oryzae' (the 16Sr XI-B genetic group of the rice yellow dwarf category) based on 16S rRNA RFLP (Lee et al., 1998).

The white leaf symptoms caused by phytoplasma can be seen in some grass species including bermuda grass, (Cynodon dactylon L.), crowfoot grass (Dactyloctenium aegyptium L.) and brachiaria grass (Brachiaria distachya L.) (Wongkaew et al., 1997). It was believed that these weeds were alternative host plants of SCWL phytoplasma. However, sequencing data showed that the phytoplasmas that infect these weeds were different from those infecting sugarcane (Wongkaew et al., 1997).

Apart from wild type sugarcane (Saccharum spontaneum L.), sugarcane commercial hybrid varieties have been identified as a natural host for SCWL phytoplasma. Up to now, the alternative host plants for pathogen reservoir has not yet been identified.

12.4 DISEASE SPREADING

The disease is spread by two ways, first from the use of infected cane stalk as planting material and second from insect vectors. The main spreading of

SCWLD is from using infected seed cane planting and it is considered to be the most rapid mean of spreading disease among farmers in different areas and causing wide spread disease outbreak. The spread of disease by insect vectors causes the disease spreading within sugarcane fields due to the slow movement of insect vectors.

Identification sugarcane white leaf symptom caused by infected cane sett or insect vectors is essential for preventing the spread of the disease during the next planting. In case of disease occurrence from using infected cane sett, the white leaf symptom can be obviously seen when the young shoots germinate and the disease plants are in a row from the same cane sett. If the disease is caused by insect vectors, no white leaf symptom show at early plant germination. The symptoms can be observed over time after the insect inoculated the phytoplasma to sugarcane plants. The disease plants show a random or aggregation (cluster) pattern within sugarcane field and also at the guard rows or edge of the field.

12.5 TRANSMISSION VECTORS

The SCWL phytoplasma is transmitted to the sugarcane plant by the phloem feeding leafhoppers *Matsumuratettix hiroglyphicus* (Mutsumura) and *Yamatotettix flavovittatus* (Homoptera: Cicadellidae) (Chen, 1974; Chen et al., 1976; Hanboonsong et al., 2002, 2006; Matsumoto et al., 1968).

Among these two vector species, the species of *M. hiroglyphicus* is a key vector contributing to spread of SCWL disease. It has high population number and its population dynamics is synchronized with monthly disease incidence (Hanboonsong et al., 2006; Pisitkul et al., 1989; Yang and Pan, 1969). Moreover, the disease transmission efficiency of *M. hiroglyphicus* (55%) is higher than that of *Y. flavovittatus* (45%), with females being more efficient than males (Hanboonsong et al., 2006). The SCWL phytoplasma is transmitted by vector insect in a persistent-propagative manner. Transmission characteristic of *M. hiroglyphicus* vector consists of the minimum acquisition and inoculation feeding periods of 3 h, and 30 min, respectively (Chen, 1979). The incubation period of SCWL phytoplasma in insect body is 25–35 days while in the host plant it is 70–90 days before the appearance of white leaf symptom (Matsumoto et al., 1968). After insect acquisition of the phytoplams from feeding on infected sugarcane plants,

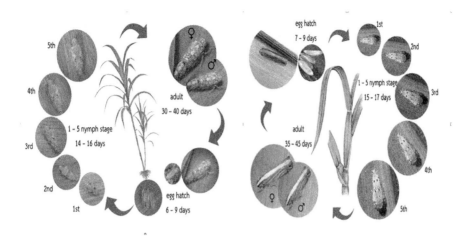

FIGURE 12.2 Life cycle of insect vectors. (A). *Matsumuratettix hiroglyphicus* (Matsumura) (B). *Yamatotettix flavovittatus* Matsumura.

the SCWL phytoplasma becomes widely distributed throughout the body of vector, and is transmitted transovarially from parents to the next generation. Not only *M. hiroglyphicus* adults but nymphs can also transmit phytoplasma to healthy plants (Hanboonsong et al., 2002, 2006). Female adults are more efficient than the males in transmission of SCWL disease (Hanboonsong et al., 2006).

12.6 ECOLOGY AND BIOLOGY OF VECTOR

The leafhopper vector, *M. hiroglyphicus*, has long been known as a key vector of SCLW disease. Thus, several research works have been carried out on its basic biology and ecology. It was reported that after mating the females lay their eggs in the soil, especially sandy soil most preferably but sometimes eggs can be found in leaf sheaths near the ground (Marcone, 2002; Promtasan, 2008). The young nymphs are usually found on the soil near the young sugarcane plants. Therefore, high rainfall may adversely affect on the survival of eggs and the newly hatched nymphs, which are very tiny and crawling around the soil. The older nymphs and adults usually stay on leaf and stem (Pisitkul et al., 1989). Moreover, insects prefer to feed on the young shoot or young tillers than older plants. This vector species usually do not move far away from its host. It was reported that the nymphs are able to move in a short distance about 1.2–1.5 meters (Hanboonsong and Boonchim, 2016). The adults dispersal distance was calculated to be about 162 meters or approximately 4.0 meters per day (Thein et al., 2011, 2012). The growth development time of vector is influenced by the temperature. The optimum temperature for the best fitness of this vector is 27–28°C. The life cycle is about 50 - 70 days, egg incubation period is about 8–9 days, nymphal periods are about 17–19 days, adult's longevity is about 49–57 days and fecundity per female is 48–50 eggs. It is estimated that the vector could produce 8 generations per year with overlapping between generations (Kobori and Hanboonsong, 2017). There is very little information on natural enemies of this vector. Only 3 parasitoids including erythraeids mite (Erythraeidae), *Pipunculus* sp (Diptera) and Drynid wasp (Dryinidae) have been reported in Taiwan (Yang and Pan, 1977).

Apart from sugarcane, other plants have not been reported as host plants for this leafhopper vector. Several grass species geminated around the sugarcane fields have also been investigated as insect host plants but the insect vector cannot complete their growth development when fed on them. For instance, adult insect vector can survive only 1–2 days on gold beard grass (*Chrysopogon aciculatus* (Retz.) Trin.), the lesser fimbristylis (*Fimbristylis miliacea* (L.) Vahl.) and the striped crotalaria (*Crotalaria striata* DC.). One exception is insect fed on the bermuda grass (*Cynodon dactylon* (L.) Pers.), which has the longest development time from eggs to adults and adults can survive for 7–8 days but female leafhoppers do not lay their eggs or produced any offspring on this grass (Anutrakunchai et al., 2012). Yangand Pan (1970) has also reported that survival of this vector after rearing from egg to the 5th instar is 13.2% on the bermuda grass, 3.7% on *Euphorbia hirta* L. and 2.7% on *Cyperus roduntus* L. but no adults could develop. It might be because of host-plant specificity between the leafhopper vector and sugarcane plants as well as phytoplasma pathogen or the unknown others host plants. Alternative host plants for insect vector still need to investigate.

The second most important insect vector of sugarcane white leaf disease is *Y. flavovittatus*. Its population occurs when the sugarcane plants are at old mature stage, therefore it may not have much effect on causing disease symptoms. The information on this species is small. The egg incubation period is 7–9 days, 5 nymphal stages and development is 15–17 days. The adult longevity is 35–45 days. The fecundity per female is 30 eggs.

(Hanboonsong, 2008). The adults' dispersal distance is approximately 387 meters (Thein et al., 2011, 2012).

12.7   POPULATION DYNAMICS OF INSECT VECTORS

Vector leafhoppers, *M. hiroglyphicus* and *Y. flavovittatus* reported in tropical climate like Thailand are found in the field during sugarcane planting from March to November, but not in December to February or after cane harvesting. The number of insects increases greatly during the rainy season from late April to August. However, the highest population peak period varies with vector species. The peaks for *M. hiroglyphicus* occur during April–May that would be during occurrence of the young tillers of the ratoon cane and late tillers of the plant cane in the field. Those for *Y. flavovittatus* occur about 2–3 months later in July-August when the sugarcane plants would be almost at full mature stage (Hanboonsong et al., 2006; Pisitkul et al., 1989; Rattanabunta and Hanboonsong, 2015). In Taiwan, the vector species of *M. hiroglyphicus* was reported to be widely distributed in central and southern parts of the country and sugarcane and *Saccharum spontaneum* L. (wild cane) are the preferred hosts. Moreover, in sugarcane fields in Taiwan, the vector, *M. hiroglyphicus* population is particularly abundant from July to October. The population declines rapidly in December and remains low until April (Yang, 1972; Yang and Pan, 1969). Overall, in both countries, disease incidence is correlated with the population trend of the vector in the field.

12.8   DISEASE PREVENTION AND CONTROL

At present, there is no effective method or any resistant sugarcane variety to control SCWL disease. Disease prevention treatment by soaking cane stalks before planting with thermotheraphy such as hot water at 50°C or dual treatment in hot and cold water as well as antibiotic treatment using tetracycline were used in the past (Pliansinchai et al., 1997). However, this method is not able to prevent or control the SCWL phytoplasma. Heat treatment affects to cane germination and antibiotic only can delay the appearance of disease symptoms for a few months but it cannot eliminate the pathogen (Kaewmanee and Hanboonsong, 2011). Therefore, the integrated disease management strategies including management of insect vectors could be one of the approaches to reduce or control the SCWL disease.

12.8.1   USE OF DISEASE FREE SEED CANE

The SCWL disease is a systemic disease and the sugarcane plant does not show symptoms when initially infected. Thus, the use of healthy or disease free sugarcane material is highly recommended. In Thailand, SCWL disease is a continuous long serious disease to the sugarcane industry, particularly in the Northeast region. The use of disease free seed cane has been promoted and progressively adopted in farmers practice. Consequentially the loss from SCWLD does not increase dramatically. The production of disease free cane stalk in Thailand is done by the research centers of the Ministry of Agriculture and Cooperative as well as some sugarcane mill companies. The disease-free cane stalks are initially screened by PCR technique using specific primers for SCWL phytoplasama then propagated through meristem tissue culture technique. The tissue cultured plants, then are planted as mother plants in isolated areas and multiplied as clean seed in propagation fields. The clean cane stalk is regularly distributed to farmers through collaboration networks between government research sectors and sugar mills. In addition, the farmers follow a participatory approach in which farmer groups are intensively trained by the researchers and extension officers in SCWL disease monitoring, management and propagation of their own clean seed cane from tissue cultured disease-free seed cane at mother plants stage. This good management of clean seed cane helps to reduce the risk of using infected seed cane and causing wide spread of disease.

12.8.2   FIELD MANAGEMENT BASED ON INFORMATION OF INSECT VECTORS ECOLOGY AND BIOLOGY

Insect vectors also play an important role in disease spreading even when disease free cane seed is used. Thus, basic information on vectors ecology and biology is essential for disease management. It was revealed that insect vector prefers sandy soil for egg laying and has short movement distance (Thein et al., 2012). Thus, it is recommended that disease free or clean sugarcane plants planted in propagation field should be in areas isolated from other planting cane by at least 300 meters and to avoid sandy soil. In addition, selective insecticide from the Neonicotinoid group including Dinothefuran and Thiamethoxam are recommended to use only in the propagation field as it was reported that it can control the insect vectors and has no harmful effect on the egg parasitoids and predators, the natural enemies of sugarcane stemborers (Poonoo and Hanboonsong, 2016). However, more studies are needed to determine the appropriate timing to apply insecticides and the length of the residue periods on sugarcane plants. Besides, adjusting of planting time to avoid planting during the rainy season (May to July) when population of key insect vector at a peak time is also an important factor for reducing the incidence of SCWL disease. At present, several works on vectors and disease management are in progress such as insect vector simulation model for disease management, sugarcane varieties resistant to insect vector feeding, microbial control and symbiotic control approach for controlling and interruption of the disease transmission in insect vectors.

12.8.3 CULTURAL CONTROL

Basic cultural practices are usually used for SCWL disease control. These are field sanitation by removing out the diseased plants which can be inoculum sources for insect vectors, campaigning for stopping burning of cane fields before harvesting, crop rotation with legume family plants such the sunhemp (*Crotalaria juncea*) or the pigeon pea (*Cajanus canja* L.) as a source of green manures. Burning of cane fields is a practice used to facilitate harvesting by manpower, which negatively affects beneficial soil microbes and soil micronutrients, making the soil infertile, thereby weakening the plants and making them prone to the disease

12.9 CONCLUDING REMARKS

SCWL disease caused by phytoplasma is a destructive disease to sugarcane crop. The disease can spread through the use of infected cane stalk for planting material and by insect vectors. So far there is no effective method or sugarcane resistant variety to control SCWL disease. Integrated pest management practices for both insect vectors and sugarcane host plants are considered to be the appropriate strategies for control the disease. The main activities including the early disease detection in cane by molecular technique, the production of clean or healthy cane stalk through tissue culture technique and multiplying clean seed in propagation field, using knowledge of insect vectors population, ecology and their movement for advising sugarcane planting timing and management to avoid the peak period of insect vectors. Moreover, general cultural practice such as removing of diseased plants and crop rotation with legume green manure plants maybe useful. However, the multiple integrated control approach described here can be used as prevention technique to reduce the incidence of the disease but not for its eradication.

Other SCWL disease control methods subject of on going research include insect vector simulation model data for forecasting disease epidemiology, using a symbiotic control approach for interruption of the disease transmission in insect vectors or microbial control of insect vectors. In addition, improving plant health by providing essential and appropriate nutrients for strengthening sugarcane plants or reducing plants stress would make them resist and survive from pathogen attack.

ACKNOWLEDGMENTS

The author thanks the Thailand Research Fund and the National Research Council of Thailand for their long-term funding support to research in sugarcane insect vectors and sugarcane white leaf disease.

KEYWORDS

- **cultural control**
- **disease free cane**
- **field management**
- **insect vector**
- **phytoplasma**

REFERENCES

Anutrakunchai, S., Wangkeeree, J., & Hanboonsong, Y., (2012). Host plant testing for leafhopper Matsumuratettix hiroglyphicus (Matsumura) vector of sugarcane white leaf disease. *Khon Kaen Agri. J.,* 40(3), 281–286.

Chen, C. T., (1974). Sugarcane white leaf disease in Thailand and Taiwan. *Sugarcane Pathologists' Newsletter, 11,* 12–23.

Chen, C. T., (1979). Vector-pathogen relationships of sugarcane white leaf disease. *Plant Protection Bulletin*, Taiwan., 21(1), 105–110.

Hanboonsong, Y. (2008). The studies of ecology, behavior, efficiency of disease transmission of new insect vector of sugarcane white leaf disease and potential of its control. Project Report, Khon Kaen University, 36 pp.

Hanboonsong, Y., Wangkeeree J., & Kobori, Y., (2016). Integrated management of the vectors of sugarcane white leaf disease in Thailand: an update. *Proceedings of the 29th International Society of Sugar Cane Technologists Congress*, pp. 1258–1263.

Hanboonsong, Y., & Boonchim, T., (2016). Movement activity of leafhopper vector of sugarcane white leaf disease. *Khon Kaen Agri. J.*, *44*(1), 73–79.

Hanboonsong, Y., Choosai, C., Panyim, S., & Damak, D., (2002). Transovarial transmission of sugarcane white leaf phytoplasma in the insect vector *Matsumuratettix hiroglyphicus* (Matsumura). *Insect Molecular Biology*, *11,* 97–103.

Hanboonsong, Y., Ritthison, W., Choosai, C., & Sirithorn, P., (2006). Transmission of sugarcane white leaf phytoplasma by Yamatotettix flavovittatus, a new leafhopper vector. *Journal of Economic Entomology*, *99,* 1531–1537.

Hoat, T. X., Bon, N. G., Quan, M. V., Hien, V. D., Thanh, N. D., & Dickinson, M., (2012). Detection and molecular characterization of sugarcane grassy shoot phytoplasma in Vietnam. *Phytoparasitica, 40,* 351–359.

Kaewmanee, C., & Hanboonsong, Y., (2011). Evaluation of the efficiency of various treatments used for sugarcane white leaf phytoplasma control. *Bulletin of Insectology*, *64*(1), S197–S198.

Kobori, Y., & Hanboonsong, Y., (2017). Effect of temperature on the development and reproduction of the sugarcane white leaf insect vector, *Matsumuratettix hiroglyphicus* (Matsumura) (*Hemiptera Cicadellidae*). *Journal of Asia Pacific Entomology*, *20,* 281–284.

Kumarasinghe, N. C., & Jones, P., (2001). Identification of white leaf disease of sugarcane in Sri Lanka. *Sugar Tech.*, *3,* 55–58.

Lee, I. M., Gundersen-Rindal, D. E., & Bertaccini, A., (1998). Phytoplasma: ecology and genomic diversity. *Phytopathology*, *88,* 1359–1366.

Ling, K. C., (1962). White leaf disease of sugarcane. *Taiwan Sugar*, *9,* 1–5.

Maramorosch, K., Kimura, M., & Chareoniridhi, S., (1975). Mycoplasma like organisms associated with white leaf disease in Thailand. *FAO Plant Pro Bullet*, *23,* 137–139.

Marcone, C., (2002). Phytoplasma diseases of sugarcane. *Sugar Tech.*, *4*(3/4), 79–85.

Matsumoto, T., Lee, C. S., & Teng, W. S., (1968). Studies on sugarcane white leaf disease of Taiwan with special reference to transmission by a leafhopper, *Epitettix hiroglyphicus* Mats. *Proceedings of the International Society of Sugar Cane Technologists*, *13,* 1090–1099.

Nakashima, K., & Murata. N., (1993). Destructive plant disease caused by mycoplasma-like organisms in Asia. *Outlook on Agriculture*, *22*(1), 53–58.

Nakashima, K., Wongkaew, P., & Sirithorn, P., (2001). Molecular detection and characterization of sugarcane white leaf phytoplasmas. In: Rao, G. P., Ford RE, Tosir, M., Teakle, T. S., (eds). *Sugarcane Pathology. Volume II: Virus and Phytoplasma Diseases.* Science Publishers, Inc., Enfield (NH), USA, pp. 157–175.

Nakashima, W., Chaleprom, P., Wongkaew, P., & Sirithorn, P., (1994). Detection. of mycoplasma like organism associated with white leaf disease of sugarcane in Thailand using DNA probes. *JIRCAS Journal*, *1,* 57–67.

Pisitkul, S., Kanta, C., Wongkaew, S., Neera, P., Chaioy, P., & Chetarash, C. (1989). Studuies on the insect vector of sugarcane white leaf disease and their control. *Khon Kaen Agri. J., 17,* 164–172.

Pliansinchai, U., Surinthu, N., Prammanee, S., Thanawon, T., & Yimsaard, M., (1997). Etiology and symptomatology of grassyshoot syndrome of sugarcane and heat therapy, [online] From http://kucon.lib.ku.ac.th/Fulltext/. KC3501048.pdf.

Poonoo, W., & Hanboonsong, Y., (2016). Effect of insecticides on insect vector of sugarcane white leaf disease. *Khon Kaen Agri. J., 44*(1), 66–72.

Promtasan, A., (2008). Studies of life history of leafhopper, *Matsumuratettix hiroglyphicus* (Matsumura) and phytoplasma transmission efficiency. MSc thesis, Khon Kaen University, 71p.

Rattanabunta, C., & Hanboonsong, Y., (2015). Sugarcane white leaf disease incidences and population dynamic of leafhopper insect vectors in sugarcane plantations in Northeast. *Pakistan Journal of Biological Sciences, 18,* 185–190.

Thein, M. M., Jamjanya, T., & Hanboonsong, Y., (2011). Evaluation of color traps to monitor insect vectors of sugarcane white leaf phytoplasma. *Bulletin of Insectology, 64*(1), S117–S118.

Thein, M. M., Jamjanya, T., Kobori, Y., & Hanboonsong, Y., (2012). Dispersal of leafhoppers, *Matsumuratettix hiroglyphicus* and *Yamatotettix flavovittatus* (Homoptera: Cicadellidae), vectors of sugarcane white leaf disease. *Applied Entomology and Zoology, 47,* 255–262.

Wongkaew, P., (2012). Sugarcane white leaf disease characterization, diagnosis development, and control strategies. *Functional Plant Science and Biotechnology, 6*(2), 73–84.

Wongkaew, P., Hanboonsong, Y, Sirithorn, P., Boonkrong, S., Choosai, C., Tinnangwattana T., Kitchareonpanya, R., & Damak, S., (1997). Differentiation of phytoplasmas associated with sugarcane and gramineous weed white leaf disease and sugarcane grassy shoot disease by RFLP and sequencing *Theoretical and Applied Genetics, 95,* 660–663.

Wu, L., Wu, R. Y., Li, H. W., & Chu, P. N., (1969). Influence of white leaf disease of sugarcane on the chloroplast development and chlorophyll biosynthesis. *Botanical Bulletin of Academia Sinica, 10,* 23–26.

Yang, S. L., & Pan, Y. S., (1969). Ecology and morphology of *Epitettix hiroglyphicus* Matsumura. *Report Taiwan Sugar Experiment Station, 48,* 25–35.

Yang, S. L., & Pan, Y. S., (1970). Bionomics of *Matsumuratettix hiroglyphicus* (Matsumura), an insect vector of sugarcane white leaf disease. II. Development in relation to host plants. *Report Taiwan Sugar Experiment Station, 50,* 73–79.

Yang, S. L., & Pan, Y. S., (1977). Parasites of *Matsumuratettix hiroglyphicus* (Matsumura). Annual *Report Taiwan Sugar Research Institute, 77,* 49–58.

Yang, S. L., (1972). Bionomics of *Matsumuratettix hiroglyphicus* (Matsumura), an insect vector of sugarcane white leaf disease. III. A study on the relationship between environmental factors and oviposition of *Matsumuratettix hiroglyphicus* (Matsumura). *Report Taiwan Sugar Experiment Station, 57,* 65–74.

CHAPTER 13

SUGARCANE RATOON MANAGEMENT

S. K. SHUKLA, LALAN SHARMA, and V. P. JAISWAL

ICAR – Indian Institute of Sugarcane Research, Lucknow–226002, Uttar Pradesh, India, E-mail: sudhirshukla151@gmail, sudhir.shukla@icar.gov.in

CONTENTS

13.1   INTRODUCTION

Ratoon cropping is an age old system, which may have begun when man first noticed regrowth of new shoots following cutting of certain crops at harvest; thus, producing a new crop without replanting. According to the Indian Sugar Committee Report (1920) ratooning of cane was not a very common practice in our country owing to whole dependency on indigenous cane varieties, which were very poor ratooner. The introductions of "Co" canes revolutionized the position and lead to the rapid increase of the area under ratoon crop. The pace of increase emphasizes the economic viability of ratoons on account of its reduced cost of cultivation owing to savings in seed material, land preparation and planting operations. In addition, the early harvesting of ratoons with relatively higher sugar recovery extends opportunity to run the sugar mills early in crushing season. In nonconventional areas, ratoon paves the way of bringing jaggery early in the market and fetches higher price to the farmers.

However, low productivity of ratoon greatly influences the national average of cane yields as ratoon occupies a sizeable share (>55–60%) in sugarcane acreage. The poor productivity level of ratoon cane include gappy stands as a result of failure of bud sprouting under low temperature conditions, accumulation of toxic exudates due to intensive monoculture and less efficient root system. The ratoons are more prone to insect-pests build up and diseases, which contribute negatively in improving the yield. Traditionally, a 4–6-months fallow (either bear, as weeds, or a sown legume) exists before the crop cycle is repeated. However, at present, adoption of intensive cropping system resulted in no fallow period. This increase in the intensity of growing monoculture cane has been accompanied by the more extensive use of inorganic fertilizers, insecticides and herbicides. While these changes have increased the efficiency of cane production, they have also been associated with the development of a plateau in productivity, which is today's big concern.

In the tropics and subtropics most sugarcane is ratooned at least once. In Hawaii, where mechanical harvesting is employed, sugarcane is usually ratooned only once because of soil problems and mechanical damage to the stools. In Mauritius, where as many as seven ratoon crops are taken, 85% of the cane milled each year is produced by ratoon cane. Therefore, for improving sugarcane production on sustainable manner, it is important to adopt appropriate soil health and crop management techniques to improve the ratoon cane yield. Ratoon or stubble crop has got some limitations and it is important to understand them for their better management.

13.2 RATOON CANE PRODUCTION CONSTRAINTS

13.2.1 ENVIRONMENTAL CONSTRAINTS

These are mainly temperature and soil moisture induced constraints. In subtropics, the harvesting of sugarcane starts during peak of the winter season in low to very low temperature. Stubble bud gets exposure to such odd ambience and usually yields gaps. Late harvesting too dries out the bottom buds of stubble and reduces plant population.

13.2.2 SOIL-BASED PRODUCTION CONSTRAINT

The compactness of the soil mainly under multi-ratooning system and build-up of weed flora and pests contribute to the loss of productive capacity of sugarcane-growing soils under long-term monoculture. It is recognized as a component of the productivity plateau and occurs in all sugarcane growing areas. The shallow root system of the ratoons reduces nutrient quenching ability and eventually makes the ratoon more sensitive to moisture stress and more demanding for available nutrients/addition of fertilizers. Because of the faulty and frequent irrigations, weeds along with their associated problem create menace to proper plant growth and reduce number of millable canes.

13.2.3 BIOTIC CONSTRAINTS

Ratoon cane is subjected to biotic stresses like red rot, wilt, smut, grassy shoot, ratoon stunting, and leaf scald diseases. The insect-pests like root borer, top borer, stalk borer, pyrilla and scale insect are the major pests causing damages to the crop in northern India.

13.2.4 CROP-BASED PRODUCTION CONSTRAINT

13.2.4.1 Short Life Span

The maturity in the ratoon crop sets in prior to that of plant crop. Under identical growing conditions, it has been observed that under subtropical climatic conditions, the normal growth and vigor in both plant and ratoons remains conspicuous till elongation phase (monsoon). Thereafter, the plant crop takes a lead and develops more vigorously than ratoon.

13.2.4.2 Root System

Sugarcane roots are short and die out quickly with little scope for substantial lateral growth, thus occupying smaller soil volume with limited foraging capacity and more prone to lodging. Moreover, ratoon crop results due to stubble buds, which are closer to the ground surface, the surface coverage shrinks further reducing the ratoons crop rendering them to become surface feeder.

13.2.4.3   Arrowing and Pith Formation

Arrowing is characterized by cessation of growth and in case of ratoons, it sets in early than plant crop. Ratoons are also reported to have more pith than plant crop and thus cause less cane weight.

13.2.5   VARIETAL CONSTRAINTS

Sugarcane varieties having good ratooning potential under low/high temperature conditions are not available in plenty. Besides, many of the prevalent varieties lack good ratoonability. Inadequate supply of seed of newly released good ratooning varieties is yet another constraint.

13.2.6   CULTURAL/OPERATIONAL CONSTRAINTS

13.2.6.1   Small Holdings

Small holding sizes and socio-economic conditions of sugarcane farmers affect the adoption of improved mechanical devices and timely completion of cultural operations. It is evident from the Table 13.1 that more than 63% of the farmers have less than five hectares holding size.

13.2.7   PLANTING AND HARVESTING METHODS

Sprouting of stubble buds occurs only at its favorable temperature. In case of sugarcane, the harvesting is mainly demand driven and procedural, without considering its impact on subsequent ratoons. An early winter harvested plant crop may result in to more gaps; similarly harvesting of over matured

TABLE 13.1   Distribution of Land Holdings in India

S. No.	% growers	Holdings size (ha)
1.	05.4	< 0.5 ha
2.	58.0	0.5 to 5.0 ha
3.	20.7	5.0 to 10.0 ha
4.	15.9	Others

canes during high temperature and desiccating winds in subtropical region render the bottom buds dry quickly and cause gaps. Piecemeal harvesting of plant cane creates problems in carrying out the cultural operations timely and the crop fails to obtain symmetry till its maturity.

The planting methods of sugarcane exhibit considerable impact in subsequent ratoon yields. Ratoons of Ring pit and trench method planted canes harvested close to the ground level by sharp edged chopper give higher ratoon yield over flat method of planting. Harvesting of cane leaving 4–6 inches above the ground portion give rise to poor ratoon crop.

13.2.8 SOCIO-ECONOMIC CONSTRAINTS

Adequate and balanced use of inputs is essential to obtain desired ratoon crop yield. Since sugarcane is grown under highly diverse situations, poor socio-economic conditions and ignorance about the actual requirement of the ratoon cane for its nutrition and plant protection and other critical inputs for specific conditions result in low ratoon cane productivity.

13.3 TECHNOLOGICAL INTERVENTIONS

In order to raise the productivity of ratoon cane, to commensurate with the future sweetener requirement, it is essential to dispense the developed agro-techniques suitable for ratoon cane under different conditions. This includes both maintenance of soil fertility and adoption of suitable agro-techniques for yield improvement.

13.3.1 ROTATIONAL BREAKS/SOIL SUSTENANCE

In addition to improved technical know-how, options include the incorporation of rotation breaks and the more general use of organic amendments in the cropping system. Other options should also focus on reducing compaction of the soils during the planting and harvesting operations, respectively, in order to reduce damage to soil structure and to improve water infiltration. Each of these options has the potential to contribute substantially to changing soil physical and chemical properties and hence influencing the balance between beneficial and detrimental organisms in the soil.

13.3.2   ORGANIC NUTRITION MODULES

For enriching rhizospheric microbial pool, maintaining soil health, enhancing cane productivity and making plant-ratoon system economically viable package of practices are recommended for plant/ratoon cane production. These are:

1. Irrigate the field 12–15 days after planting in spring and 18–20 days in autumn.
2. Blind hoeing at optimum field moisture to ensure adequate germination.
3. The irrigation and hoeing operations may be repeated at 15–20 days interval.
4. Earthing-up by mid of July for autumn and by mid of August in spring planted crop.
5. Thereafter fixing and propping may be done to avoid lodging.
6. Autumn planted crop is harvested in February and spring planting in March.
7. Trash is aligned in the wind rows to facilitated operation for ratoon initiation.
8. Field is irrigated and inter-culture operation done at optimum field moisture.
9. Bio-manures viz., well composted SPM 10 t/ha + FYM 10 t/ha or SPM 10 t/ha + *Gluconacetobacter* proves ideal organic farming module.

13.3.3   ADOPTION OF SUITABLE VARIETIES

Cane varieties possessing high regeneration capacity and free from sett borne diseases are preferred. A mid-late sugarcane variety Colk 94184 (Birendra) developed by Indian institute of sugarcane Research, Lucknow has good ratooning potential and performs well and moderately waterlogging/drought conditions. The optimum temperature for stubble bud sprouting ranges between 25–30°C, therefore, harvesting of plant crop is to be scheduled accordingly. The characteristic features of varieties with better ratooning potential include- uniform and good bud sprouting and tillering in ratoon cane, non lodging tendency, less tiller mortality, adequate number of millable cane formation, resistant to insect pest and diseases, higher plant *vis-a-vis* ratoon yield and good cane juice quality. In All India Coordinated Research Programme on Sugarcane, the varieties are evaluated on the performance of two sugarcane plant and one ratoon crop. The recommended sugarcane varieties through AICRP(S) during last 15 years are given in Table 13.2.

13.3.4 PLANTING TECHNIQUES AND METHOD OF HARVESTING

Although for subtropical region, there is general recommendation of sugarcane planting at 90 cm, but there is a positive interaction between row spacing and soil fertility. Under low level of soil fertility, closer row spacing is better while in case of well fertile soils, wider row spacing is more remunerative (Singh et al., 1984). Tiller mortality is reported to be more in cases of 60 cm and 75 cm than 90 cm. Ring pit method of planting fetched the highest ratoon yields under multiple rationing system at Lucknow (Table 13.3).

TABLE 13.2 Sugarcane Varieties Released by AICRP(S) During 2000–2016

Recommended Zone	Early group	Mid late Group
Peninsular Zone	Co 85004, Co 94008, Co 0403, Co Snk 05103	Co 86032, Co 87025, Co 87044, Co 8371 CoM 7714 (Krishna), CoM 88121, Co 91010, Co 99004, Co 2001–13, Co 2001–15, Co 0218, Co 06027, CoSnk 05104
East Coast Zone	CoC 01061, CoOr 03151	Co 86249, Co 06030
North West Zone	CoH 92201, CoS 95255 Co 98014, CoS 96268, Co 0118, Co 0238, Co 0239 Co 0237, CoPK 05191, Co 05009	CoS 91230 CoPant 90223, CoS 94270
		CoH 119, CoPant 97222, CoJ 20193
		CoS 96275, Co 0124, CoH 128, Co 05011
North Central and North Eastern Zones	Co 87263, Co 87268	BO 128, CoSe 92423, CoSe 96436, Co 0232, Co 0233, CoP 06436 (CoP 2061)
	Co 89029, CoSe 95422, CoSe 96234, CoLk 94184, CoSe 01421	

TABLE 13.3 Effect of Planting Methods on Cane Yield and Sucrose Content in Ratoon/ Ratooning Potential

Planting method	Plant crop	I ratoon	II ratoon	III ratoon	IV ratoon
Cane yield					
Ring pit method	157.0	112.7	106.3	99.3	94.7
Space transplanting	86.3	89.0	85.0	76.3	67.7
Flat planting	79.0	74.7	66.0	57.3	50.3
Sucrose% juice					
Ring pit method	17.3	17.6	18.7	18.8	19.2
Space transplanting	17.2	17.3	17.4	17.6	17.7
Flat planting	17.2	17.3	17.4	17.6	17.7

Sugarcane ratoon yield depends on planting methods. The methods of planting of plant crop give rise to different levels of ratoon cane yield. In Mujaffarnagar and Gorakhpur locations of subtropical region trench planting and ridge level of harvesting (Table 13.4) gave higher yield of subsequent ratoon than that obtained from the flat method of planting harvesting at ground level (Chauhan, 1993). At jalandhar trench planted and earthed up plant cane gave better sprouting subsequent ratoon crop.

13.3.5 CLIMATIC CONDITIONS DURING RATOON INITIATION

Proper development of ratoon crop is essentially dependent on sprouting of underground buds. Bud sprouting and germination is the function of optimum temperature, moisture and general vigor of the bud itself. Research evidences suggest that the ratoon obtained after winter harvesting fetched 33.5% loss in yield as compared to spring harvested cane under Pantnagar

TABLE 13.4 Effect of Plant Cane Harvesting on Number of Millable Canes and Yield of Ratoon Cane (Mean of Two Years) at Lucknow (Subtropical India)

Treatment	Millable canes	Cane yield (t/ha)
Harvesting of plant crop		
At ground level	90.3	64.4
At 5 cm above ground level	82.9	57.7
At 1o cm above level	81.2	54.8

conditions of Northern India (Verma, 2005; Verma et al., 1988) (Tables 13.5 and 13.6)

13.3.6 STUBBLE SHAVING

If the crop has been harvested close to the ground after flattening of ridges, then there is no need for stubble saving. But under farmers' practice where cane is harvested at 5–10 cm above ground level, stubble saving is essential because bud sprouted at apical portion dries up in due course of time for want of support, nutrients and water from the soil. This give rise to gappy stands in the field. Generally a sharp edged stainless steel chopper is used for harvesting or stubble saving. Precautions are taken that that the stubble remains intact while practicing stubble saving operation mainly in shallow planting otherwise plant population may reduce (Singh et al., 2005). The experiments carried out at Indian Institute of Sugarcane Research, Lucknow on multiple ratooning indicated that the highest ratoon yield was obtained when stubble shaving coupled with gap filling operations were employed (Table 13.7).

13.3.7 OFF-BARRING

In order to modulate rhizospheric environment, practice of off barring which connotes cutting older roots and creating space for own root system to support the clump for its nutrition needs and provide proper anchorage during elongation is an essential part of ratoon culture. This involves cutting older roots from both sides of the cane row followed by fertilizer application near root zone when the soil attains proper tilth after irrigation. Due precautions are taken against approaching the cultivator beneath the stubble which may eventually disturbing it.

TABLE 13.5 Effect of Time of Ratoon Initiation on Ratoon Cane Yield in Lucknow Conditions

Time of ratoon initiation	Gap (%)	Millable cane (000/ha)	Yield (t/ha)	
			Cane	Sugar
November–December	22–30	78.10	50.10	5.01
February–March	9–17	90.39	60.15	6.02
April–May	10–15	80.14	55.36	5.54

TABLE 13.6 Comparative Economic Evaluation of Different Ratooning Systems at Lucknow

Ratooning systems	Cane yield (t/ha)	Net profit (Rs/ha)		
		Ratoon	Intercrop	Total
December started ratoon (sole)	48.0	5480	–	5400
December started ratoon+ potato	62.0	8420	2460	10,880
December started ratoon + wheat	43.5	4535	4390	8925
February started ratoon (sole)	59.0	7790	–	7790
CD (0.05)	4.7	–	–	–

13.3.8 PLANT POPULATION MANAGEMENT

A close look at the field for observing gaps and there by taking appropriate steps to nullify their negative impact on overall performance of the crop is inevitable to maintain desired plant population. If the gaps are more than

TABLE 13.7 Effect of Different Management Practices on Yield and Quality Under Multiple Ratooning System at Lucknow

Treatments	Ratoon yield (T/ha)				Average ratoon yield
	I	II	III	IV	
T_1	55.1	49.3	42.6	33.3	45.0
T_2	66.6	60.3	53.0	43.6	55.6
T_3	63.7	52.8	48.0	39.0	50.8
T_4	67.0	60.0	53.0	43.0	55.7
T_5	68.2	61.0	52.0	44.0	56.3
T_6	65.4	55.0	49.0	39.2	52.1
T_7	71.0	66.3	59.0	57.6	63.4
T_8	71.0	65.6	60.1	58.6	63.8
T_9	73.5	68.0	61.0	59.0	65.0
CD (0.05)	3.80	6.25	4.38	3.30	–

Symbols used: T_1 – control (no cultural practices), T_2 – trash mulching in alternate rows, T_3 – trash burning, T_4 – stubble shaving+ T_2, T_5 – phorate insecticide (15 kg/ha), T_6 – T_3 + stubble shaving, T_7 – T_4 + gap filling, T_8 – T_5 + gap filling, T_9 – T_8 + 20% more seed rate.

15% of the total clump population (27,000–29,000/ha) the productivity of the ratoon is adversely affected (Chauhan, 1992). Gaps could be reduced by putting sprouted stubbles of the same variety or through polybag raised sugarcane settlings. Material for gap filling (sprouted setts) should be 30–35 days old (Table 13.8).

In order to sustain ratoon cane yield under multi-ratooning system, combination of various agro-techniques viz; trash mulching, gap filling, phorate application (15 kg/ha), stubble shaving and 20% extra seed to plant crop enhanced ratoon yield to the tune of 34, 38, 43, and 77% over control in Ist, IInd, IIIrd, and IV ratoons, respectively. As an individual component technology, trash mulching and gap filling contributed 21–28 and 10–29% increase in yield, respectively indicating that these are the key management issues for multi-ratooning in sugarcane.

Use of healthy seed material for planting, adequate nutrition/water supply and adoption of appropriate plant protection measures in time are known to reduce the gaps in subsequent ratoons. Ratoons have shallow root system and cannot tap nutrients from deeper soil layers unlike the plant crop, hence application of an extra 25% nitrogen over the plant cane and specific

TABLE 13.8 Effect of Gaps and Gap Filling Materials on Yield and Quality of Sugarcane Ratoon

Treatments	Cane yield (t/ha)	% decrease/increase over normal	Sucrose% in juice
A. Gaps in ratoon			
Normal crop (27777 clumps/ha)	63.0	-	18.3
15% gaps (23611 clumps/ha)	60.0	4.7	18.2
30% gaps (19445 clumps/ha) 47.6	46.0	26.9	18.1
40% gaps (15279 clumps/ha)	33.0	47.6	18.2
CD (0.05)	4.30	-	-
B. Gap filling materials			
Control (no filling	47.0	-	18.4
Ungerminated 3 bud setts	51.30	9.1	18.4
Polythene bag raised 1 bud sett	57.5	22.3	18.1
Nursery raised 3 bud settelings	59.8	27.2	18.2
Stubble ratoon	60.0	27.6	18.3
CD (0.05)	4.30	-	NS

micro-nutrients as per deficiencies exhibited by the crop are essential components for raising the productivity of the ratoon.

13.3.9   USE OF RATOON MANAGEMENT DEVICE-AN INTEGRATED APPROACH

For raising a good ratoon crop above mentioned cultural operations *viz.* stubble saving, off-baring, fertilizer application and deep tilling in inter-row spaces are inevitable. To perform all these operations in time is an arduous task and labor intensive. The situation is more alarming for availability of manual workforce owing to Government aided different employment guarantee programs in the countryside. In order to provide a viable option to lend a technical support to the peasantry, the Indian Institute of Sugarcane research, Lucknow has developed a machine called ratoon management device (RMD), which can be operated by 35 HP tractor. The machine performs all the operation viz. stubble shaving, deep tilling, off-barring, placing of fertilizer/bio-agents and earthing-up simultaneously in a single pass. Its field capacity is 0.40 ha/h. The machine costs about Rs. 75,000.

13.3.10   INTERCROPPING WITH RATOON SUGARCANE

Under intensive cropping system, keeping fallow is practically not feasible. In order to harness the benefits of crop diversification, intercropping with ratoon sugarcane (Figure 13.1) can be successfully adopted. In an intercropping of winter initiated ratoon with potato and wheat at Lucknow, Potato proved to be superior. At Mujaffarnagar a cash crop of guar in ratoon cane yielded 9.2 tonnes of green fodder/ha with 10% higher yield of ratoon. In the similar field trial in Delhi region, cowpea intercropped with spring-raised ratoon gave additional income without affecting the ratoon yield. Onion was found to be a successful inter crop of spring raised ratoon crop at Lucknow.

13.3.11   NUTRIENT MANAGEMENT

Various technologies have developed to provide nutrient requirements during sugarcane cultivation. These are:

FIGURE 13.1   Intercropping with ratoon (sugarcane + potato).

- High C: N ratio at initial stage and less efficient root system of ratoon cane responds well with 25% more nitrogen than plant crop. Split application improves N use efficiency. Balanced nutrition is essential for both cane productivity and juice quality. Secondary and micronutrient are becoming critical in monoculture systems and must be supplemented wherever noticed.
- An integrated approach of nutrient management using organic manures, crop residues, green manures and bio-fertilizers with inorganic fertilizers is more effective in maintaining soil fertility and sustaining crop productivity.
- Intercropped green manure crop in plant cane and its residues in first ratoon increase the N use efficiency in terms of cane yield by 5.2% in the former and 13.4% in the latter case.
- Application of *Gluconacetobacter diazotrophicus* based bio-fertilizer saves 25–50 kg of chemical N/ha. *Berseem* (green manure) is reported to have increased the availability and uptake of water-soluble phosphorus (P) in the soil.
- The severity of chlorotic symptoms appear to be associated with: low protein, P, NO_3 contents, higher%age of Na and reducing sugars, low K/Na ratio and high N/P, lower Zn, Cu, Fe contents and lower Fe/Mn ratio.
- Manganese deficiency in sugarcane can be corrected by spray of 125 ppm Mn, which also improves cane quality. Amelioration of lime-induced chlorosis could be achieved through application of $FeSO_4$ and $MnSO_4$.

13.3.12   WATER MANAGEMENT

Ratoon cane transpires less water than plant crop but the amount of water required to produce per unit of dry matter is more in case of ratoons. The shallow root system and relatively more compact soil medium prefers light but frequent irrigation scheduling in ratoon crop. In subtropics the crop responds well when irrigated at 12–15 days interval depending upon the nature and type of the soil and crop stage. In tropical region the water requirement is about three times more than that of its subtropical counterpart. Water is a precious commodity and scarce natural resource that needs to be used in judicious manner (Prasad et al., 1987). Following are some of the water saving techniques, which can be adopted without significant loss in cane yield (Table 13.9).

13.3.12.1   Water Saving Irrigation Methods

Water saving irrigation methods includes:

- selection of suitable varieties;
- trash mulching;
- trench planting;
- applying irrigation at most critical stages of crop water need;
- sub-tropical belt (UP): Skip furrow irrigation saves 36% water;
- tropical belt (Maharashtra): Drip irrigation economizes 45% water;
- water use efficiency (WUE): Skip furrow, FIRB and drip methods increase WUE.

TABLE 13.9   Cane Yield, Water Saving and WUE in Plant and Ratoon Crops of Sugarcane Under Skip Furrow Irrigation Lucknow

Irrigation method	Cane yield (t/ha)	Water used (ha-cm)	Water saving (%)	WUE (t/ha-cm)
Plant crop				
Skip furrow	66.9	22.5	25.0	2.97
Check basin	66.3	30.0	-	2.31
Ratoon crop				
Skip furrow	75.9	30.0	33.3	2.53
Check basin	71.2	45.0	-	1.58

Water scarcity sometimes creates a drought like situation. In the areas of inadequate water availability, the losses in yield could be minimized by adopting some useful agronomic tools viz. Setts and crop hardening, closer planting, deep trench planting and using soil amendments/moisture absorbents.

13.3.13 WEED MANAGEMENT

In ratoons, the *rapid-close-in* is much earlier than plant crop as the tillers emerge quickly just after harvesting as the apical dominance is removed. So, desirable weed free environment in ratoon fields is the first 10–12 weeks after its initiation.

Various methods, generally employed for controlling weeds in ratoon sugarcane include cultural, chemical, mulching and integrated approach. Among cultural methods use of trash mulch has been found quite effective. Mulching in inter rows with cane trash holds promise as it is easily available in the field for ratoon crop. It has been observed that a uniform 10 cm thick layer of trash gives desired level of weed control. Integration of chemical methods with trash mulching adequately controls the weeds. Triazines particularly atrazines and metribuzin (2 kg a.i./ha) applied at ratoon initiation stage as pre-emergence treatment (in relation to weeds) have been found quite effective in controlling weeds in ratoon fields (Annual Report, IISR, 2003–2004). However, *Cyperus rotundus* is least affected with these treatments. A post-emergence application of 2,4-D (0.5 kg a.i./ha) could yield desired results (Table 13.10).

13.3.14 RESIDUE MANAGEMENT

The crop residues (trash) left in the field after sugarcane harvesting accounts for about 10–15% of the weight of the cane harvested and left over root biomass in the soil constitute 5% of the cane harvest. Upon decomposition, this trash may supplement 30 kg N, 9 kg P, and 24 kg K/ha. Conventionally the trash is burnt in situ to prevent the harboring of insects and pathogens as well as weeds and to clear the field at one go. The experiments have shown that for sustainable productivity under multiple ratooning systems, trash mulching is more useful than to burn it out. For accelerating decomposition of trash, application of *Trichoderma*

TABLE 13.10   Effect of Weeds and Weed Free Conditions during Initial Stage of Ratoon Cane at Lucknow

Treatments	Millable canes (000/ha)	Cane yield (t/ha)
Weedy up to 30 DARI*	110.2	67.6
Weedy up to 60 DARI	97.2	58.4
Weedy up to 90 DARI	80.4	49.4
Weedy through out	71.7	46.3
Weed free up to 30 DARI	75.8	47.2
Weed free up to 60 DARI	95.2	61.4
Weed free up to 90 DARI	108.2	68.7
Weedy free through out	110.4	69.0
CD (0.05)	5.8	5.4

*DARI- Days After Ratoon Initiation.

is suggested. Trash mulching is done on alternate rows of ratoon cane to facilitate the irrigation and other cultural operations without interruption (Table 13.11).

13.3.15   HARVESTING

13.3.15.1   Scheduling by Variety

Sugarcane varieties are designated into early, mid-late and late, not exactly on the basis of crop-duration but on the prospective period of harvest during the harvest season. Thus, an early variety is required to be harvested first in the crushing season, which also indirectly implies that it has attained maturity in terms of commercial sugar recovery. Similarly, Ratoons mature

TABLE 13.11   Amount (kg) of Organic Matter and Major Nutrients in Per Tonne of Sugarcane Trash and FYM

Constituent	Trash	FYM
Organic matter	390.0	900.0
Nitrogen	5.30	5.40
Phosphorus	1.10	3.50
Potash	5.80	15.20

*10 cm thick layer of trash mulch applied during April.

earlier than plant crop so their harvesting schedule is be prepared keeping in view the nature of variety and time of ratoon initiation.

13.3.15.2   Scheduling by Date of Ratoon Initiation

When only one variety is predominant, it is desirable to schedule harvesting according to the date of planting ratoon initiation or harvesting of plant crop for ratoon. This practice is widely used in Maharashtra and Tamil Nadu.

13.3.15.3   Contingency Plans

Crop affected by floods, fire, frost or flowering should be given priority in harvesting over the normal crop.

13.3.16   HARVESTING METHODS AND TRANSPORT

Predominantly, harvesting of sugarcane is done manually. It is advisable to use sharp steel blade mounted on wooden base with a handle. It cuts the cane stalks at an angle. Cane should be cut flush to the ground. The canes should be de-trashed thoroughly after topping with sickle or a stripper. If the transport of the cane is likely to be held up for a period exceeding 48 hours, proper care should be taken as to heap the bundles under shade, cover them with trash and occasionally sprinkle water to prevent fast deterioration.

13.3.17   RATOON MANAGEMENT UNDER SPECIFIC CONDITIONS

13.3.17.1   Under Optimal Conditions of Input Availability

- At the time of harvest of plant crop, stubble shaving flush to the ground level is essential to promote a uniform sprouting.
- The inter row spacing should receive deep cultivation (off-barring) to improve the soil physical condition and also to prune the stubble roots. It should be irrigated soon after.

- The ridges should be dismantled after harvesting of plant crop.
- Fertilizer nitrogen may be applied in two equal splits within 60 days of ratoon initiation. The requirement is 25% higher than for the plant crop. The ratoon crop generally requires phosphorus and potash fertilization at the rate of 60 kg/ha. The actual requirement may, however, better be ascertained by soil test.
- Gap filling, using pre-germinated setts or by utilizing the stubble of the same variety is essential if gaps are more than 15%.
- Malathion @ 1 kg a.i./ha is required to protect the crop where black bug incidence is high.

13.3.17.2   Under Conditions of Piece-Meal Harvesting of Plant Crop

In Uttar Pradesh, Bihar, Haryana and Punjab where the numbers of suppliers are too many and quantity of cane supplied per harvest order is rather small, whole field cannot be harvested in one lot even if it is smaller in size. In such a case stubble shaving may be deferred until the last harvest. This will ensure a uniform crop stand. However, it should not be postponed beyond April.

13.3.17.3   Ratoon with Late Shoots

These are called water shoots or late shoots (lalas), which are often thick, succulent and appear later when early tillers are fully developed. The retention of such shoots during late harvesting of plant crop give good ratoon crop. Heavy earthing up, fertilizer application and frequent irrigation are essential for satisfactory ratoon yields.

13.3.17.4   Ratooning Under Low Temperature Conditions

Plant crop harvested during very low temperature (below 10°C) produces less ratoon yield due to poor sprouting. This should be avoided, however, if it is essential, a thick trash blanket (20 cm) may be spread over the stubble to prevent them from sprouting until the season becomes warm (February) when the trash may be removed and spread in the inter-row spaces. Inter-cropping of berseem in winter-initiated ratoon also protects the buds from extreme winter and encourages sprouting.

- Though farmers do get much information from various authentic sources that some machines are available for certain practices. The irony is that they do not get them on time owing to limited financial resources. It would be, therefore, appropriate if the Sugar Industry comes forward and takes a lead in dispensing the technology/suitable machines in their respective command areas by providing the services to the farmers on hire basis.
- When small farmers with minimal physical resources or financial assets attempt to improve productivity, they have a limited choice. The only resource they can maximize is knowledge, which they can obtain from research institutions. Sugar factories and NGOs active on rural developmental activities may sponsor for Farmers' training organized at research institutions time to time.
- Sugarcane (plant/ratoon crop) cultivation is an arduous work and physically straining. But in our country still a major part of the field operations is done manually as compared to foreign countries. Use of Ratoon Management Device is a viable option and can be tried successfully to minimize manual labor and complete the necessary agronomic practices in one go.

13.4   CROP PROTECTION MEASURES

Efficient and effective insect-pest and diseases management of sugarcane crop principally involves the selection disease resistant sugarcane varieties (where available and appropriate), selection of healthy seed cane and the employment of sound agronomic practices.

13.4.1   MAJOR INSECT-PESTS AND THEIR CONTROL MEASURES

Sugarcane is attacked by at least 125 species of different insect pests. In sub-tropics, root borer, stalk borer, top borer, Gurdaspur borer, pyrilla, and scale insect are the common insect pests while in the tropics, shoot and internodes borers are more prevalent (Table 13.12).

TABLE 13.12   Integrated Pest Management Schedule Against Major Insect Pests of
Sugarcane/Ratoon Crop

Crop stage	Pest covered	Management practice
At the time of ratoon initiation	Pyrilla, black bugs, armyworm	Stubble saving, removal of water shoots, burning of all remainants of previous crop residues.
March–June	Termites	Application of chlorpyriphos @5.0 L/ha through irrigation water.
	Borers (root, shoot and top borer)	Collection and destruction of egg masses and infested shoots
	Scale and mealy bugs	Drench spraying with 0.1% Dimethoate @ 0.06% after removal of lower dried leaves at 4–5 internode stage.
	Black gugs	Drench spraying with chlorpyriphos/quinalphos @ 0.2 kg a.i./ha.
	Pyrilla	Conserve *Epiricania malanoleuca* cocoons in the area @ 4000 to 5000/ha and 4–5 lakh eggs/ha.
	White fly*	Fields prone to water logged by sprayed with Dimethoate with 0.06% solution in fields of susceptible varieties in case of heavy build up of the pest.
July–August	White grubs	• Collection of adult beetles through light and sex pheromones and their destruction.
		Soil application of chlorpyriphos 5 G granules @ 25 kg/ha.
July–September	Gurdaspur borer plassey borer	Destruction of infested shoots in gregarious stage of the rests.
September–October	Stalk borer*	• Detrashing of dried foliage.
		• Removal of late shoots at 15 days interval till harvest of the crop.
	For pests in general	Pest control strategies recommended against the different insect pests in plant crop may also be adopted as the need based in ratoon crop.

* In areas where pests are seen regularly.

13.4.1.1   Root Borer

Root borer (*Emmalocera depressella*) is main pest of sugarcane crop but it has host sorghum and bajara also (Figure 13.2a). It is not only found in India but also in other nearby countries like Pakistan and Bangladesh. Sugarcane infected plant with this pest causes critical symptoms "Dead heart" and general yellowing of the leaves (Figure 13.2b). Due to infestation results in poor tillering in mature plants is observed. Studies have been conducted on its economic damage assessment which varies productivity loss ranging between 1.3–10%, and even severe infestation it leads up to 66.2 to 73.0% with reduced of cane length and weight, respectively (Gupta et al., 1952). There are some reports available that increased incidence of wilt disease of sugarcane (caused by *Sephalosporium sacchari)* in some varieties recorded but this association was not proven in all cases. Simultaneously occurrence both problems may lead to 100% crop loss.

13.4.1.2   Stalk Borer

Sugarcane stalk borer (*Diatraea saccharalis*) is a common pest of rice, paspalum, sorghum, johnson grass and maize (Figure 13.3). It caused variety of

FIGURE 13.2   (a) Root borer; (b) Dead hearts symptom.

FIGURE 13.3   Sugarcane stalk borer.

symptoms like dead heart, rot, wilting, discoloration of bark and fasciation. But tunneling by larvae inside sugarcane stems or stalks can result in major damage. Generally, dead hearts is more common on young cane but dead hearts in older plants sometimes occur. Commercial growing sugarcane varieties are more susceptible to this insect and causes up to 33% damage to the crop (David et al.; 1986). Stalks extensively damaged by borer may be unfit to use as seed cane and reduce the profitability of growing cane to produce juice for alcoholic fermentation.

13.4.1.3   Top Borer

Top borer (*Tryporyza nivella* F.) is a major pest of sugarcane in all cane growing regions of India (Figure 13.4). Young larva after hatching moves on the leaf surface for 3–4 hours infect at suitable site and penetrate 'mid-rib' from the underside of the leaf and bores its way towards the axil. The tunnel made in the mid-rib forms the characteristic white streak, which changes to red later. Severe infestation is seen when it infects crop at early stage (before the cane formation) results in dried top portion may sprout and develop into small-sized side-canes.

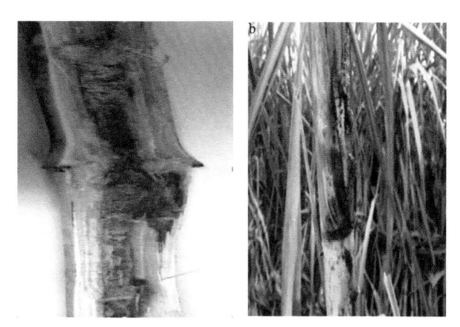

FIGURE 13.4A,B   (a) Sugarcane top borer; (b) Top borer infested cane.

13.4.1.4   Gurdaspur Borer

Gurdashpur borer (*Acigona steniella*) (Figure 13.5) is a serious pest of sugarcane. It infects sugarcane crop during July–October. It causes great losses to sugarcane especially ratoon crop. The female insect lies about 90–300 flattened scale like eggs in clusters on the leaves along the midrib in 3–4 rows and hatch in 4–9 days. The young larvae enter the top portion of a cane through a single hole just above the node. It feeds gregariously by making spiral galleries, which run upwards. After 7–10 days the larvae enter the adjoining canes resulting in wilting of the tops. In case of severe infestation 70–75% losses have been seen.

13.4.1.5   Pyrilla

Pyrilla (*Pyrilla purpusilla,* Walker) is the most destructive foliage-sucking pest of sugarcane. Both, adults and the nymph suck leaf sap from the under surface of the lower leaves results leaves turn yellowish white and wither from the tip. Due to continuous desapping by large number of hoppers, top

FIGURE 13.5   Gurdashpur borer.

leaves in the affected canes dry up and lateral buds germinate. Heavy rainfall followed by 75–80% humidity, intermittent drought periods, high temperature (26–30ºC) and wind movement favors the rapid build-up of the of pest population. Economic damage is caused by it as estimated, is up to 28% in cane crop while 1.6% unit reduction in sugar content. The hoppers exude a sweet sticky fluid known as "honeydew" which promotes quick and luxuriant growth of the fungus, *Capanodium* species and as a result the leaves are completely covered by the sooty mould (Jayanthi et al., 2009).

13.4.1.6   Scale Insect

Scale insect (*Melanaapsirs glomeata*) is serious pest of cane and ratoon crop. It enters into inter node and establish itself where internode is covered with leaf sheath. The sugarcane leaves of infested show signs of tip drying and unhealthy pale green color. Continuous sucking of leaf sap leads to non-opening of leaves, which turn yellow and finally dry up. Infested crop losses vigor, canes shrivel and inhibited growth results in poor quality of cane production. It is reported in literatures that waterlogged condition, high temperature and humidity favors tremendous build-up of scale population. High-speed wind assist in fast dispersion of pest population and covers large area within short period of time. Economic damage by this pest varies with pest population and susceptibility of the varieties, but up to 20% damage is reported.

13.4.2 MAJOR DISEASES AND THEIR CONTROL MEASURES

At present, sugarcane is being cultivated over 5 mha area in India. A total of 55 diseases of sugarcane caused by fungi, bacteria, viruses, phytoplasmas and nematodes have been reported from India. The economic importance and causing huge losses caused by them to the crop are red rot, wilt, smut, ratoon stunting, grassy shoot, pokkah boeng and yellow leaf disease (Table 13.13).

13.4.2.1 Red Rot

Red rot (*Colletotrichum falcatum* Went, Perfect stage – *Glomerella cingulata*) is serious disease of sugarcane crop and spread all over states of India except Maharashtra (Figure 13.6). This disease drastically retards the crop yield (25%), considerably deteriorates the juice quantity and quality. Many good varieties have gone out of cultivation due to red rot incidence. The red rot pathogen infects both leaves and stalks. On the leaves upper surface, initially dot shaped reddish lesion develop and it increases up to 2 to 3 mm in length and 1 mm in width. Later, these spots collapsed with each other and make it large symptoms. It can be easily seen in field conditions. It is reported that 3rd and 4th leaf of the crown at margins becomes dry and later entire crown drops down. It appearance in the stems can be seen on splitting them. Infected cane, mainly internal tissue shows red patched with transverse bands and on smelling it emit alcoholic like sour (Saksena et al., 2013).

13.4.2.2 Wilt

Sugarcane wilt (*Cephalosporium sacchari*) is also economic disease of sugarcane crop (Figure 13.7). It occurrence is reported in all the sugarcane

TABLE 13.13 Integrated Disease Management Against Major Diseases of Sugarcane/Ratoon Crop

Diseases	Management practices
Red rot	• Grow disease resistant
	• Healthy looking seed cane
	• Uproot the infected plants and burn them
	• Avoid replanting sugarcane crop at least 3 years in the same field
	• Treat seed with Aretan for 2–3 minutes during sowing
Wilt	• Grow disease resistant varieties
	• Select healthy looking seed cane
Smut	• Growing resistant sugarcane varieties
	• Disease-free seed cane
	• Disease-free planting material can usually be obtained by subjecting seed to a Hot Water Treatment
	• Roguing diseased clump and burn them
Ratoon stunting disease	• Disease-free seed cane can be obtained by tissue culture (Meristem Tip Tissue Culture), or after Hot-Water Treatment of sugarcane or Moist Hot Air Treatment (3 hours at 50°C)
Grassy shoot of sugarcane	• Disease-free seed cane can be obtained by tissue culture (Meristem Tip Tissue Culture), or after Hot-Water Treatment of sugarcane or Moist Hot Air Treatment (3 hours at 50°C)
Pokkah Boeng Disease	Cultivation of disease tolerant sugarcane varieties /use of fungicides
Yellow Leaf Disease of Sugarcane	Use of disease tolerant sugarcane varieties

growing states of India but more prevalent in south India. Because of more favorable climatic conditions are exist throughout the crop year. The first symptoms of the disease become apparent only when the plant has grown for about 4–5 months. The infected leaves of affected clumps gradually turn yellow and dry up. The infected canes show gradual withering. On examination of affected clumps, the pith will be seen discolored purple or dirty reef and with longitudinal streaks. A cottony white mycelium can also be seen in the pith region during moist condition, sometimes in dry condition. The pathogen is soil born in nature and survives up to 3–5 years in soil conditions.

FIGURE 13.6   Red rot infected cane.

13.4.2.3   Smut

Sugarcane smut disease is caused by the fungus *Sporisorium scitamineum (Ustilago scitaminea)*. The disease spread is worldwide covering most of the sugarcane producing countries. The most recognizable diagnostic feature of sugarcane infected with smut is the emergence of a long, elongated whip from the infected crop. The whip morphology differs from short to long,

FIGURE 13.7   Wilt infected sugarcane crop.

twisted, multiple whips, etc. (Figure 13.8). Affected sugarcane plants may tiller profusely with spindly and more erect shoots with small narrow leaves "grass-like" with poor cane formation. Others symptoms are leaf and stem galls and bud proliferation can be observed. It causes variable level of crop loss with respect to susceptibility of variety.

13.4.2.4   Ratoon Stunting of Sugarcane

Ratoon stunting of sugarcane is caused by coryneform bacterium inhabiting the xylem cells, named *Clavibacter xyli subsp, xyli*. It is probably present in almost all sugarcane growing countries in the world. It has considered one of the major diseases of sugarcane. The crop loss caused by it varies from 2 to 20%. Yield loss is generally associated with sugarcane variety as well as environmental conditions prevalent there.

FIGURE 13.8   Whip-like symptoms.

13.4.2.5   Grassy Shoot of Sugarcane

Grassy Shoot disease is caused by Phytoplasma (Figure 13.9). It is one of the serious diseases of sugarcane and affecting crop throughout crop growth stages. It is reported in almost all the states of India. Disease causes severe damage to the ratoon crop.

13.4.2.6   Pokkah Boeng Disease

Sugarcane pokkah boeng (Figure 13.10) is caused by the *Fusarium* species complex (*F. sacchari, F. proliferatum,* and *F. andiyazi*). It is emerging disease of sugarcane. The disease symptom occurs at 3 to 7 month-old sugarcane crop, more susceptible stage to infection than plants in later growth stages. After infection, the leaves become crumpled, twisted and shortened. Irregular reddish stripes and specks develop within the chlorotic tissue. The most serious injury is infection of the growing tip of the plant which results in the loss of the entire top of plant referred "top rot." In severe infection, it causes 5 to 25% economic damage to the crop.

FIGURE 13.9   Grassy shoot infected crop.

FIGURE 13.10   Pokkah Boeng disease symptoms.

13.4.2.7   Yellow Leaf Disease (YLD)

Yellow leaf disease is caused a virus and its symptoms initially develop at mid rib during 6–8 months crop stage (Figure 13.11). If disease severity increases, the yellowing spreads to laminar region and later there may be drying of affected mid rib and adjoining laminar tissue from leaf tip downwards along the mid rib. Another important symptom would be bunching of leaves in the crown. Highly susceptible variety may exhibit severe foliage drying during maturity stage. In place of yellow discoloration, purple or pinkish purple discoloration may also be seen on the mid rib and lamina. Canes of the affected plant do not dry.

13.5   CONCLUDING REMARKS

Sugarcane is important cash/industrial crop and grown worldwide. The demands of sugarcane farming continually require additional costs and inputs. Ratoon crop is lagging behind the sugarcane plant crop because of poor crop

FIGURE 13.11   Yellow leaf disease infected crop.

yield and less income. But nowadays, scenario is changing, and farmers are getting equal or more yields per unit area per day through ratoons than plant crop. In western Uttar Pradesh, ratoon yields are about 30–40% higher than plant crop. Productivity can be improved by adoption of advanced methods of sugarcane production as well as protection (viz., insect-pests, diseases and weed management) technologies. Sugarcane farmers can double their income by proper ratoon management. Good ratoon starts after planting of seed cane. A good crop of sugarcane gives good ratoon as well.

KEYWORDS

- **sugarcane**
- **crop diversification**
- **diseases**

- **harvesting and transport**
- **insect-pests**
- **management**
- **production constraints**
- **ratoon cane**
- **ratoon management device**
- **technological interventions**

REFERENCES

Annual Report, (2003). Indian Institute of Sugarcane Research, Lucknow, 04.

Chauhan, R. S., (1992). Effect of gaps and gap filling materials on yield and quality of ratoon crop of sugarcane. *Indian Sugar XLI, 10*, 743–45.

Chauhan, R. S., (1993). Effect of height of plant cane harvesting, stubble shaving and earthing up on ratoon crop of sugarcane. *Bharatiya Sug., 18*(3), 57–59.

David, H., Easwaramoorthy, S., & Jayanthi, R., (1986). *Sugarcane Entomology in India* (Eds.). Sugarcane Breeding Institute, Coimbatore, 564p.

Gupta, B. D., & Avasthy, P. N., (1952). Biology of sugarcane borer, *Emmalocera depressella* in UP. *Journal of Agriculture and Animal Husbandry,* UP 2, 19–25.

Jayanthi, R., Mukunthan, N., Salin, K. P., Srikanth, J., Geetha, N., Singaravelu, B., & Sankaranarayanan, C., (2009). Sugarcane Pests and their Management. Course Material for M.Sc. (Sugarcane Technology), Directorate of Open Distance Learning, *Tamil Nadu Agricultural University*, Coimbatore. 88p.

Magarey R. C., Lonie, K. J., & Croft B. J., (2006). BSES Limited, Indooroopilly, *Sugarcane Disease – Field Guide*. BSES Limited, p. 109.

Prasad, S. R., Alam, M., & Singh, R. N., (1987). Role of skip furrow irrigation in increasing water use efficiency in sugarcane under staggerd row planting. *Indian Sugar, 36*(11), 569–573.

Ricaud, C., Egan, B. T., & Gillaspie, A. G. Jr., Hughes, C. G. (Eds.). (1989). *Diseases of Sugarcane: Major Diseases*. Elsevier Publishing Company, Amsterdam, pp. 189–210.

Saksena, P., Vishwakarma, S. K., Tiwari, A. K., Singh, A., & Kumar, A., (2013). "Pathological and molecular variation in *Colletotrichum falcatum* Went isolates causing red rot of sugarcane in the Northwest zone of India." *Journal of Plant Protection Research. 53*(1), 37–41.

Singh D., & Singh S. M., (2005). Agro-techniques for multiple ratooning in sugarcane. National Seminar on improving sugar recovery in early crushing season held at IISR, Lucknow. Sept. 29–30.

Singh, K., Prasad, R., & Yadav, R. L., (1984). Agrotechnology for efficient use of irrigation water in sugarcane. Paper presented in a seminar on Science and Technology Development in U.P. held at Roorki University, Roorki, March 9–10.

Verma, R. P., (2005). *Unnat krishi taknik se peri ganne ki adhik padawar-Aarthik vishlesha*n. Presented in National Seminar on improving sugar recovery in early crushing season held at IISR, Lucknow, Sept. 29–30.

Verma, R. S., & Yadav, R. L., (1988). Intercropping in sugarcane for improving stubble sprouting under low temperature conditions in subtropical India. *Bhartiya Sug.*, *13*(3), 45–48.

CHAPTER 14

MECHANIZATION FOR SUSTAINABLE SUGARCANE PRODUCTION

AKHILESH KUMAR SINGH

Division of Agricultural Engineering, ICAR-Indian Institute of Sugarcane Research, Lucknow–226002, India, E-mail: aksingh8375@gmail.com

CONTENTS

14.1   INTRODUCTION

India is predominantly an agricultural economy country with 60–65% of her population living in villages and earns their livelihood through agriculture and allied activities. Rural population of India was 91% in 1901 and may reach to 50% by 2020. Rural people migrate to urban areas for employment and better amenities as such opportunities are not adequately available in rural areas. With the implementation of Mahatma Gandhi National Rural Employment Guarantee Scheme (previously known as NAREGA), there is further division of rural labor resulting into scarcity of labors for agricultural operations and increased labor wages. Indian agriculture is characterized by small and scattered holdings and sugarcane cultivation is no exception. Sugarcane crop remains in the field for almost a year. There is a heavy demand of labor and machinery throughout its crop cycle right from land preparation to harvesting of the crop and its timely supply to the mill. Sugarcane accounts for 60–70% of the cost of sugar production and thus has a vital role to make sugar industry a commercially viable venture.

Agricultural mechanization is a crucial input to agricultural crop production. It is frequently very capital intensive, compared to other (usually annual) inputs and it has repercussions on the efficiency of all other inputs used in crop production, including seeds, fertilizer, water, and time/labor. Farm mechanization is a crucial input for improving agricultural production. Without farm power and the appropriate tools, implements and machinery farmers would struggle to emerge from subsistence farming. With increasing demand for food and agricultural products being exerted on the planet's natural capital base, the essential role for sustainable mechanization in production systems development becomes increasingly obvious.

The state of agricultural mechanization in the country is characterized by large variations in power availability, which in 2001 varied from 0.6 kW/ha of agricultural land in some states to 3.5 kW/ha in Punjab. The average farm power available country-wide was about 1.91 kW/ha which comprised about 88% from mechanical and electrical sources and 12% from animal power and human labor. There is a strong linear relationship between the farm power available and agricultural output per ha. This underscores the emphasis on the growth and development of power machinery systems in Indian agriculture.

Mechanization of sugarcane agriculture aims at timeliness of operation, reduced cost of unit operations, reduced human drudgery and increasing productivity of other critical inputs such as labor, fertilizer and insecticide, etc. Sugarcane cultivation requires various operations like land preparation, planting, irrigation, interculture, earthing up, plant protection, harvesting including detrashing and detopping, and ratoon management. Most of the cultural operations are labor intensive. Use of suitable agricultural implements may play an important role towards achieving sustainable productivity level and inculcating cost efficiency in sugarcane production system. Number of useful equipments have been designed, developed and demonstrated at the farmer's fields with a fair degree of success. There is a need for concerted efforts on the part of researchers, sugar industry, private entrepreneurs and manufacturers of agricultural implements by way of providing commercially viable prototypes to the cane farming community. Operation wise equipment have been described in the article for mechanization of sugarcane agriculture.

14.2   SEED CANE TREATMENT

Moist Hot Air Treatment is performed to administer thermotherapy to seed cane for controlling seed piece transmission diseases of sugarcane. It was recommended by ICAR-IISR, Lucknow. Moist hot air treatment plant (MHAT unit) was developed for performing moist hot air treatment. It consists of a thermally insulated mild steel chamber, a cylindrical drum for loading of seed cane to be treated, heaters for heating and blower for circulation of air and auxiliary chambers with 2 kW industrial immersion heaters for generation of steam. Treatment is imparted at 54°C for uninterrupted duration of 4 hours at in an atmosphere saturated with moisture. Inside humidity of the chamber is increased by injecting steam from the auxiliary chambers. Output is 600 kg of seed per batch of treatment. Hot water treatment has also been recommended for seed treatment.

14.3   SEEDBED PREPARATION

Two types of machineries are used for seedbed preparation prior to planting of sugarcane. Primary tillage machinery like moldboard and disc plough and secondary tillage machinery like disc harrow and cultivator. One operation of primary tillage machinery and two operations of secondary tillage machinery is generally sufficient to achieve good soil-tilth for planting of sugarcane. Use of subsoiler is also recommended once in four years for breaking the hard pan 35–40 cm underneath the soil surface.

14.4   SUGARCANE PLANTING

The planting operation of sugarcane is an energy and labor-intensive operation. Human drudgery is also involved when the operation is performed manually.

Several methods and techniques of planting sugarcane have received attention of researchers from time to time. These methods include flat method, trench method, furrow method, spaced transplanting technique (STP), cane node method, etc. However, flat method of planting is much prevalent among cane growers of North Indian Plains. Planting of sugarcane comprises many unit operations such as opening of furrows, cutting of cane into pieces known as seed setts, placement of setts, fertilizer and insecticide in the furrows and providing soil cover over the setts. Furrows are opened with the help of animal or tractor drawn ridgers. Forty to forty five man-days are required in one hectare to carry out other operations. Since, arranging such a huge number of labor in a day is very difficult, the planting operation prolongs resulting into moisture loss of soil as well as seed setts. A lot of efforts have been made at ICAR-Indian Institute of Sugarcane Research, Lucknow to mechanize sugarcane-planting operations. Brief accounts of these machines are presented below;

14.4.1 MACHINERIES FOR MECHANIZING FLAT METHOD OF SUGARCANE PLANTING

Various models of sugarcane planters viz. Animal or tractor drawn semi-automatic (billet) planters and later tractor operated sugarcane cutter planters suiting to different agro-climatic and soil conditions have been developed at IISR. Different variants of tractor operated sugarcane cutter planters are either tractor PTO or ground wheel driven. Sett cutting is continuous and uninterrupted in PTO driven planters but proper sett metering is achieved at a particular combination of forward speed and PTO rpm. Sett metering remains same in ground wheel driven planters but precaution is required that ground wheels do not skid and remain in firm contact with soil. Tractor operated planters take four to five hours to cover one hectare. Four to five laborers are needed to operate the planter. There is saving of more than 50% in the cost of planting operation by using sugarcane cutter planter as compared to traditional method.

14.4.2 MACHINERIES FOR MECHANIZING FURROW METHOD OF SUGARCANE PLANTING

Recently, there is awareness of water shaving in sugarcane cultivation. In north India, it is being recommended to plant the cane in furrow method to save irrigation water. Planting of sugarcane in furrow method needs machine for deep furrow opening. For this purpose tractor operated deep furrower, deep furrower-cum-fertilizer applicator and deep furrow sugarcane cutter planter have been developed at IISR during last two years. Deep furrow sugarcane cutter planter is a multitasking machine, which performs all the unit operations involved in sugarcane planting including sett cutting, in single pass of the machine. It facilitates planting of sugarcane in deep furrow (20–25 cm) and maintains 5–7 cm loose soil bed underneath the planted seed setts. Planter has been field tested at IISR farm and on-farm trials also going on at farmers field of western, central and eastern U.P. and Bihar.

14.4.3 MACHINERIES FOR MECHANIZING TRENCH METHOD OF SUGARCANE PLANTING

Planting of sugarcane in deep and wide trenches under wide spaced paired row geometry (30:120 cm) has shown promising results on cane yield, water saving, reduced lodging and better ratooning. In order to reap the benefit of trench method of planting tractor operated trencher and trench planter were developed at IISR. While trencher performs opening of deep and wide furrow for paired row planting of sugarcane manually whereas, trench planter performs all the unit operations involved in cane planting including sett cutting, like earlier developed sugarcane cutter planters, in single pass of the machine.

14.4.4 MACHINERIES FOR MECHANIZING RING-PIT METHOD OF SUGARCANE PLANTING

The ring pit planting technique is very good from the point of view of increased cane productivity but digging of large number of pits over the entire field was found to be very cumbersome and labor intensive. Therefore, the technique could not be pushed for large-scale adoption by the farmers. Efforts were made at IISR to develop tractor drawn pit digger for mechanization of pit digging operation. The developed pit digger was able to dig one pit at a time. There was a problem of excessive vibrations and dynamic instability during the operation. Design refinements were made and modified prototypes of pit digger was developed. The equipment digs two pits simultaneously at a time. The developed equipment was tested and evaluated in sandy loam soil at IISR farm. With the help of the equipment approximately 150 pits (75 cm diameter x 30 cm depth) at a spacing of 30 cm were dug per tractor-hour operation. Cost of pit digging operation was saved by 70% by using the pit digger.

14.4.5 MACHINERIES FOR MECHANIZING PLANTING OF INTERCROPS WITH SUGARCANE

Equipment for planting of inter crop like wheat or pulses with sugarcane has been developed at IISR. Two types of machineries have been developed for inter cropping on the raised bed with sugarcane (i) raised bed seeder-cum-fertilizer applicator (RBS), and (ii) raised bed seeder-cum-sugarcane planter (RBS cane planter). The raised bed seeder is used for making three furrows and sowing of companion crop like wheat on the two raised beds. Sugarcane is planted in the furrows at a later stage manually. With the help of raised bed seeder-cum-sugarcane planter, planting of sugarcane in the furrows and sowing of companion crop like wheat on the raised beds, are accomplished simultaneously in a single pass of the equipment. Recently, sugarcane-cum-automatic potato planter, deep furrow sugarcane cutter planter-cum-multi-crop bed seeder, sugarcane trench planter-cum-multicrop bed seeder have also been developed for planting/sowing of intercrop simultaneously with sugarcane. These equipments are performing well during field trials at IISR and other locations.

14.5 INTER-CULTURING OPERATIONS

About 4–5 inter culture operations are quite common in sugarcane and each operation, if carried out manually, requires 25–30 man-days/ha. During early stage of crop growth (up to 50 cm of crop height) intercultural operations can easily be mechanized by using conventional 9-type cultivators, engine operated walking type rotary weeders and tractor operated rotary weeders. These equipments are commercially available. It is advisable to use sweep shovels in place of reversible shovels. Sweep shovels completely cover the spacing and no weed is left in the covered space. A tractor operated inter-culturing equipment with sweep shovels for conventional as well as wide spaced paired row planted cane crops has been developed at IISR. It covers 0.50 ha/h. Of late, tractor operated sugarcane manager has also been developed which performs interculturing as well as band application of fertilizer near to root zone of cane crop. Effective field capacity of this machine is 0.40 ha/h.

14.6 SUGARCANE HARVESTING

Harvesting of sugarcane and its transportation to sugar mills is an energy and labor intensive operation. Harvesting of cane involves cutting of cane stalks, detopping and detrashing of cane stalks, making bundles of 10 to 12 stalks and finally loading of clean cane bundles into transport vehicles. The harvested cane is transported to sugar mills for its processing using different modes of transportation viz., bullock carts, tractor operated trailers and trucks. Basically, two types of harvesting systems are prevalent world wide, i.e., green cane harvesting and burnt cane harvesting. The green cane harvesting is performed without burning the standing crop. The green cane harvesting is preferred due to superior quality of juice. Also, this method allows green top to be used as fodder and incorporation of trash into the soil. In case of burnt cane harvesting the cane field is first burned. The cane is harvested once temperature goes down to the operating comfortable level. In India, generally green cane harvesting is practiced. As the availability of labors is precarious and varies with the season, the availability of cane to the mills becomes uncertain. This necessitates the adoptability of mechanization of harvesting operation of sugarcane along with suitable transportation system. The mechanization efforts in the country have been basically limited to the development of whole stalk harvesters for the partial mechanization of sugarcane. These harvesters were basically designed to cut the cane and in some cases to detop the green top also. The remaining operations such as detrashing, bundle making and loading were to be performed manually. At few places such as in Andhra Pradesh and Maharashtra whole stalk harvesters were imported and evaluated for partial mechanization of sugarcane. So far these harvesters are still under development and evaluation stage. Of late, combine chopper harvesters have been imported at few places in Tamilnadu and Maharashtra. These harvesters are intended to mechanize the whole operation of sugarcane and found acceptability at few places at limited scale. Combine chopper harvesters are capable of harvesting green as well as burnt crops and also heavy yielding and lodged crops unlike whole stalk harvesters. Whole stalk harvesters are suitable for harvesting of erect and medium tonnage crops.

The available sugarcane harvesting options can be grouped into manual harvesting and mechanical harvesting

14.6.1 MANUAL HARVESTING

In India, harvesting of sugarcane is done manually using conventional harvesting tools. The productivity of manual cutters can be improved by giving them proper training, the use of correct cane knife, balance diet, etc. On an average an out put of 0.8 to 1.0 tonne per man-day is obtained. Human drudgery involved in cane cutting, detrashing, detopping and carrying head loads to the transport wagon together with shortage of labor during harvest season, is compelling the cane growers to look for alternate viable options for mechanizing of harvesting operation.

14.6.2 MECHANICAL HARVESTING

Attempts have been made at different places to develop and adopt mechanical harvesters for sugarcane. Mechanical harvesters can be grouped into (i) whole cane; and (ii) combine chopper harvester.

14.6.2.1 Whole Cane Harvester

A tractor operated side mounted whole harvester was developed for cutting of single row of cane. Power to the cutting blade was provided from P.T.O. of the tractor through chain sprocket and P.T.O. pulley. Stationery guider was provided to guide the cut cane stalks for the windrowing. The equipments needed for space for the tractor and could be operated only in one direction. The harvested cane needed to be lifted prior to cutting of next row of the cane. Due to these limitations the equipment could not be taken for further trial.

A tractor front mounted sugarcane windrower harvester was developed at IISR. With the help of this harvester two rows of sugarcane was cut simultaneously and windrowed at the center of the rows. Removal of green tops and dry trash, bundle making and loading into transport vehicles, were performed manually. The major components of the harvester were mainframe, drive system, base cutters and rotating crop dividers and front reels. The drive from tractor P.T.O. was through double chain and sprockets. Base cutters consisted of a disc carrying three cutting blades. Crop dividers having spirals, rotating in opposite directions were provided to facilitate harvesting of lodged crop and proper windrowing. Limitations were the lifting of harvested and linearly windrowed crop.

The performance of imported sugarcane harvesters, viz Bonnel, Carib, Cameco and indigenous model VSI-Merado were also done in the country. Bonnel sugarcane harvester consisted of a front mounted and three-point link mounted, side harvesting cum windrowing unit on a standard tractor chassis. The detopped cane were cut at the base by twin disc blades and windrowed with the help of a conveyor along the direction of travel. Carib whole stalk harvester is a single row harvester on a standard tractor base, having two spiral scrolls to separate standing canes which are either lodged or bent and for gathering canes towards the center of the cutting row for base cutting. The base cutter is a twin blade cutter having adjustable depth wheels to take care of field undulation to maintain a constant height of cut. The cut cane stalks are windrowed beneath harvester along the travel direction thereby overlaying harvester canes over each other. Cameco whole cane harvester is a single row machine having a single spiral scroll at the left to separate entangled and lodged cane stalks from the adjoining rows of standing crops. It has upper and lower conveyors to gather and convey standing cane stalks for base cutting along row. Height of these conveyors can be adjusted independently. Base cutter height can be adjusted by hydraulic power and is always inclined to the ground. Windrowing unit has a conveyor which can discharge detopped and base cut canes at any angle, to a maximum, perpendicular to the direction of travel. The topper detops and shreds immature green tops and leaves. All the functional components and sub systems are hydraulically powered. The VSI-Merado prototype was similar to Carib model having a detopper at the front to detop immature green tops. It has twin base cutter powered by tractor power take off. A front detrashing roller was mounted to remove dried and loose leaves adhered to the cane stalks. The base cut canes was windrowed underneath the harvesting machine along forward travel. It has twin spiral scrolls to separate entangled stalks and for gathering lodged or bent cane stalks towards base cutter. All these whole cane harvesters has limitations of functioning well in the cane crop which was low yielding and erect or partially lodged. Due to these limitations these harvesters could not be adopted on commercial scale and phased out from the scenario.

A tractor front mounted sugarcane harvester was developed for cutting and windrowing of two rows of cane. Windrowing of one row is in transverse direction while the other row is windrowed linearly to the direction of travel of the tractor. Attachments, consisting of a M.S. frame and hydraulically controlled arms through hydraulic cylinders, were developed for raising and lowering of harvester during transportation as well as field operation. Power to the cutting blades was provided through tractor P.T.O. Attachments were also provided with the main frame for windrowing of harvested cane by guiding the cane towards cutting blades during harvesting operation and up to some extent for raising the partially lodged canes. The harvester was tested at IISR farm at different speeds of cutting blades (300–350, 450–500 and 600–650 rpm). The cutting was clean without any splitting and rupture of cane stubbles at cutting blade speed of 450–500 rpm. The cutting and windrowing was satisfactory for crops, which were not lodged and where the canopy was not intermingled with the other rows. Equipment needed free space towards right hand side to facilitate transverse windrowing of cut cane stalks. Due to this it could be operated in one direction only. During return equipment is not engaged in cutting. Due to these limitations, the equipment could not be commercially exploited.

For removal of green top as well as dry trash from the harvested sugarcane a power operated detrasher was developed. Equipment consisted of mechanisms for cane feeding, detrashing and delivery. It separates the top from the cane by breaking it from the natural weak point at the joint of immature top with mature cane stalks. It can be transported on three-point linkage of the tractor and operated by a electric motor, diesel engine or tractor P.T.O. Performance of the equipment was evaluated by feeding different varieties of harvested canes, with their tops first, to the detrashing rollers through the feeding chute. The trash left on the cane after passing through the detrasher varied from 1.5 to 6.6%. Trash removal efficiency varied from 77.5 to 94.5% depending upon the variety. The output of the detrasher was 2.4 tonnes/h for feeding of 2–3 cane stalks at a time. There was a saving of about 17% in cost of operation and 84% in labor requirement using the detrasher as compared to manual method.

Self-propelled whole stalk harvesters were imported in Tamilnadu from China. It was a Hansen make whole stalk harvester. It performs cutting and detrashing. The cut and detrashed cane stalks were delivered to the rear side of the harvester which is needed to be collected manually or could be collected in a bin attached with the rear of the harvester in the later models.

The field-testing of these harvesters have been conducted in Tamilnadu. There was problem in handling the recumbent and lodged cane crops hence it could also not be popularized.

14.6.2.2   Self-Propelled Chopper Harvester

Austoft 7000 sugarcane harvesters were imported from Australia at Sakthi Sugars Limited, Sakthinagar (Tamilnadu). These were combine chopper harvesters. With the help of this harvester cane is cut, cleaned and loaded into an articulated infield transporter. The harvested cane was transported to the mill using trucks. It has been reported that the harvester is commercially used. The average quantity of sugarcane harvested in a day was reported as 200 tons per day. The maximum quantity harvested in a day was 405 tons. The net cane output of combine chopper harvester (Austoft 7000) was found to be 24 to 30 tonnes per hour. The minimum working row space for this harvester was 150 cm. The cost of harvesting varied from Rs. 157 to 200 per tonne in case of mechanical harvesters, which could be lowered by increasing the total tonnes of cane harvested per day by operating the harvester for longer hours per day.

Presently few firms in India (New Holland, John Deere, Shaktiman) are manufacturing the self-propelled chopper harvesters. These harvesters are suitable for cane row spacing of 150 cm. They are also manufacturing the harvesters for 120 cm row spacing. These harvesters are also needed to be tested for harvesting of wide spaced paired row sugarcane crop.

14.7   RATOON MANAGEMENT

About more than 50% of the total sugarcane area is occupied by ratoon crop in India. It is an integral part of sugarcane cultivation being a profitable proposition. Raising ratoon crop of sugarcane has economic benefits not only for cutting down the cost of land preparation, seed material and cost of planting, but also ensure an economically high recovery in the initial phase of the crushing season because of early maturity than the plant cane. In the countries having sugar export oriented economy, taking 3–5 ratoon is quite common, but in India only 1–2 ratoon is common. Keeping a good ratoon crop is always a problem and it is often less cared for. On an average yield of conventionally grown sugarcane ratoon crop is lower than the sugarcane plant crop. Investigations reveal the fact that the productivity of sugarcane ratoon crop could be improved by applying crop inputs orderly in time and by executing cultural operations like (i) shaving stubbles close to the ground surface, (ii) off-barring or cutting old roots on either side of the stubbles, (iii) interculturing, (iv) deep tilling or breaking soil hard pan, (v) applying fertilizer, manure, insecticide or pesticides, and finally (vi) by providing desired amount of soil cover over stubbles or to the plantlets as the case may be. These operations are not only difficult and arduous but also far too uneconomical to be carried over by using conventional tools like spades, cultivators, ridgers, etc.

Efforts were made to develop independent tools like stubble shaver, stubble shaver with off-barring and fertilizer application, chemical applicator, power weeder, cultivators, earthing up equipment, and rippers with limited success. In the present context of increased global competition in world sugar market, there is need to increase the sugar productivity at reduced cost of operation. These objectives could be achieved by breaking the myth that ratoon is a free gift of nature. Due attention is needed to be paid for raising ratoon by performing requisite cultural operations in time in cost effective manner. Mechanical alternatives for accomplishing the cultural operations may be explored to achieve these objectives. Adoption of appropriate mechanical tool to undertake majority of the cultural operations simultaneously in a single pass, will be a major step towards increasing the productivity of ratoon at reduced unit cost of operation. Concerted efforts have been made at IISR and three prototypes were developed to undertake most of the cultural operations simultaneously in a single pass.

14.7.1   STUBBLE SHAVER-CUM-FERTILIZER APPLICATOR

During eighties a tractor operated two-row multi-purpose stubble shaver was developed at IISR. The equipment performs stubble shaving, off-barring and interculturing, and fertilizer application simultaneously in its single pass. The equipment consisted of M.S. Angle Iron frame with three point linkage system for mounting with hydraulic system of the tractor, gear drive unit taking drive from the tractor P.T.O. shaft through a telescopic shaft and universal joints, rotating disc with blade holders, shaving blades (2 nos.) and auxiliary frame with fertilizer metering system and 7 times with reversible double point shovels for off-barring/interculture. After harvesting of the plant crop, the trash remained in the field is handled properly either for using as mulch or through burning, as the case may be. For obtaining the best performance from the machine, management of trash was essential and the moisture in the field should be optimum so that shaving is proper without uprooting of the clump. Planting in straight rows is desirable to avoid uprooting while using the equipment. After ascertaining the field condition, the equipment is checked for the sharpness of the shaving blades and the wearing of the off-barring/interculture shovels. In the last, but not of least importance is the skill of the operator. The driver should be properly trained because the operation requires simultaneous operation of P.T.O. shaft while driving the tractor. If used, properly, two rows stubble are shaved, old root system chopped off so as to promote new rooting at a faster rate, interspace between the rows is tilled and fertilizer is also applied beside the clump in the close proximity of the root zone.

14.7.2   RATOON MANAGEMENT DEVICE (RMD)

Equipment namely ratoon management device (RMD) was developed at IISR. The equipment performs all the recommended cultural operations viz. stubble shaving, off-barring & deep tilling, fertilizer, manure and chemical application, interculturing and soil-covering in its single pass. It consisted of units namely stubble shaving, off-barring including old root pruning, Manure, fertilizer, liquid chemical dispensing and earthing up units for performing all recommended cultural operations independently or in a single pass of the tractor. It is a two-row tractor mounted type equipment that requires a minimum of 35 hp to execute operations in field. The performance of the equipment was satisfactory and output of equipment was 0.25 ha/h.

14.7.3   DISC TYPE RATOON MANAGEMENT DEVICE (DISC RMD)

Disc type ratoon management device (Disc RMD) was developed at IISR for performing cultural operations in ratoon field even having surface trash. It was equipped with stubble shaving serrated blades mounted on a disc, two tillage discs for off-barring (pruning of old roots) on either side of the stubbles and device for application of fertilizer at root zone. The effective field capacity of the equipment was 0.28 ha/h.

14.8   TRASH MANAGEMENT

In the present scenario where manual harvesting is in vogue, handling of trash is another area requiring attention of the researchers. Researches conducted have indicated that application of vinasse and filter cake to the residues, promotes decomposition of the dry matter so that resulting compost can be harrowed into the soil within 30 days. Nutrients derived from the trash may include 32 kg N/ha, 6 kg P_2O_5/ha and 30 kg K_2O/ha. Plant residue shredder has been developed at IISR for trash shredding in the field. The equipment is mounted with the tractor and is operated by P.T.O. shaft. The system picks up trash, passes it on to the chopping unit where trash is chopped into small bits. Provision has also been made for applying chemical/other substances for quick decomposition of trash. To start with, burning of trash may be avoided in select areas where insect pests are not a major problem and this precious material can be put to effective use either as a mulch to conserve soil moisture or as organic matter, thereby improving the soil health.

14.9   CONCLUSION

Most of the tools and machinery described in the article have been demonstrated at different places. These have been found effective in saving of time and cost and have potential to make sugarcane cultivation more profitable besides reducing human drudgery. It can be summarized that many useful and innovative machineries have been developed for mechanization of sugarcane crop at IISR Lucknow and other places in the country. With the increase in the availability of tractor power at Indian farms, the emphasis was on to the development of tractor-operated equipment. In the beginning the individual machines were developed for mechanizing the unit operations such as sett cutting machine, semiautomatic drop type planter, fertilizer applicator, stubble shaver, raised bed seeder, etc. Thereafter, equipments such as sugarcane cutter planter, stubble shaver-cum-fertilizer applicator, raised bed seeder-cum-sugarcane planter, ratoon management device, multi-harrow, sugarcane-cum-potato planter, were developed for performing more than one unit operations by integrating different units on a common framework. There is need for mass multiplication and adoption of these useful equipments. IISR has signed memorandum of agreement (MoA) with few manufacturers for manufacturing of these equipments on commercial scale. Few more are in the pipeline for signing the MoA for commercial production of these equipments. There is need for the close liaison and linkages among the researchers, manufacturers, sugar mills, entrepreneurs, and the extension agencies, cane department government officials for popularization of useful sugarcane machineries for inculcating the cost efficiency and timeliness of cultural operations thereby achieving sustainability in increased sugarcane production.

KEYWORDS

- **combine chopper harvester**
- **cutter planter**
- **intercropping**
- **labour efficiency**
- **mechanization**
- **production**
- **ratoon management device**
- **sugarcane**
- **sustainable**

REFERENCES

Singh, A. K. (2017). Advances in sugarcane mechanisation research in India. Compendium of Research Papers, National Symposium on Sugarcane Mechanisation: Challenges and Opportunities, BAIT Sathyamangalam, Mar 17–18, 8–13.

Singh, A.K., Sharma, M. P., & Gupta, R. (2017). Development of tractor operated double bottom pit digger for mechanizing ring pit method of sugarcane planting. *Sugar Tech*, *19*(5), 510–516.

Singh, A. K., & Singh, P. R. (2017). Development of a tractor operated sugarcane cutter planter for mechanisation of sugarcane planting in deep furrows. *Sugar Tech*, *19*(4), 416–423.

Singh, A. K., Singh, P. R., & Solomon, S. (2017). Design and development of a tractor operated disc type sugarcane ratoon management device. *Sugar Tech*, *19*(5), 501–509.

Singh, A. K., Gupta, R., Singh, S., & Singh, R. D. (2017). Mechanization of stubble shaving, off barring and fertilizer application during initiation of ratoon after sugarcane harvesting. Technical Compendium, 51st Annual Convention of ISAE and National Symposium on "Agril. Engg. for Sustainable and Climate Smart Agriculture" held at HAU Hisar during 16–18 Feb, 2017, p. 10.

Singh, A. K., Gupta, R., Singh, S., & Singh, R. D. (2016). Mechanization of sugarcane for sustainable sugarcane production. Souvenir, National Symposium on Challenges, Opportunities and Innovative Approaches in Sugarcane: Agriculture, Bio energy and Climate Change, UPCSR Shahjahanpur, Dec 21–23, 65–66.

Singh, A. K. (2016). Mechanization of sugarcane cultivation. Souvenir, Seminar on Imroving Productivity and Quality of Hybrid Rice and Sugarcane in Uttar Pradesh, held at IISR Lucknow during 18 Dec, 47–50.

Singh, A. K., & Singh, S. (2016). Mechanization of sugarcane cultivation in North India-status, problem and prospects. Proceedings, Annual Convention, NISSTA held at IISR Lucknow during 29–30 April, 49–56.

Singh, A. K. (2016). Recent developments in mechanization of sugarcane planting at IISR. Proceedings, 74th Annual Convention of STAI held at Delhi during 28–30 July, 123–129.

Singh, A. K., Singh, S., Gupta, R., & Singh, R. D. (2016). Development of agricultural machineries at IISR for mechanization of sugarcane cultivation. Abstracts, International Conference and Exhibition on Sugarcane Value Chain, held at VSI Pune during 13–16 Nov, ASP70.

Singh, A. K., & Solomon, S. (2015). Development of a sugarcane detrasher. *Sugar Tech,* *17*(2), 189–194.

Singh, A. K., Singh, P. R., & Gupta, R. (2012). Mechanization of sugarcane harvesting in India. *Journal of Sugarcane Research, 2*(2), 9–14.

Singh, P. R., Gupta, R., & Singh, A. K. (2012). Culti-harrow: A time and energy saving land preparation implement for sugarcane farming. *Journal of Sugarcane Research, 2*(2), 70–72.

CHAPTER 15

ALTERNATIVES TO INCREASE THE SUSTAINABILITY OF SUGARCANE PRODUCTION IN BRAZIL UNDER HIGH INTENSIVE MECHANIZATION

H. C. J. FRANCO, S. G. Q. CASTRO, G. M. SANCHES, O. T. KÖLLN, R. O. BORDONAL, B. M. M. N. BORGES, and C. D. BORGES

Brazilian Bioethanol Science and Technology Laboratory – CTBE, Brazilian Center for Research in Energy and Materials – CNPEM, Giuseppe Maximo Scolfaro Street 10000, Cidade Universitária, Campinas, 13083–100, Brazil, E-mail: henrique.franco@ctbe.cnpem.br

CONTENTS

15.1   INTRODUCTION

Brazil has the largest productive agricultural area on the planet, which consists of approximately 200 million hectares and plays a major whole in food security worldwide (FAO, 2015). Agriculture is of great importance to the Brazilian economy and accounts for 11% of the GDP in 2014 (96.7 billion USD) (MAPA, 2015). This includes cereal, sugar-energy (sugarcane-based), livestock and fruit sectors, which are important to the generation of economic gains for the entire nation. Sugarcane and the industrial processing of its raw material to obtain sugar, ethanol and bioelectricity represents the Brazilian sugar-energy sector and is the largest in the world (FAOSTAT, 2015). At present, the planted area surpasses nine million hectares, representing 40% of the worldwide area. Projections indicate that by 2024, sugarcane in Brazil will reach 11.5 million hectares, allowing an increase of nine million tons of sugar and an increase of 12.5 billion liters of ethanol (OECD, 2015).

Sugarcane production in Brazil has been more substantial in the last four decades (Figure 15.1) from 1975 to 2014 (FAOSTAT, 2015; Otto et al., 2016). In the agricultural year 2015/2016, 665 million tons were processed yielding an average 77 Mg ha^{-1} (CONAB, 2016). The main producing area was the South-Central region, representing 88% of the national production; this region includes São Paulo State, which owns 55% of all sugarcane cultivated in Brazil. This progress was fostered by the National

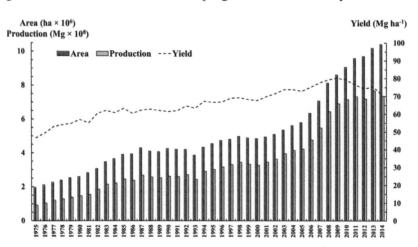

FIGURE 15.1   Evolution of sugarcane crop in Brazil from 1975 to 2014 according to FAO (Reprinted with permission from Otto, R. et al. (2016). Nitrogen use efficiency for sugarcane-biofuel production: what is the next? Bioenergy Research, doi: 10.1007/s12155-016-9763-x. © 2016 Springer Nature.)

Alcohol Program (*Proálcool*), which was created in 1975 by the Brazilian government to stimulate the consumption of ethanol to replace oil and its derivatives (Goldemberg, 2007). In this context, Brazil represents one of the first countries to produce clean and renewable energies, either by the production of ethanol or, more recently, by the cogeneration of electric energy. In 2014, the Brazilian sugar and ethanol agroindustry produced 20,815 GWh, which was 18% more than in 2013. The co-generation of electricity by the sugarcane sector accounted for 14% of all production from hydroelectric plants. For 2015, the National Electric Energy Agency (Aneel) forecasts an increase of 531.4 MWh from biomass, which is an increase of 4% from 2014 (UNICA, 2016).

There is no doubt that sugarcane is an excellent source of energy and food that contributes to the generation of employment and income. However, a substantial drop in agricultural productivity has been observed since 2008 (Figure 15.1), when the mechanization of the crop was intensified. Alternative management activities are and will continue to be necessary to overcome this problem, and they are vital to raise crop productivity back to levels of 100 Mg ha^{-1}.

15.2   SUGARCANE PRODUCTIVITY AFTER INTENSIVE MECHANIZATION

Since the first reports of sugarcane crops in Brazil, planting and harvesting operations were all manual, including loading and transporting (Belardo et al., 2015). From 1930 to 1950, the first prototypes of sugarcane loaders appeared (Figure 15.2) as well as the first attempt to mechanize sugarcane harvesting (Braunbeck et al., 2008). However, manual harvesting continued to dominate until the mid-1990s, when the advance, development and expansion of sugarcane harvesting began (Belardo et al., 2015).

Over 16 years (1990–2006), mechanized harvest was practiced with the previous burning of the straw, aiming to facilitate the operation. In 2006, 66% of the sugarcane area in the south-central region of Brazil utilized the previous burning management of sugarcane plantations. However, the impact of burning on human health and the environment in addition to legislation established deadlines for the gradual elimination of burning. Deadlines, procedures, rules and prohibitions were established to regulate burning in agricultural practices; these actions related to agronomic and energy issues

FIGURE 15.2 Development of machinery for mechanized harvesting of sugarcane from 1930 to 2016 (*Source*: www.macmanutencaoagricola.com.br / www.deere.com.br / www. lexicarbrasil.com.br).

involved making a substantial change in the harvest procedure. The gradual increase in mechanized harvesting without previous burning of the straw was inevitable, and in the south-central region of Brazil, the percentage of mechanized harvesting (green cane) increased to 84% of all harvested area in 2014 (UNICA, 2015).

In this new scenario, in which mechanized agricultural practices were consolidated in producing sugarcane crops, there was a great reduction in sugarcane yield (Figure 15.3). A decrease of 15% at the end of the 2010/2011 agricultural year led to an average yield of 68 Mg ha^{-1}, even with a 5% increase in crop area during the same period (CONAB, 2015).

The rapid development of mechanization was the main reason for the decrease in productivity. The soil compaction of the ratoon is one of the main reasons that are responsible for the average fall of 33% in stalk yield (Otto, 2015). The crops in Brazil were not adapted to the traffic of large and heavy agricultural machines (average weight of a cane harvester: 20,000 kg – Segato et al., 2006), and there was a mismatch between machine wide-span and planting spacing of sugarcane plantations.

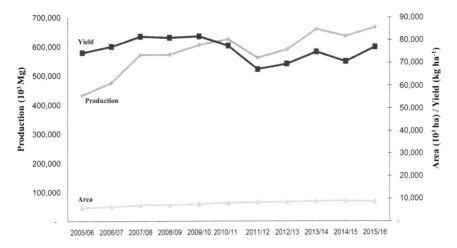

FIGURE 15.3 Production, yield and cultivated area with sugarcane in Brazil in the last 10 years (Source: CONAB, 2015).

In most recent years (2013–2015), there was a small increase in Brazilian productivity (Figure 15.3), but this increase was still lower than that obtained in the period prior to intensive mechanization (CONAB, 2015). The adoption of practices that optimize the return of productivity above 80 Mg ha^{-1} is vital. Along with intensive mechanization, the adoption of new practices is necessary to guarantee the sustainability of sugarcane yield in Brazil. Current practices have led to an early renewal of sugarcane plantations (lower longevity), increasing the total agricultural cost (R$ ha^{-1}) to an average value of R$ 11,813.16 (2015), which is 12% greater compared to 2014 (R$ 10,555.19) (IDEA, 2016).

15.3   YIELD DECLINE PROJECT: AUSTRALIAN LESSONS THAT SHOULD BE FOLLOWED BY THE BRAZILIAN SUGARCANE INDUSTRY

The stagnation and decrease of sugarcane yield in Brazil that occurred over the last decade is not something new when comparing the management of this crop to other countries. Australia, for example, is a country where mechanization has been traditionally used for years. Along with a major technological advance in agricultural operations, stagnation and decline in yield occurred from the 1960s to the 1990s (Figure 15.4). Australia overcame this

scenario by adopting a set of measures that promoted yield increases; nowadays, Australian productivity (excluding years with climatic adversities) is
higher than the period of stagnation (Figure 15.4).

Australian researchers aimed to put an end to the stagnation of sugarcane
yield and have proven that appropriate management can have a significant
impact in boosting crop yield (Sugar Yield Decline Joint Venture). From
this study, the sustainable pillars for the cultivation of sugarcane emerged as
follows: (i) maintenance of the straw on the soil surface (coming from the
mechanized harvest without burning); (ii) control of traffic and systematization of the crop (adequacy of machine wide-span to planting row-spacing
and use of autopilot and GPS); (iii) adoption of crop rotation during renewal
of the sugarcane field; and (iv) adoption of minimum tillage during sugarcane field reforms.

Analogous to the situation of sugarcane production in Brazil, the impacts
of a monoculture (cane over cane) and the improvement of mechanized harvesting are highly significant. Given the magnitude of the advancements in
agricultural machinery, the producers saw the necessity to systematize the
crops to carry out traffic control, aiming to increase operational efficiency

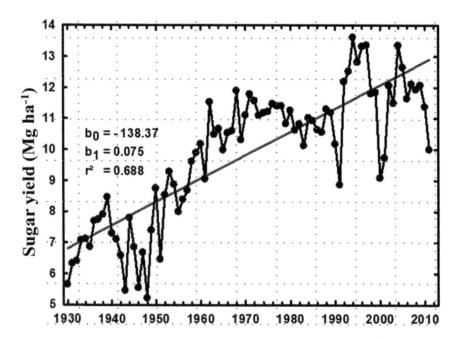

FIGURE 15.4   Sugar yield in Australia (*Source*: Alan L. Garside – CTBE /2015).

and yield recovery. On the other hand, the soil preparation practices and minimum cultivation as well as the adoption of crop rotation are still not widely used in Brazil; therefore, there is a clear need for diffusion of these technologies.

15.4 FRIENDLY AGRICULTURAL PRACTICES TO MAXIMIZE YIELD OF SUGARCANE FIELDS SUBMITTED TO INTENSIVE MECHANIZATION

It is possible for sugarcane areas in Brazil to adopt the four Australian sustainability pillars that contribute to the increase of sugarcane longevity and the sustainable development of the productive sector. In this sense, four topics will be addressed (soil preparation and crop rotation, planting systematization and row-spacing, mechanized harvesting and cultivation) with the objective of exposing friendly agricultural practices and the productive returns of the adoption of these steps under Brazilian conditions.

15.4.1 SOIL TILLAGE AND CROP ROTATION

Traditionally, soil tillage performed at plantations has a high cost since it involves several operations, such as harrowing, plowing and/or subsoiling. These operations are known as conventional tillage. The adoption of mechanized harvest without previous burning (green cane) deposits a large amount of plant material on the soil (10–20 Mg ha^{-1} – Fortes et al., 2012; Franco et al., 2007; Vitti et al., 2007), which is often difficult since re-harrowing is needed to incorporate these crop residues into the soil, thus facilitating plowing and/or subsoiling operations that are carried out in the sequence.

The adoption of conventional tillage causes excessive mobilization of the superficial layers of the soil, which could generate changes in soil physical properties when performed in inadequate conditions (excess moisture). This fact results in compaction and restriction of water infiltration, damaging sugarcane root development (Souza et al., 2014) located mainly in the 0.40 m soil layer (Otto et al., 2009). The growth of the root system is directly related to soil physical conditions (Otto et al., 2009) as well as its distribution along the soil profile (Smith et al., 2005).

During the last five years (2010–2015), there was a 47% increase in the cost of soil tillage operations (IDEA, 2016), reaching an average value of

US$540 ha^{-1}. This value corresponds to ~24 tons of sugarcane per hectare (considering US$22 per ton of sugarcane). There is an evident need for alternatives to soil management in order to reduce costs without hampering the establishment, development and yield of sugarcane plantations.

Based on the technological advancements in the production of grains, the use of the no-tillage system emerges as an alternative to sugarcane cultivation (Pereira et al., 2011). The premise of no-till in sugarcane follows the same guidelines of no-till for grains. Once the replantation area has been defined, a chemical desiccation is performed, followed by the application of soil amendments. After performing a soil acidity correction, the rotation crop is sown to later plant the sugarcane on straw (Tavares and Zonta, 2010). In cases when the occurrence of soil pests is evident (e.g., *Sphenophorus levis*), the adoption of a no-tillage system becomes impracticable, requiring mechanical removal of the ratoon.

By assessing conventional tillage versus no-till in loamy soils in the region of Ribeirão Preto (São Paulo – Brazil), Castro and Franco (2016) reported that there was no difference in yield during four consecutive seasons (Table 15.1). In a conventional plantation, a sequence of operations was adopted: chemical desiccation of ratoon, heavy harrow, plowing and grader harrow. In the treatment with no-tillage, only the desiccation of the old ratoon was done, and the land was left fallow until the planting of the sugarcane. The lack of response in the productivity of the different methods of preparation suggests that it is possible to obtain a reduction of the agricultural costs of implantation on a sugarcane field. Moreover, no-till favors the adaptation of cane plants to new soil conditions, benefiting its growth (Tavares and Zonta, 2010). The reduction of soil preparation with maintenance of the previous

TABLE 15.1 Effect of Soil Tillage on Sugarcane Yield in Mg ha^{-1} (Castro and Franco, 2016)

Tillage	2013	2014	2015	2016	Sum	Average
Conventional tillage	131 a*	106 a	91 a	79 a	407 a	102 a
No-tillage	130 a	98 a	94 a	86 a	408 a	102 a
LSD (5%)[†]	5	6	5	4	7	5
CV (%)[††]	10	8	4	6	8	9

[†] Least significant difference.

[††] Coefficient of variation.

[*] Mean values followed by different letters within each column indicate differences between treatments.

crop residues presents a more stable microclimate, maintaining greater soil moisture, decreasing temperature and reducing the risks of laminar erosion and nutrient leaching (Silva Junior et al., 2010).

In this scenario, the Brazilian Bioethanol Science and Technology Laboratory (CTBE/CNPEM) started a project in 2013 to evaluate the effects of soil tillage on the productivity of sugarcane. Four experiments were carried out in contrasting soils (sandy, clayey and very clayey) using two conventional methods (desiccation, subsoiling, plowing and harrowing) and no-tillage (desiccation and planting). The results (Figure 15.5) demonstrate that there were no differences in productivity, regardless of soil type and preparation systems; these results were similar to those observed in Castro and Franco (2016). The results of the CTBE/CNPEM project are presented in Barbosa (2015), and in the later harvests this same trend was observed, in which there was no difference between soil tillage systems adopted (data not shown herein).

Results from Otto et al. (2009) showed that 87% of the crop root system is distributed up to 0.40 m soil depth and 0.30 m on each side of the crop lines. These findings were also obtained in the works conducted by CTBE

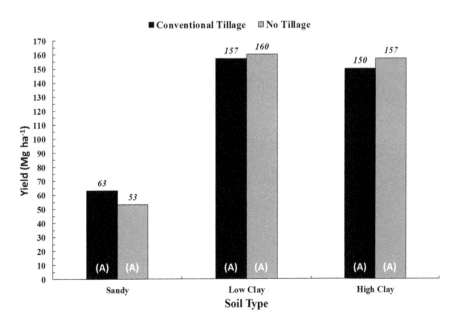

FIGURE 15.5   Effects of soil tillage on sugarcane yield in different soils (Adapted from Barbosa (2015). *Same letters indicate no difference between the methods of preparation within the same soil type.).

(Barbosa, 2015) in soils with different textures (Figure 15.6). It can be seen from the data that, regardless of soil texture (sandy or clayey) and type of tillage (conventional or no-till), most of the root system was concentrated to a depth of 0.40 m (black line; Figure 15.6), as highlighted by Otto et al. (2009). Consequently, deep preparation operations did not provide the development of the root system in layers deeper than 0.40 m and laterally to the inter-row. In this sense, the adoption of localized tillage, or even direct notation with deep furrowing, can be adopted by the producers and do not cause reductions in the productivity of the sugarcane-plant and subsequent ratoons. This practice will allow a significant reduction in planting costs by not using soil tillage operations in the pre-planting of the crop.

It is important to highlight that the grooving in the planting should be deep (Figure 15.7) to ensure that the stalks are deposited farther from the soil surface. This allows the plants to be less susceptible to the pull-off of their clumps promoted by harvesters and contributes to a greater longevity of the sugarcane field (Magalhães and Braunbeck, 2010). It is also interesting to couple a "swallow-wing" (Figure 15.7) to ensure that the region below the planting groove is also mobilized (Campanhão, 2003). The implementation of deep grouting with suitable equipment ends up being a local tillage operation that removes existing soil compaction, and conditioning positively determines the point of the greatest root development (Vasconcelos et al., 2008).

In order to have a conservationist management associated with high yields, the use of crop rotation appears to be an easily adoptable practice that can create prompt economic returns, but these management practices

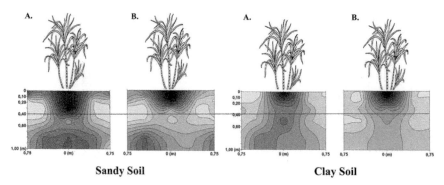

FIGURE 15.6   Distribution of the root system of sugarcane plants in different soils with (A) no-tillage and (B) conventional tillage (Adapted from Barbosa (2015). Note: darker regions designate higher root concentrations in the soil profile.).

FIGURE 15.7   Deep grooving (0.50 m) carried out in the (a) planting of sugarcane and (b) adaptation ("swallow-wing") made in the furcation rod.

are oftentimes not used by producers. The adoption of this management results in benefits for physical, chemical and biological soil attributes, weed, diseases and pest control. It also enhances the recycling of nutrients and protects the soil from the effects of climate, reducing issues related to compaction (Miranda et al., 2011).

When crop rotation is conducted with sugarcane, it is recommended to use nitrogen fixing species that have a deep root system. In comparative studies between plants used as soil cover, Duarte Júnior and Coelho (2008) found the highest dry matter yield using *Crotalaria juncea*, which provided a high coverage rate and potential for soil protection in addition to biological nitrogen fixation. Also, soil cover has the capacity to improve soil physical attributes by the addition of organic matter, reduction of density and increase of macropores and total soil porosity (Andrade et al., 2009).

Results presented by Duarte Júnior and Coelho (2008) showed a 37% increase in sugarcane yield when cultivated in a crop rotation area compared to a regular plantation area. When compared to the conventional production system, Ambrosano et al. (2011) observed an increase in sugarcane stalk yield with the use of *Crotalaria juncea*. The yield of the sugarcane plant and the average of three cycles of the area with rotation and without rotation were, respectively, 145/130 Mg ha^{-1} and 88/68 Mg ha^{-1}; that is, green manure yielded a gain of 15 tons already in the first cut and 20 tons per year.

In addition to *C. juncea*, other options are soybean and peanut that can provide extra income in the off season and reduce costs of sugarcane planting. However, it must be considered that these crops often do not break the cycle of pests and diseases, in addition to exporting most of the nitrogen fixed by the grain harvest.

The definition of which crop best fits the rotation system becomes difficult mainly in economic uncertainty frames, where extra gains from economic crops can lower the total costs of production. One of the first studies that assessed which crop rotation promoted the greatest financial return to sugarcane was carried out by Cáceres and Alcarde (1995); they identified that in the first harvest and in the sum of three cycles, the species *C. juncea* and *C. spectabilis* presented higher productivity than other species, besides the significant increases in sugarcane stalk yield when compared to the areas without rotation (Table 15.2).

15.4.2 SYSTEMATIZATION OF AREA AND ROW-SPACING

With the advent of mechanization in agricultural operations for sugarcane (planting and harvesting), it was necessary to develop a new strategy to be used in areas of replanting. The crop rows should be planned prior to planting, considering the total dimensions of the area in order to achieve greater efficiency of agricultural operations (Santos et al., 2010). In addition to the

TABLE 15.2 Effects of the Green Manure Species on the Increase of Sugarcane Stalk Yield

Species	Harvests			Cumulative	Difference from no-rotation
	First	Second	Third		
C. juncea	133 a*	90 ab	63 ab	286	19
C. spectabilis	134 a	92 a	66 ab	292	25
Cajanus cajan	127 ab	85 ab	61 ab	273	6
Mucuna aterrima	120 b	88 ab	60 ab	268	1
Mucuna pruriens	126 ab	88 ab	58 b	272	5
Lablab purpureus	127 ab	86 ab	60 ab	273	6
Canavalia ensiformis	126 ab	85 ab	60 ab	271	4
No-rotation	119 b	84 b	64 ab	267	

*Same lowercase letters do not differ means within column.

slope of the terrain, the concavity and rippling must be considered. In general, systematization starts at the field plot, which are considered to be the basic unit for sugarcane. Its area and shape are variable according to the soil type and topography as well as with the intrinsic conditions of each location (roads, currencies, among others). Usually the plots present maximum areas of 20 hectares and may vary from region to region (Santos et al., 2010).

The average length of a sugarcane row within the field plot should be planned based on the expected yield and operating income expected to be obtained during harvesting. In general, the average length of a row varies from 400 to 700 m (Figure 15.8a). Another important feature is the allocation of trucks to the traffic of trucks and trans-shipments, which should be between three and five meters wide and preferably five meters where truck traffic occurs (Figure 15.8b). It is also convenient to allocate discharge yards for the trans-shipments to transfer the load to the road assembly, eliminating truck traffic on the crop. Generally, discharge yards should be located in flat, easily accessible locations, with a width of 7 to 10 meters and based on a yard for every 70–80 hectares (Figure 15.8).

An important aspect to note is that the systematization of the area begins with the use of GIS (Geographical Information System) software. Along with the area of the georeferenced plot, software can be used to simulate the planting rows as well as their length. Once this planning is completed, it should be checked whether the project is suitable for implementation in the field. In many cases, adjustments are necessary due to relief conditions and topography. The main objective of this work of systematization and planning the rows is to obtain more sugarcane rows per hectare. Spekken and

FIGURE 15.8 Adequate length of planting line for high crop yield (a) and allocation of transverse carriers (b) in order to direct traffic in the crop.

Bruin (2012) and Spekken et al. (2016) highlight a series of considerations regarding planning and optimization of the systematization for plantations. Considering that the planting will be done with the aid of GPS, there being traffic control and the maintenance of the exact spacing between rows, the georeferenced groove generates gains of ~10% compared to when GPS is not used as a tool.

On the other hand, considering that the systematization seeks a higher operating income, the result obtained in most cases is the reduction of the number of sugarcane rows, increasing their length without decreasing the planted area. There is an increase in the operating income to the detriment of the number of maneuvers, since more rows need more maneuvers. This reasoning is justified by the work carried out by Sanches et al. (2012), in which the systematization carried out in two agricultural areas provided a reduction of ~7% in the number of grooves and an increase of ~2% in the average length of grooves. Similar results are presented by the Santa Fe Sugar Mill in the systematization of an area of 186.6 hectares (Figure 15.9). The systematization of the area increased the average length of grooves from 499 to 570 meters with ~9.5 km of ridges in total (Table 15.3). Considering an average yield of 80 Mg ha^{-1} with 1.5 m row-spacing, this total increase in sugarcane lines translates into a gain of 114 Mg of stalks. One of the great advantages is also observed in the reduction of the number of maneuvers (~10%, 280 maneuvers), which generates great operational efficiency for all agricultural operations. As reported by Spekken et al. (2014), the maneuvers are a small fraction of the production costs, but they have a major impact on the final economic revenue. According to the same authors, sugarcane rows with lengths of less than 50 meters do not generate sufficient revenues to pay the maneuver costs.

Traditionally, what always dictated sugarcane-planting row-space were machines, including both tractors and harvesters. This fact substantially

TABLE 15.3 Use of Systematization to Adapt the Area to Mechanization with High Operational Efficiency

	Area (ha)				Man.	Length (m)	
	Crop	Curves	Aisle	Total		Mean	Total
Stand.	172.7	7.9	6.1	186.6	2384.0	499.1	1189868.0
System.	177.4	2.1	7.2	186.6	2104.0	570.1	1199441.0

Man: maneuvers; Stand.: Standard procedures; System.: Systematized procedure; Source: Santa Fé Sugarcane Mill.

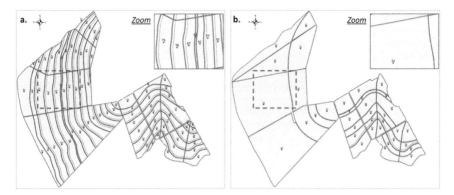

FIGURE 15.9   Conventional (a) and systematized (b) plantation of sugarcane
(Source: Santa Fé Sugarcane Mill.).

reduces the longevity of sugarcane plantations, since with the available machines there is a compaction of inter-row between the fields at approximately 60% of the soil surface. Environmental legislation and the agro-environmental protocol, which restrict the prior burning of crop before harvesting, contributed to a very quick shift from manual harvest to mechanical harvest during the first decade of the 21st century. Taking this into account, the Brazilian Bioethanol Science and Technology Laboratory (CTBE) has developed a traffic control structure (TCE) aimed at alleviating many of the constraints that the sugarcane sector faced, such as a reduction in soil compaction of inter-row and an increase in the sustainability of the energy balance of ethanol. Moreover, this new concept aims at eliminating the constraints imposed by mechanization in terms of planning row-space, plant configuration at harvest, and soil compaction in inter-rows. With the advent of this structure, what could be the best row-space for sugarcane production? This option is not restricted only to row spacing but also to plant spacing, aiming at less competition for nutrients, water and solar radiation, which are determinant factors of productivity. Based on this questioning, there were five experimental areas distributed in the central region of Brazil: Iracemápolis/SP, Guaíra/SP, Teodoro Sampaio/SP, Chapadão do Céu/GO and Lençóis Paulista/SP. In the experiment areas, 6 row spaces were tested and are listed as follows: Standard – Conventional row-spacing with 1.5 m between crop rows (one row); T1 – Double spacing (dual row) with 0.90 m ×1.5 m between rows; T2 – Triple spacing (triple row) with $0.75 \times 0.75 \times 1.5$ m between rows; T3 – pp. 1.0 m (1.0 m precision planting) with 1 m between rows and between plants; T4 – pp. 0.75 m (0.75 m precision planting) with 0.75 m between rows and between plants; T5 – pp. 0.5 m (0.5 m precision planting) with 0.5 m between rows and between plants (Figure 15.10).

This project presented savings with regard to seedlings used for planting. With the adoption of precision planting, there is an average saving of 27% or 14 Mg ha^{-1} of sugarcane seedlings (average of three experiments – Figure 15.11). Thus, considering the potential of the renovation areas in Brazil is around 1,100,000 ha year^{-1} × 14 Mg ha^{-1} × US\$ 20 Mg^{-1}, a potential savings of US\$ 308,000,000 per year is obtained only saving planting material.

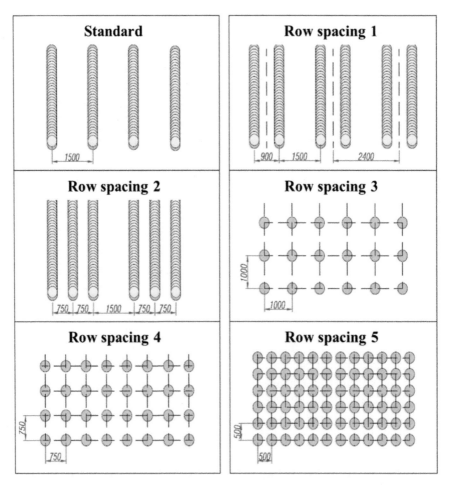

FIGURE 15.10   Tested planting in 6 row-spacing by the Brazilian Bioethanol Science and Technology Laboratory (CTBE) research team.

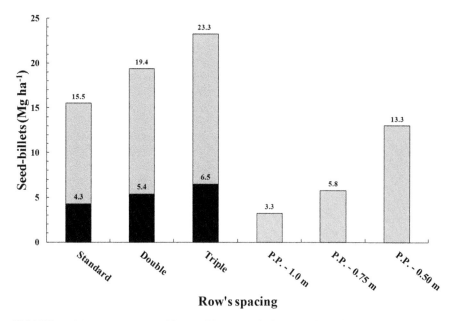

FIGURE 15.11 Expenses with seedlings used for planting in different planting configurations. Black bars represent expense with seedlings with precision planting at conventional spacing.

Results of yield gain due to reduced plant spacing and precision planting were also significant. An average gain of 20 tons of stalks per ha per year was obtained. Even in less favorable environments, such as Teodoro Sampaio/SP and Lençóis Paulista/SP, there was a yield gain with reduced row-spacing and precision planting (Figures 15.12 and 15.13).

15.4.3 MECHANIZED HARVEST

The process of mechanized sugarcane harvesting without preliminary burning of residues involves 10 operations: (1) cutting of tops; (2) lifting, aligning the stalks; (3) cutting the stalks; (4) lifting the stalks' base; (5) transport stalks with separation of part of the soil derived from the base cut; (6) chop stalks in small parts; (7) ventilation and primary straw cleaning; (8) transport stalk with the aid of elevator for discharge; (9) ventilation and secondary cleaning; and (10) stalk deposition in transport vehicle. The combination of these operations allows the chopped cane

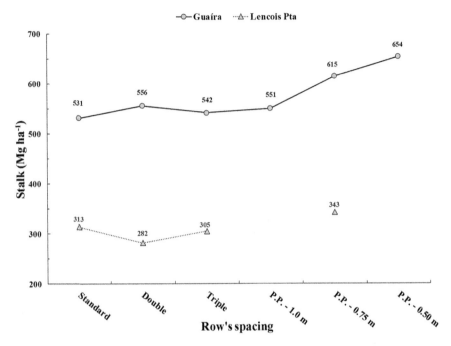

FIGURE 15.12 Yield of stalks with different planting configurations, sum of four harvests, Guaíra/SP and Lençóis Paulista/SP.

system to detrash different sugarcane fields, thereby contributing to the expansion and adoption of this system (Magalhães and Braunbeck, 2010). However, there are restrictions related to quality and loss of raw material, soil compaction and harvesting on sloping terrain. In the 2014/2015 harvest season, 632.6 million tons of stalks were processed, of which around 60 million were related to the presence of mineral impurity adhered to the stalks during the harvest process (Datagro, 2015). This is attributable to the type of cutting, loading and crop productivity, which affect the amounts of soils adhered in sugarcane stalks: clayey soils adhere more than sandy soils; rain and high moisture also facilitate the adhesion of soil in the stalks (Segato et al., 2006). The impact of mineral impurities is highly significant. There exists a loss of 2.6 kg of sugar per ton of sugarcane for each 1% of mineral impurities (IDEA, 2016), which means that the Brazilian sugarcane sector could have an additional production of 156,000 Mg of sugar in the 2015/2016 harvest season, thereby generating a loss of approximately US$ 31,200,000.

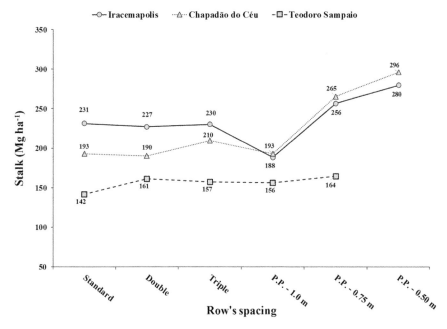

FIGURE 15.13 Stalk yield with different planting configurations, sum of two harvests, Iracemápolis/SP, Chapadão do Céu/GO and Teodoro Sampaio/SP.

The advancement of mechanized harvesting in Brazil was faster than the systematization of cropland, and this led to a great reduction of the Brazilian sugarcane yields and, consequently, total production (see Section 15.2). One of the major factors is the trampling on ratoons caused by the harvester, mainly because of a discrepancy between the wide-span of the machines (generally 1.85 m) in relation to the spacing of the planting rows used in Brazil (1.5 m). In order to lessen trampling and increase productivity and longevity of sugarcane fields, the use of autopilot during harvesting is mandatory. This technology enables to control traffic in the inter-rows (Figure 15.14), with reductions of up to 78% of trampling on ratoons (Campanelli, 2013).

The reduction of trampling in the mechanized harvest is critical in order to have better yields and increase longevity in the current scenario of sugarcane mechanization. As reported by Echeverry (2014), the cultivated areas under traffic control presented higher yields during five consecutive harvests compared to those without traffic control; average reductions of 17% (i.e., 19 Mg ha^{-1}) in cane yields were observed in areas without traffic control (Table

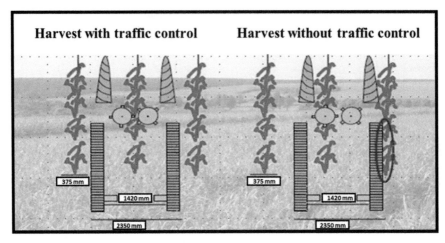

FIGURE 15.14 Effect of traffic control using autopilot in the trampling on ratoons during the harvest operations (*Source*: Campanelli, 2013).

15.4). There was no difference between areas in the first harvest, since they were not under machinery traffic. In the following harvests, the area without traffic control had lower yields than that under traffic control, especially because of trampling on ratoons as well as their pulling off.

As the implementation of mechanized harvesting is irreversible, the sustainability of sugarcane production in Brazil relies on the adoption of autopilot mechanisms for controlling traffic either in harvesters or in towrope operations. A proper use of this technology will certainly prevent trampling of ratoons in any agricultural operation by avoiding soil compaction

TABLE 15.4 Effects of the Adoption of Traffic Control on Yields and Longevity of Sugarcane

Harvests	With traffic control	Without traffic control
1st	120 a	115 a
2nd	120 a	95 b
3rd	110 a	90 b
4th	100 a	80 b
5th	95 a	70 b
Average	109 a	90 b

* Mean values in line followed by the same letter do not differ.

Adapted from Echeverry (2014).

in the cane row and concentrating the traffic in the inter-rows, which in turn increases the longevity and profitability. Efforts to optimize the basis harvest of the existing harvesters may also promote significant increases in crop longevity, since this process is one of the main causes of ratoons pulling off in sugarcane fields (Magalhães and Braunbeck, 2010).

15.4.4 CULTURAL PRACTICES ON INTENSIVELY MECHANIZED SUGARCANE FIELDS

A "straw mulching" composed of an average of 46% of tops and 54% of dry leaves came to exist with the advance of the green mechanized harvest (Franco et al., 2013). This forms a thick layer on the soil surface with 8 to 30 Mg ha^{-1} on a dry basis (Carvalho et al., 2013), depending on the variety and stage of cutting (Landell et al., 2013) as well as the agricultural practices and edaphoclimatic conditions (Santos et al., 2014). Because of its high C:N ratio (e.g., 100:1; Robertson and Thorburn, 2007), the decomposition of this residue is a slow process, so that less than 20% of the N contained in the straw (e.g., total N contained in a crop year of 40–80 kg N ha^{-1}; Trivelin et al., 2013) is mineralized after an agricultural cycle (Ferreira et al., 2015). On the other hand, the potassium (K) contained in the straw is quickly mineralized during an agricultural cycle (e.g., 85% of the total; Oliveira et al., 1999). This makes it possible to reduce the application rates of this nutrient in the crop management (Vitti et al., 2010), thereby reducing the fertilizers application costs.

The maintenance of straw on the soil surface also contributes to the retention of soil moisture that allows greater development and production of stalks, especially because of higher water availability to the plants. Evaluating soil moisture and sugarcane yields in clayey soils, Castro et al. (2012) found soils under straw retention with 25% more humidity compared to those without straw, which in turn promoted a 15% increase in stalk yield. In a literature review regarding the moisture retention in clayey soil with different straw fractions (0, 50, and 100%), Carvalho et al. (2017) reported that the maintenance of 50 and 100% of straw on the soil led to moisture retention of 392 and 556 mm along the crop cycle, respectively. On average, there was an increase of 39 Mg ha^{-1} of stalks in areas under straw retention, highlighting the importance of straw on soil with regards to moisture retention, soil conservation and weed control.

Contrarily, the synergistic effects of the larger amounts of water and increased straw on the soil surface have increased populations of pests, such as spittlebug (*Mahanarva fimbriolata*), sugarcane borer (*Diatrea saccharalis*) and sphenophorus (*Sphenophorus levis*). The latter has caused serious damages to the crop with significant implications for longevity and productivity of sugarcane fields (Arrigoni, 2011; Dinardo-Miranda and Fracasso, 2013). In addition to promoting the incidence of pests, sugarcane straw also hinders the agricultural operations carried out in the field, especially in the fertilizers incorporation during the scarification and fertilization of the inter-rows.

The presence of straw on the soil surface and the cropping season (wet or dry season) are important aspects that must be taken into account for agricultural management of sugarcane ratoons, especially associated with applications of N fertilizers. According to IPNI (International Plant Nutrition Institute), it is crucial to answer four questions during fertilizer management in order to have a sustainable and efficient agriculture: (i) Which source to use? (ii) What rate to apply (how much)? (iii) How to apply (form)? and (iv) When to apply (season)?

Despite the high losses through volatilization of ammonia (NH_3) when fertilizers are applied on the straw, urea is still the most prevalent source in Brazilian sugarcane fields, followed by ammonium nitrate ($NH_4^+NO_3^-$) and ammonium sulfate [$(NH_4)_2SO_4$]. Generally, the N is applied superficially without incorporation into the soil, and therefore the volatilization losses of NH_3 from urea would be significantly reduced or even eliminated if incorporation was performed in addition to increasing crop yields.

With regard to the N-fertilizer application form (Figure 15.15), six different approaches are used: (a) application incorporated into the soil through inter-row cultivation (triple operation); (b) banded superficial application; (c) banded applications on straw; (d) application incorporated on both sides of the cane row; (e) surface application in total area; and (f) aerial application by airplane.

The applications of N-fertilizer by means of the inter-rows (Figure 15.15a) or superficially in strip (Figure 15.15b) are still the most widely used practices in sugarcane fields and have been since when burning was prevalent. The superficial application is more often used in areas under green cane management because of its greater operational ease. The thick layer of straw may hinder the scarification of the inter-rows when such cultivation is chosen. Thus, there is a need to evaluate which methods of application

FIGURE 15.15 Application methods of nitrogen fertilizers in sugarcane ratoons. (*Source*: Castro and Franco (2016); Image E (*Source*): Henrique Coutinho Junqueira Franco; Image F (*Source*): www.controlservicerio.com.br).

are the most viable and those that present the highest productive returns in the management of N fertilization, taking into account the environmental impacts that this nutrient may cause to the natural resources. To evaluate if there is a need to perform the management of the inter-rows in a green cane system, Castro et al. (2014) carried out an experiment in a clay soil during four agricultural seasons in the Sales Oliveira region and observed no effect on crop productivity by adopting this management (Table 15.5). Other studies also reported the lack of response to the management of the inter-rows in areas under green harvest system (Campanhão, 2003; Manechini, 2000), indicating the ineffectiveness of this agricultural operation. In addition, such management is unfeasible because sugarcane roots do not exploit the region of the inter-rows (Otto et al., 2009), even in areas where there is no machinery traffic between crop rows (Rossi Neto, 2015).

Comparing the methods of N fertilizer application under green cane system, Castro et al. (2016) found that the banded superficial application and

TABLE 15.5 Sugarcane yield (Mg ha⁻¹) Associated with Mechanical Cultivation of the Ratoons (triple operation) under Green Cane System during Four Agricultural Season (Castro et al., 2014)

Treatments	2008	2009	2010	2011	Cumulative
With cultivation	125 b*	112 a	85 a	81 a	404 a
Without cultivation	131 a	112 a	88 a	81 a	412 a
LSD (10%)†	5.5	4	3.8	2.5	10
CV (%)††	11	9	11	7	6

† Least significant difference.

†† Coefficient of variation.

* Mean values followed by different letters within each column indicate differences between treatments.

the incorporated method in the inter-rows were substantially lower when compared to the application performed on both sides of the row (0.30 m next to the row with incorporation to 0.08 m depth) (Table 15.6). In general, the incorporation of N-fertilizer showed an increase in cane yields of 14% (28 Mg ha⁻¹) and 19% (38 Mg ha⁻¹) compared to the applications on surface and inter-rows zone, respectively, which did not differ between them. The gains in sugarcane yields in the incorporated method may be attributed to the fertilizer that is allocated directly to soil and close to the root system, thereby transposing the layer of straw. Moreover, 80% of the crop root system is located up to a 0.40 m depth and with a horizontal distance of the cane row of 0.30 m (Otto et al., 2009), which justify the greater efficiency of this operation. The application by triple operation, although the fertilizer is

TABLE 15.6 Effects of N Application Methods on Sugarcane Yields (Mg ha⁻¹) in Two Agricultural Seasons in an Area Located in Sales de Oliveira (Castro et al., 2016)

Methods of Application	2010	2011	Cumulative
Mechanical cultivation	115b*	80ab	195b
Surface	129a	76b	205b
Incorporated	139a	94a	233a
Control	84c	61c	145c
LSD (10%)†	5	14	18
CV (%)††	9	15	8

† Least significant difference.

†† Coefficient of variation.

* Mean values followed by different letters within each column indicate differences among treatments.

incorporated into the soil, presents a lower efficiency in terms of yields due to the distance in which the fertilizer is placed in relation to the ratoons and root system, which is lesser in the inter-row zone (Otto et al., 2009).

Although a higher increase in cane yield is observed, the incorporated application of N fertilizer in large areas may be limited because of its lower operating efficiency compared to other options. An alternative is to apply N-fertilizer granules with high pressure on soil covered with straw, so that they can cross this physical barrier and reach the soil surface (Figure 15.15c). Borges et al. (2016) compared three application methods of N fertilizer in ratoons under a green cane system, including banded application below straw, banded surface application and incorporated application into the soil. The authors found that sugarcane yield was higher when N fertilizers were incorporated or placed below the straw in relation to the surface application (Table 15.7). Considering the sugarcane yields over two crop seasons, the applications incorporated into the soil and placed below the straw provided increases in productivity of 16% (26 Mg ha^{-1}) and 13% (21 Mg ha^{-1}) compared to surface application, respectively. Therefore, the choice of the application methods of N fertilizer is one of the fastest and most efficient ways to increase productivity in sugarcane fields. Furthermore, the application of this nutrient incorporated into the soil contributes to reduce greenhouse gas emissions (Borges et al., 2016; Liu et al., 2005; Van Kessel et al., 2013) and ensures the sustainability of sugarcane-based ethanol.

It should be emphasized that the sugarcane harvest season in the central-south region of Brazil varies from April to November, during which time

TABLE 15.7   Effects of the Application Methods of N Fertilizer on Sugarcane Yields (Mg ha^{-1}) During Two Agricultural Seasons (Borges et al., 2016)

Methods of Application	2015	2016	Cumulative
Control	72 c*	77 b	149 c
Surface	84 b	81 b	165 b
Below straw	96 a	90 a	186 a
Incorporated	100 a	90 a	191 a
LSD (10%)†	10	9	13
CV (%)††	8	7	5

† Least Significant Difference.

†† Coefficient of variation.

* Mean values followed by different letters within each column indicate differences among treatments.

fertilizations are performed (Otto et al., 2016). As previously described, the maintenance of straw on the soil surface contributes to the retention of soil moisture over time. However, it is advisable to apply N fertilizer to rain-fed crops (more than 90% of Brazilian sugarcane areas) in periods when there is greater water availability, thereby increasing the nitrogen use efficiency (NUE) by the plant. Such statement can be justified by analyzing the average productivity (Figure 15.16; Castro, 2016) over two crop seasons, in which the same variety was harvested at different times (beginning: April or mid-season: August) and fertilized with the same dosage (100 kg N ha^{-1}) in the wet or dry seasons.

15.5 FUTURISTIC APPROACH

The proper management of sugarcane fields is one of the key aspects that can ensure greater economic returns from agricultural production. This management should take into account the intrinsic spatial variability of the soil and crop, managing the crop according to its real needs. Precision agriculture (PA) is an increasingly popular practice among sugarcane producers in Brazil. With all technologies and techniques advocated by the PA (e.g., yield monitors, geographic information systems – GIS, variable rate technologies,

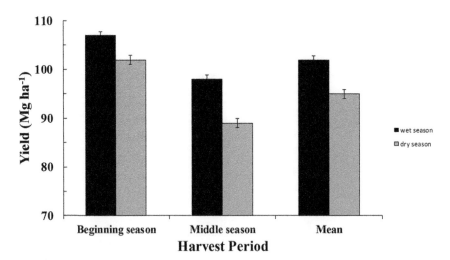

FIGURE 15.16   Productivity of sugarcane fertilized with N (dosage of 100 kg N ha^{-1}) under rain-fed regimes and harvested at different seasons. Adapted from Castro (2016).

soil and plant sensors, among others), these tools seek to maximize crop yields and minimize production costs, aiming to identify and eliminate possible causes of reduced productivity. The PA assists the producer to make localized management of the agricultural production, involving the spatial and temporal variability of the crop fields as well as allowing a greater economic and sustainable return of the production. PA is not an end in itself but an innovative, integrated and internationally standardized approach towards sustainable agriculture in which resource efficiency is increased, thereby reducing risks and uncertainty of decision-making in the management process.

One of the fundamental tools in this technological package is the yield monitors that are essential for the monitoring and identification of the problems in the fields for proper decision making. Widely developed and used in grain crops, yield monitors are still rarely used in commercial sugarcane fields in Brazil. The application examples come mainly from the academy with the first works of Magalhães and Cerri (2007); these examples are used by other authors to investigate the correlation of productivity with soil attributes (Rodrigues et al., 2013; Souza et al., 2010), delineate management zones (Zacharias et al., 2013), among other scientific applications (Menegatti and Molin, 2004; Rodrigues et al., 2012). Interesting results are presented by Magalhães et al. (2014) using the SIMPROCANA® yield monitor (ENALTA, São Carlos, Brazil) for accompaniment of two agricultural seasons. The authors found that in a terrain slope greater than 10% (western part of the experimental field), there was an abrupt change in productivity from one year to another (Figure 15.17). This fact is attributed to the ratoons pulling off, which is induced by harvesters. The results revealed that the absence of this tool could bring serious economic losses to the producers, since it was possible to identify the place where the abrupt reduction in productivity took place only through a yield monitor. Therefore, it becomes possible to take mitigating measures, such as the reform of the site or even the non-application of fertilizers, which increases the efficiency use of resources. The application of N fertilizer may also benefit from this tool through application of fertilizers according to the variability of the crop's agricultural productivity, thereby increasing the profitability of production in addition to lesser environmental impacts on the natural resources.

On the other hand, with estimates of the need to increase sugarcane production (OECD-FAO, Agricultural Outlook 2015–2024, 2015) and a growing concern regarding the environment issues (COP-21, Paris, France,

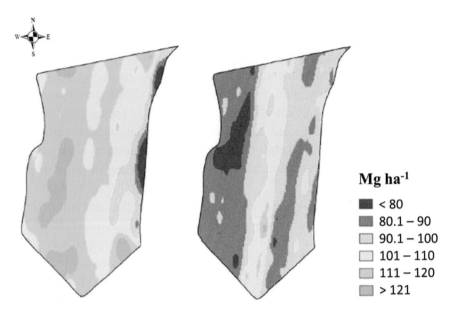

FIGURE 15.17 Sugarcane productivity map for plant-cane (left) and first ratoon (right) (*Source*: Magalhães et al. (2014)).

2016), the consumption of fertilizers to meet these demands will play a key factor for ensuring the sustainability of the entire production process. Fertilizer management should be increasingly efficient and rational, managing the crops not as homogeneous entities (Rossato, 2011) but according to the intrinsic spatial variability of soil and crop. Within this context, the application of fertilizers at a variable rate will be fundamental in sugarcane fields, following 4R IPNI premise, i.e., apply at right place, right amount, right source and right time.

Currently, in order to have an adequate mapping of the physical and chemical soil attributes to apply fertilizers at variable rates, it is necessary to perform a dense sampling of the area. This involves sampling, pretreatment and laboratory analysis of the sample; therefore, the activity becomes impractical both physically and economically (Peets et al., 2012). Within the PA technologies that are accessible for acquisition of high-quality information and that aim at the adequate management of the soil spatial variability, the apparent electrical conductivity (ECa) of the soil has been highlighted as an efficient method to quickly evaluate with high resolution and low cost the general condition of soil fertility (Sudduth et al., 2005). In addition to being intrinsically related to moisture content, research data show that ECa is able to detect variations in soil properties, such as salinity, clay content, cation exchange capacity, pore size and distribution and organic matter (Corwin and Lesch, 2003; Kaffka et al., 2005; Kitchen et al., 2003; Sudduth et al., 2001). Along with the information supplied by ECa, statistical and geostatistical techniques are able to provide tools that can assist the producer in the establishment of an intelligent sampling (Sanches et al., 2016). It is characterized by soil sampling in representative locations of the area that can lead to an accurate description of the soil spatial variability by means of few sampling points that is economically feasible. Magalhães and Sanches (2015) evaluated five different scenarios to verify the benefits of limestone application through targeted sampling and geostatistical techniques, such as: scenario 1 – ordinary kriging in a regular grid (1 sample per 0.25 ha); scenario 2 – average by field in a regular grid (1 sample per 2.5 ha); scenario 3 – inverse distance weighting in a regular grid (1 sample per 2.5 ha); scenario 4 – ordinary kriging in a targeted sampling (1 sample per 2.5 ha); scenario 5 – kriging with external drift in a targeted sampling (1 sample per 2.5 ha). Considering scenario 1 as the most real (dense sampling) and scenario 2 as the most practiced scenario by sugarcane producers in Brazil, the authors found that the total application of limestone would be 30 and 40 tons (e.g., difference of 10 tons), respectively (Figure 15.18). By means of targeted sampling (scenarios 4 and 5), the authors showed that it is possible to apply limestone in the correct locations and in the quantities needed to correct soil acidity (Figure 15.19), demonstrating a better approach to manage the real needs presented by the crop. This methodology may contribute significantly to the application of other important fertilizers for sugarcane cultivation, such as potassium and phosphorus, thereby improving crop yields and producer's profits as well as contributing positively to natural resources.

15.6   CONCLUDING REMARK

Because of the size of the sugarcane production area in Brazil and the need for intensive agricultural mechanization, it is determined that new sugarcane management systems are adopted to maximize crop productivity and reduce production costs. Currently, the Brazilian sugarcane sector goes through periods of productivity stagnation, but there is a disposition for the agricultural sector to follow techniques that are easy to adopt (e.g., crop rotation, minimum cultivation, plots systematization, use of GPS and traffic

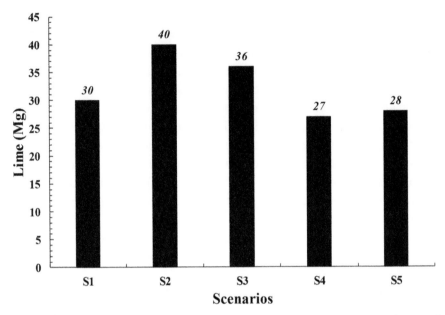

FIGURE 15.18 Total lime application in different evaluated scenarios by Magalhães and Sanches (2015).

control) and that exhibit proven efficiency in increasing productivity while reducing production costs and environmental impacts. These techniques associated with PA technologies will certainly generate a modernization of the Brazilian sugarcane fields, rationalizing the use of agricultural inputs and ensuring efficient management of the physical, financial and environmental resources of sugarcane production. It should be pointed out that

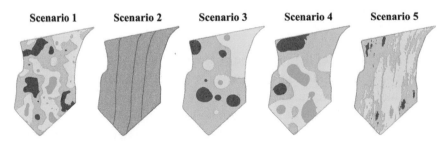

FIGURE 15.19 Lime application maps at the different evaluated scenarios. Dark green (no application); light green (0.0–0.5 Mg ha⁻¹); orange (0.5–1.0 Mg ha⁻¹); red (>1.0 Mg ha⁻¹). *Source*: Magalhães and Sanches (2015).

Brazilian sugarcane farming is almost totally based on rain-fed production, and agricultural practices that help to maintain soil moisture (e.g., minimum cultivation and maintenance of straw on the soil after harvesting without burning) for longer periods are crucial and should be prioritized.

Therefore, there are fast adoption methods to overcome the productivity problems that have arisen with the intensification of mechanization in the Brazilian sugarcane sector. Producers need to pursue these mechanisms, and this learning can be practiced in other producing regions worldwide under intensive mechanization of agricultural operations, noting that these principles are not new and have already been deployed for some years in sugarcane fields in Australia.

ACKNOWLEDGMENTS

The authors thank to Brazilian Bioethanol Science and Technology Laboratory (CTBE) for the infrastructure provided, Santa Fé mill for the data and information availability, and the National Council for Scientific and Technological Development (CNPq) for the research and financial support.

KEYWORDS

- **biomass production**
- **conservationist management practices**
- **crop rotation**
- **precision agriculture.**
- **reduced tillage**
- *Saccharum* **spp.**
- **straw management**
- **traffic control**

REFERENCES

Ambrosano, E. J. et al. (2011). Produtividade da cana-de-açúcar após o cultivo de leguminosas. *Bragantia, 70,* 810–818.

Andrade, R. I., Stone, L. F., & Silveira, P. M. (2009). Culturas de cobertura e qualidade física de um Latossolo em plantio direto. *Revista Brasileira de Engenharia Agrícola e Ambiental, 13,* 411–418.

Arrigoni, E. B., (2011). New Pests of the New System. *Revista Opiniões, 28,* 65–67.

Barbosa, L. C., (2015). *Atributos físicos do solo e desenvolvimento radicular à cana planta em diferentes sistemas de manejo.* Dissertação (Mestre em Engenharia Agrícola). Faculdade de Engenharia Agrícola, Feagri – Unicamp. 79p.

Belardo, G. C., Cassia, M. C., & Silva, R. P., (2015). *Processos Agrícolas e Mecanização da Cana-de-Açúcar.* Sociedade Brasileira de Engenharia Agrícola SBEA, Jaboticabal.

Borges, C. D. et al. (2016). Formas de aplicação de N-fertilizante em cana-de-açúcar. *In: 10° Congresso Nacional da STAB,* Ribeirão Preto – SP, *Anais.,* 1, 84–87.

Braunbeck, O. A., Magalhães, P. S. G., & Garcia, M. O., (2008). Colheita e recuperação de biomassa. In: Cortez, L. A. B., Lora, E. S., Gomez, E. O., (Eds.). *Biomassa para energia.* Campinas, Editora Unicamp, 1, pp. 63–90.

Cáceres, N. T., & Alcarde, J. C., (1995). Adubação verde com leguminosas em rotação com cana-de-açúcar (*Saccharum spp.*). *STAB, 3,* 16–20.

Campanelli, V. P. C., (2013). Uso de tecnologias para a produção de cana-de-açúcar. In: VI Simpósio tecnologia de produção de cana-de-açúcar. *Escola Superior de agricultura "Luiz de Queiroz" Universidade de São Paulo, On-line.*

Campanhão, J. M., (2003). *Manejo da soqueira de cana-de-açúcar submetida à queima acidental da palhada remanescente da colheita mecanizada.* Dissertação (Mestrado em Produção Vegetal). Faculdade de Ciências Agrárias e Veterinárias Universidade Estadual Paulista "Julio de Mesquita Filho", Jaboticabal, 75p.

Carvalho, J. L. N. et al. (2013). Input of sugarcane post-harvest residues into the soil. *Scientia Agricola, 70,* 336–344.

Carvalho, J. L. N. et al. (2017). Agronomic and environmental implications of sugarcane straw removal: a major review. *Global Change Biology Bioenergy, 9,* 1181–1195.

Castro, S. G. Q. et al. (2016). Best practices of nitrogen fertilization management for sugarcane under green cane trash blanket in Brazil. *Sugar Tech.* DOI: 10.1007/s12355–016–0443–0.

Castro, S. G. Q. et al. (2012). Aplicação Diferenciada de Nitrogênio em Soqueira de Cana-de-Açúcar. In: Reunião Brasileira de Fertilidade do Solo e Nutrição de Plantas, 30 ed., Maceió-AL, SBCS/UFAL. *Anais.*

Castro, S. G. Q., & Franco, H. C. J., (2016). Em busca da produtividade perdida: I. Manejo da adubação nitrogenada em soqueira de cana-de-açúcar. *Revista Canavieiros, 12,* 61–65.

Castro, S. G. Q., (2016). *Manejo da Adubação nitrogenada em cana-de-açúcar e diagnose por meio de sensores de dossel.* Tese (Doutorado em Engenharia Agrícola) Faculdade de Engenharia Agrícola – Feagri/Unicamp, 129p.

Castro, S. G. Q., Franco, H. C. J., & Mutton, M. A., (2014). Harvest managements and cultural practices in sugarcane. *Revista Brasileira de Ciência do Solo, 37,* 299–306.

CONAB – Companhia Nacional de Abastecimento, (2015). Acompanhamento da safra Brasileira. Cana-de-açúcar, Safra 2014/15. *Quarto Levantamento,* Brasília, *2,* 1–20.

CONAB – Companhia Nacional de Abastecimento, (2016). Acompanhamento da safra brasileira. Cana-de-açúcar, Safra 2015/16. *Quarto levantamento,* Brasília, *2,* 1–76.

Corwin, D. L., & Lesch, S. M., (2003). Application of soil electrical conductivity to precision agriculture: Theory, principles, and guidelines. *Agronomy Journal, 95,* 455–471.

Datagro – Consultoria de Etanol e Açúcar. Acompanhamento e índices da safra de cana-de-açúcar no Brasil. Available at: www.datagro.com.br. Accessed: September 4, 2015.

Dinardo-Miranda, L. L., & Fracasso, J. V., (2013). Sugarcane straw and the populations of pests and nematodes. *Scientia Agricola.*, *70,* 305–310.

Duarte Júnior, J. B., & Coelho, F. C. A., (2008). Cana-de-açúcar em sistema de plantio direto comparado ao sistema convencional com e sem adubação. *Revista Brasileira de Engenharia Agrícola e Ambiental, 12,* 576–583.

Echeverry, C. H. I., (2014). Cosecha mecanizada y agricultura de precisión. In: 9° Congreso de Técnicos Azucareiros de Latinoamérica y el Caribe. ATALAC, *San Jose, Costa Rica, Anais.*, pp. 207–207.

FAOSTAT. Food and agriculture organization of the United Nations. Statistics Division. Available at: http://faostat3.fao.org/home/E. Accessed: April *18,* 2015.

Ferreira, D. A. et al. (2015). Contribution of N from green harvest residues for sugarcane nutrition in Brazil. *Global Change Biology Bioenergy, 8,* 859–866.

Food and Agriculture Organization of the United Nations (FAO-2015). Available at: http://fao.org.br/home. Accessed: September *02,* 2015.

Fortes, C., Trivelin, P. C. O., & Vitti, A. C., (2012). Long-term decomposition of sugarcane harvest residues in Sao Paulo state, Brazil. *Biomass and Bioenergy, 42,* 189–198.

Franco, H. C. J. et al. (2007). Estoque de nutrientes em resíduos culturais incorporados ao solo na reforma de áreas com cana-de-açúcar. *STAB.* Sociedade dos Técnicos Açucareiros e Alcooleiros do Brasil, *25,* 32–36.

Franco, H. C. J. et al. (2013). Assessment of sugarcane trash for agronomic and energy purposes in Brazil. *Scientia Agricola, 70,* 305–312.

Goldemberg, J., (2007). Ethanol for a sustainable energy future. *Science*, 3*15,* 808–810.

Idea. A evolução dos custos de produção e ganhos possíveis de ATR na colheita. *In:18° Seminário de Mecanização e produção de cana-de-açúcar. Ribeirão Preto.* 2016. *CD-ROM.*

Kaffka, S. R. et al. (2005). Relationship of electromagnetic induction measurements, soil properties, and sugar beet yield in salt-affected fields for site-specific management. *Computer and Electronics in Agriculture, 46,* 329–350.

Kitchen, N. R. et al. (2003). Conductivity and Topography Related to Yield for Three Contrasting Soil-Crop Systems. *Agronomy Journal, 95,* 483–495.

Landell, M. G. A. et al. (2013). Residual biomass potential of commercial and pre-commercial sugarcane cultivars. *Scientia Agricola, 70,* 299–304.

Liu, X. J. et al. (2005). Tillage and nitrogen application effects on nitrous and nitric oxide emissions from irrigated corn fields. *Plant Soil, 276,* 235–249.

Magalhães, P. S. G. et al. (2014). Precision agriculture in sugarcane production: a key tool to understand its variability. In: 12 International Conference on Precision Agriculture, Sacramento. *Proceedings, 1,* 1–16.

Magalhães, P. S. G., & Braunbeck, O. A., (2010). Avaliação tecnológica da mecanização da cana-de-açúcar. In: Bioetanol de Cana-de-Açúcar: *P&D para produtividade e sustentabilidade, 14.*

Magalhães, P. S. G., & Cerri, D. G., (2007). Yield Monitoring of Sugar Cane. *Biosystems Engineering, 96,* 1–6.

Magalhães, P. S. G., & Sanches, G. M., (2015). Improving the kriging of soil attributes using soil electrical conductivity as external drift. In: First Conference on Proximal Sensing Supporting Precision Agriculture, Torino. *Proceedings*, 1, 1–4.

Manechini, C. Cultivo mecânico da soqueira de cana colhida sem queimar (condensado da experiência acumulada). Relatório Interno Coopersucar – Projeto Cana crua, Piracicaba. *1,* pp. 1–10.

Menegatti, L. A. A., & Molin, J. P., (2004). Remoção de erros em mapas de produtividade via filtragem de dados brutos. *Revista Brasileira de Engenharia Agrícola e Ambiental,* 8(1), 126–134.

Ministério da Agricultura Pecuária e Abastecimento. Balanço do Agronegócio do Brasil, ano de 2014. Available at: http//agricultura.gov.br. Accessed: September 02, (2015).

Miranda, J. M., Rigone, M. V., & Silveira, F. T., (2011). Associação da crotalaria com adubação orgânica e mineral na produtividade da cana-de-açúcar. *Bioscience Journal, 27,* 948–953.

OECD. Food and agriculture organization of the United Nations. OECD-FAO agricultural outlook 2015. OECD Publishing, Paris. doi:10.1787/agr_outlook-2015-en Accessed: January 06, 2016.

Oliveira, M. W. et al. (1999). Decomposição e liberação de nutrientes da palhada de cana-de-açúcar em campo. *Pesquisa Agropecuária Brasileira, 34,* 2359–2362.

Otto, R. et al. (2016). Nitrogen use efficiency for sugarcane-biofuel production: what is the next? *Bioenergy Research,* doi: 10.1007/s12155-016-9763-x.

Otto, R., (2015). Desafios para aumento da produtividade e longevidade do canavial In: VII Simpósio tecnologia de produção de cana-de-açúcar. *Escola Superior de agricultura "Luiz de Queiroz"* Universidade de São Paulo.

Otto, R., Trivelin, P. C. O., Franco, H. C. J., Faroni, C. E., & Vitti, A. C., (2009). Root system distribution of sugar cane as related to nitrogen fertilization, evaluated by two methods: Monolith and probes. *Revista Brasileira de Ciência do Solo, 33,* 601–611.

Peets, S. et al. (2012). Methods and procedures for automatic collection and management of data acquired from on-the-go sensors with application to on-the-go soil sensors. *Computers and Electronics in Agriculture, 81,* 104–112.

Pereira, F. S. et al. (2011). Qualidade física de um latossolo vermelho submetido a sistemas de manejo avaliado pelo índice. *Revista Brasileira de Ciência do Solo, 35,* 87–95.

Robertson, F. A., & Thorburn, P. J., (2007). Management of sugarcane harvest residues: consequences for soil carbon and nitrogen. *Australian Journal of Soil Research, 45,* 13–23.

Rodrigues, F. A. et al. (2013). Correlation between chemical soil attributes and sugarcane quality parameters according to soil texture zones. *Soil Science, 178,* 147–156.

Rodrigues, F. A., Magalhães, P. S. G., & Franco, H. C. J., (2012). Soil attributes and leaf nitrogen estimating sugar cane quality parameters: Brix, pol and fiber. *Precision Agriculture, 14,* 270–289.

Rossato, O. B., (2011). Metodologias de avaliação e aplicação de nutrientes nas culturas de cana-de-açúcar e algodão. Tese (Doutorado Agronomia – Energia na Agricultura) Faculdade de Ciências Agronômicas, *Universidade Estadual Paulista,* Botucatu.

Rossi Neto, J., (2015). Desenvolvimento do sistema radicular e produtividade da cana-de-açúcar em diferentes espaçamentos. Dissertação (Mestrado em Engenharia Agrícola). *Faculdade de Engenharia Agrícola* – Feagri/Unicamp.

Sanches, G. M. et al. (2016). Determinação de grade de solo por meio de ferramentas de agricultura de precisão baseada na variedade espacial da argila. In: 10º Congresso Nacional da STAB, Ribeirão Preto – SP, *Anais., 1,* 150–154.

Sanches, G. M., Magalhães, P. S. G., & Rodrigues, F. A., (2012). O planejamento da área de reforma para o plantio da cana-de-açúcar com o uso do piloto automático pode contribuir para melhorar o rendimento? In: Congresso Brasileiro de Agricultura de Precisão. *Anais., 1,* 1–4.

Santos, F., Borém, A., Caldas, C. (Ed.), (2010). Cana-de-açúcar: bioenergia, açúcar e álcool – tecnologias e perspectivas. Viçosa/MG: Universidade Federal de Viçosa. 577p.

Santos, F. A. et al. (2014). Otimização do pre-tratamento hidrotérmico da palha de cana-de-açúcar visando a produção de etanol celulósico. *Química Nova, 37,* 56–62.

Segato, S. V. et al. (2006). Atualização em produção de cana-de-açúcar. Piracicaba: CP 2nd ed. 415p.

Silva Junior, C. A. et al. (2010). Alterações nos atributos físicos de um Latossolo Vermelho sob diferentes métodos de preparo para o plantio da cana-de-açúcar. *Revista Agrarian, 3,* 111–118.

Smith, D. M., Inman-Bamber, N., & Thorburn, P. J., (2005). Growth and function of the sugarcane root system. *Field Crops Research, 92,* 69–183.

Souza, G. S. et al. (2014). Effects of traffic control on the soil physical quality and the cultivation of sugarcane. *Revista Brasileira de Ciência do Solo, 38,* 135–146.

Souza, Z. M. et al. (2010). Análise dos atributos do solo e da produtividade da cultura de cana-de-açúcar com o uso da geoestatística e árvore de decisão. *Ciência Rural, 40,* 840–847.

Spekken, M. et al. (2016). Planning machine paths and row crop patterns on steep surfaces to minimize soil erosion. *Computers and Electronics in Agriculture, 124,* 194–210.

Spekken, M., & Bruin, S., (2012). Optimized routing on agricultural fields by minimizing maneuvering and servicing time. *Precision Agriculture, 14,*1–2.

Spekken, M., Molin, J. P., & Romanelli, T. L., (2015). Cost of boundary manoeuvres in sugarcane production. *Biosystems Engineering, 129,* 112–126.

Sudduth, K. A. et al. (2005). Relating apparent electrical conductivity to soil properties across the north–central USA. *Computers and Electronics in Agriculture, 46,* 263–283.

Sudduth, K. A., Drummond, S. T., & Kitchen, N. R., (2001). Accuracy issues in electromagnetic induction sensing of soil electrical conductivity for precision agriculture. *Computers and Electronics in Agriculture, 31,* 239–264.

Tavares, O. C. H., Lima, E., & Zonta, E., (2010). Crescimento e produtividade da cana planta cultivada em diferentes sistemas de preparo do solo e de colheita. *Acta Scientiarum Agronomy, 32,* 61–68.

Trivelin, P. C. O. et al. (2013). Impact of sugarcane trash on fertilizer requirements for Sao Paulo, Brazil. *Scientia Agricola, 70,* 345–352.

União da Industria de Cana-de-Açúcar (UNICA – 2016) Notícias – Valorização do Etanol da cana-de-açúcar. Available at: http://www.unica.com.br/noticia/2778178692031116810́5/california-valoriza-etanol-brasileiro-de-cana-de-acucar-para-manter-seu-ar-mais-limpo/. Accessed: February *02,* 2016.

União da Industria de Cana-de-Açúcar (UNICA – 2015) Histórico de produção e moagem. Available at: www.unica.com.br. Accessed: September *03,* 2015.

Van Kessel, C. et al. (2013). Climate, duration, and N placement determine N_2O emissions in reduced tillage systems: a meta-analysis. *Global Change Biology, 19,* 33–44.

Vasconcelos, A. C. M., & Casagrande, A. A., (2008). Fisiologia do sistema radicular. In: Dinardo-Miranda, L. L., Vasconcelos, A. C. M., Landell, M. G. A., (Eds.). Cana-de-açúcar. Campinas: *Instituto Agronômico,* pp. 79–97.

Vitti, A. C. et al. (2010). Utilization of nitrogen from trash by sugarcane ratoons. *Sugarcane International, 28,* 249–253.

Vitti, A. C., et al. (2007). Produtividade da cana-de-açúcar relacionada à localização de adubos nitrogenados aplicados sobre os resíduos culturais em canavial sem queima. *Revista Brasileira de Ciência do Solo, 31,* 491–498.

CHAPTER 16

TRANSFER OF TECHNOLOGY APPROACHES FOR SUSTAINED SUGARCANE PRODUCTIVITY

T. RAJULA SHANTHY

Extension, ICAR-Sugarcane Breeding Institute, Coimbatore–641007, Tamil Nadu, India,
E-mail: rajula.sbi@gmail.com

CONTENTS

16.1   INTRODUCTION

Indian farming had been on conventional lines till the first waves of Green Revolution in the 60s, which gave a swift improvement to the production and productivity of main crops. Rapid diffusion of technical knowledge from the Agricultural Research System to the farmers and reporting of farmer's response to the research system is one of the decisive inputs in transfer of farm know-how. Extensive research over the last numerous years have made technologies available to enhance cane production in India. However, the effect of such massive research works could not be completely reflected in terms of cane output. In many developing countries, more than the technology, the major concern today is the rate of diffusion of technology from the points of production to the ultimate users.

Our nation has grown by leap and bounds in sugarcane cultivation from 1930–31 to 2015–16. The area under sugarcane was just 1.19 m.ha in 1930–31 with 36.35 m.tons of cane production and the area increased to 5.20 m.ha area with a record sugarcane production of 370 m.tons in 2015–16. In the same way, productivity increased from 30.9 t/ha in 1930–31 to 72.0 t/ha in 2015–16. Nevertheless, there is a broad discrepancy in productivity amid the cane growing states of the country. This gap has to be bridged so as to capitalize on per unit production and reduce the expenditure of sugarcane farming.

Brazilian sugar industry is quite flexible, wherein the dealing out of cane for sugar or alcohol relies on the existing market trends of these supplies, which ensures sustained growth and profitability of the industry; Indian sugar industry lacks this suppleness (Solomon, 2011). India ranks third in the domestic sugar price and is among the lowest in the globe. The production price of Indian sugar is in the medium range and is higher than that in Australia and Brazil but lower than in the USA. Probably, we may face taut rivalry from African countries, whose cost of production is lower than that in India (Nair, 2014).

Sugarcane growers in India have their own convincing grounds for non-adoption of scientific sugarcane technologies, which could be overlooked in 'top-down' approach. This includes:

- technical know-how is too intricate;
- non-trialability of the recent technologies;
- incompatibility with farm and individual intentions;
- highly rigid and unrefined;
- not lucrative;
- too high principal expense;
- extra learning as well is required;
- risk and vagueness is high;
- contradictory information;
- non-availability of infrastructure; and
- lack of communal infrastructure.

In 'top–down' approach, scientists decide research priorities; produce innovations they believe are superior for the growers and provide research outcomes to extension agents. This is typically referred as linear adoption or diffusion model (Rogers, 1983). Outreach initiatives, which focus on growers are to be supported and this should be a primary task in achieving the sugarcane targets for 2025 AD and beyond.

16.2 A SYSTEMS PERSPECTIVE

The worth of a systems viewpoint for accepting and analyzing technology generation and propagation is already recognized. Even though this approach is criticized because they are so abstract, systems analyses propose holistic vantage points for considering the factors that impede or augment the two-way flow of technology and scientific information between farmers and the public organizations that compose the system.

Any agricultural knowledge system consists of four components set in a larger context (Figure 16.1). The components are technology generation, technology transfer, technology utilization, and support system. The organizations that constitute the components, as well as others in the system environment influence each other in complex ways.

Systems standpoint in agricultural research requires a total turnaround of the technology generation approach *viz.,* a farmer first or farmer participation approach where farmers are involved in all stages of technology

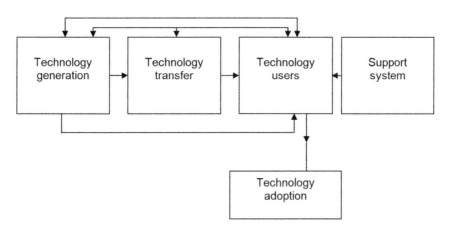

FIGURE 16.1 Agricultural knowledge system.

development (Chambers and Ghildyal, 1985; Van den Ban and Mkwawa, 1997). Farmer participatory research is a methodical procedure involving people in the analysis of their own situations. The term is increasingly used in agricultural research and development circles. In the context of dwindling resource base, in future, agriculture can be sustained only by producing more food, fuelwood, fiber and other commodities from less land, water and energy per unit of output without auxiliary ecological damage. Earlier approaches to on-farm research recognized the importance of farmers, but were not able to effectively incorporate farmers' skills and experimental practices into the research process. This has laid the way for the growth of 'farmer participation'.

Food and Agriculture Organization has developed the Strategic Extension Campaign (SEC) methodology by giving weightage to people's participation in strategic planning, systematic management and field implementation of extension programs.

16.3   CHALLENGES FACED

16.3.1   IN SUGARCANE AGRICULTURE

Sugarcane crop currently faces severe challenges on several fronts in our country. This includes:

- Increasing cane production to meet the ever-increasing demand for sugar.
- Diminishing interest in agriculture due to decreasing net return from the crops.
- Diversion of labor from agriculture to other sectors. The fact here remains that adequate farm labor is unavailable and the available labor is unaffordable.
- Sugarcane despite being the singular source for green power and ethanol is often accused as a water guzzler. With growing ecological concerns, cane crop can no longer remain oblivious of the need to conserve water.
- Monocropping and excessive usage of fertilizers and pesticides have horrendously harmed our soil productivity. We hence need to put in place appropriate alternative and ameliorative measures to restore soil productivity for sheer sustainability of cane cultivation.

- As the sugar industry moves from control to competitive era, conflicts of role must necessarily give way to convergence of interests. We need to build in synergies between diverse stakeholders so that we are ready to take on global competition.

16.3.2 IN SUGARCANE EXTENSION

The dynamic situation in sugarcane farming poses several challenges to the extension system. Some of the challenges with which the extension system has to cope up include:

- How to reach individual farm families involved in sugarcane cultivation and deal with risk prone complex scenario?
- How to improve technology standards of farmers through effective and optimal technology use, especially when the pressure on land is increasing and common property lands are slowly fading out.
- How to face the emerging agricultural development situation as a sequel to the technological and development interventions?
- How to take cognizance of the changes that are taking place in the society in terms of Shift from farming to industry, rural to urban (migration), focus from social to economic issues.

16.4 WAY FORWARD

Sugarcane agriculture is faced with several challenges, as also uncommon opportunities. Being a crop of a well-organized sector, namely the sugar factories, public private partnership as a governance strategy can very well be utilized to converge the efforts of varied organizations at the grass root level of a community. The sugar industry is a major player to the rural economy because the sugar mills are situated in rural areas and provides substantial service to rural population. A SWOT analysis of the extension approaches till late 1980s reveal a lacuna of trickle down approach rather than a democratic one. Change thus became inevitable in all developmental spheres of steering India into a developed country. In this era of globalization, mankind has become assertive and needs a rationalization of any developmental efforts being undertaken.

16.4.1   TECHNOLOGY GENERATION

It is imperative to develop critical technologies that could serve as panacea of a contemporary constraint faced by users in their situation. Always, technologies should be developed in terms of the capabilities, resources, opportunities, social and economic considerations that are both intrinsic and extrinsic to the technology highlighting on positive and negative consequences for the users (Gupta, 2009). Farmers need to be involved in the planning, executing, monitoring, evaluation and follow-up actions in technology generation and dissemination. Nonetheless, the attributes of technology *viz.,* relative advantage, compatibility, complexity, trialability, and observability have to be kept in mind when making explicit recommendations.

16.4.2   PUBLIC PRIVATE PARTNERSHIP (PPP)

PPP through participatory mode has been acknowledged as one of the successful mechanisms to upgrade the country's infrastructure services. There is a possibility to influence PPPs as a significant means in the agricultural sector also. Improved yield and per unit production is a critical need, with India still battling food uncertainty and poverty. Participatory approach in farming is normally found at the community level where the strengths of the public and private sectors supplement each other in providing information and advisory services that cater to the needs of growers and agrarian communities. The public sector's obligation for providing technical knowledge and services can be achieved through tapping the potential of the private sector to add local context in a commercial setting.

16.4.3   CONVERGENCE OF EXTENSION SERVICES

There are a number of service providers in the field, providing different useful services like information and service support to farmers. This includes state and central government organizations, agribusiness companies, agri-based entrepreneurs, input dealers, manufacturing firms, Non Governmental Organizations, farmers' cooperatives and progressive farmers. Duplication of efforts with multiplicity of agents attending extension work without convergence is witnessed often times. A coordinated attempt is needed to synergize and converge these efforts at the district level through Agricultural

Technology Management Agency and below to improve the performance of various stakeholders.

16.4.4 INCREASED ROLE OF EXTENSION AGENTS

Along with the government efforts in stepping up the allocations for sugarcane development programs, the sugar factories/department of agriculture are expected to play an increasing role in improving the productivity levels in their respective zones through their extension agents. Such initiatives would help the growers in each area to improve their sugarcane yields apart from assisting the sugar factories to get large quantities of better quality cane at appropriate time, thus ensuring higher sugar recovery levels with increased crushing duration. Extension agents can use the following ways to help the cane growers to reach their goals:

- giving advice in the right time to create awareness of a problem;
- providing wide range of alternatives enabling the growers to make better choice;
- letting them know about the expected outcomes of each alternative;
- making them realize which goal is the most important;
- assisting them to take decisions in a planned way, be it as an individual or as a group;
- make them to learn from experience, from experiments; and
- stimulating them to exchange information with fellowmen and act upon it.

16.4.5 FRONTLINE DEMONSTRATIONS

'Seeing is believing' is an adage that holds good even today. Realizing this, the Indian Council of Agricultural Research, New Delhi introduced the concept of 'frontline demonstrations.' These are also referred as first-line demonstrations conducted by the scientists to demonstrate proven technologies in the farmers' fields under different agroclimatic regions and farming situations. The objective is to convince extension functionaries and growers about the potentialities of technologies for wide scale diffusion and adoption. Frontline demonstrations are generally conducted in at least one hectare to have better impact of the demonstrated technologies. The special features of frontline demonstrations are:

- Conducted under close supervision of the scientists.
- Newly released or technologies likely to be released in near future are demonstrated.
- Critical inputs and training alone are provided from the scheme budget; remaining inputs are to be borne by the growers themselves.
- Training of farmers associated with frontline demonstration is a pre-requisite.

ICAR-Sugarcane Breeding Institute, Coimbatore is involved in conducting frontline demonstrations in Coimbatore and Tirupur districts in Tamil Nadu since 2001–02. Considerable improvement was obtained in yield in all the demonstration plots compared to the farmers' practice. As the demonstrations are conducted in growers' fields, they serve as a motivation for the farmers in the neighborhood. During the course of conducting such demonstrations in farmers' fields, the factors contributing to high crop production and field constraints of production were studied to get feedback information and discussed in scientific meetings (Rajula Shanthy, 2011).

16.4.6 PLURALISTIC EXTENSION

Development practitioners have coined an "Integrated, Multi- Disciplinary and Holistic approach to Rural Development (IMDH-RD)," reiterating the need to address both agricultural and rural problems to make development effective and environmentally sustainable (Davidson, 2007). To overcome the challenges created by a multisectoral approach, pluralistic agricultural extension mechanism is required to coordinate the various stakeholders. Pluralism well realizes the heterogeneity of farming population and thus the need for diversity in extension delivery systems. This approach requires farmers and other rural inhabitants, extension agents, input supply dealers, researchers, policy makers, and so on to focus their efforts on advocating sustainable agriculture, which is inclusive rather than exclusive.

16.4.7 COOPERATIVE APPROACH

In states like Uttar Pradesh, and to a limited extent in Andhra Pradesh, Bihar, Haryana, Madhya Pradesh and Punjab sugarcane is procured by the sugar mill through cooperative societies, whereas, factories deal directly with sugarcane growers in all the other major sugarcane producing States like Tamil Nadu and Karnataka. It is advocated to have a direct link between the sugar mills and cane growers in terms of computerized operations for preparation of cane supply calendars, issuing supply orders to the farmers, cane price payment through banks and maintenance of grower-wise records, etc. This in turn creates an opportunity for the farmer to interact with the cane development personnel on scientific crop management. Tripartite agreement of sugar mills with lending banks/credit agencies and growers can be promoted for procurement of sugarcane to smooth the progress of use of *Kisan* Credit Cards, availability of loans to farmers and custom hiring of machineries.

16.4.8   YIELD GAP ANALYSIS

The notion of yield gaps in agricultural crops originated from constraint studies carried out by International Rice Research Institute during the late seventies. Yield gap comprises two major components; the first component, yield gap I is the difference between research station yield and potential farm yield and this component is not exploitable. The second component is referred as yield gap II and is the difference between potential farm yield and actual average farm yield. Yield gap II is exploitable and can be minimized by deploying research and extension approaches and government interventions.

Yield gap in sugarcane in the different states of India during 2015–16 are given in Table 16.1.

This unexploited reservoir of yield as revealed by the difference in productivity among research institutions, demonstration plots, crop yield competition plots and farmers fields in general form the future source of output growth. To achieve utmost sugarcane production and meet the growing needs, the breach between the potential and actual farm yields should be bridged.

The means to bridge the yield gap in sugarcane cultivation include:

- Extension agencies should concentrate to improve the yield per unit area.
- Need based technological interventions have to be introduced at field level to increase the adoption rate thus it reduces the technological gap in the actual field condition.
- Farmer participated 'service providing activities' have to be upscaled to ensure the timely operations of field activities.

TABLE 16.1   Yield Gap in Sugarcane in the Different Indian States During 2015–16

States	Potential yield* (t/ha)	Existing yield# 2015–16 (t/ha)	Yield gap (%)
Andhra Pradesh	169.01	78.0	53.8
Bihar	127.16	56.8	53.3
Gujarat	188.37	65.0	65.5
Haryana	181.27	73.0	60.0
Karnataka	183.00	84.6	54.0
Madhya Pradesh	167.25	42.2	74.8
Maharashtra	192.33	77.4	60.0
Odisha	158.70	66.3	58.2
Punjab	182.60	70.0	61.7
Tamil Nadu	203.70	107.0	47.5
Utter Pradesh	162.72	62.4	61.7
West Bengal	165.07	105.0	36.40
Average	173.43	73.97	57.24

*Highest cane yields level (potential productivity) obtained in Zonal Varietal Trail plots in the respective state sugarcane research station.

#Provisional productivity obtained by the farmers in different states (State wise yield of sugarcane in India - Cooperative sugar).

- Extension personnel can classify the farmers into lower, medium and higher yield categories and should concentrate on first two categories to bring up them into third category.
- Extension personnel should undertake farming system research and extension approach to narrow down yield gap.
- Before initiating various measures to reduce yield gap, one should prepare 'technological mapping' prevailing in the area and then the need based technological intervention should be promoted.
- In technology development programme, farmers' practices have to be taken care for inclusive development of sugarcane production system.

16.4.9   CONSTRAINT ANALYSIS

Constraints are very specific to each location and hence there is a need to identify the farm level constraints that hinder realizing the untapped yield reservoir. Constraints are generally knowledge and information based, technology related, socioeconomic, infrastructure oriented and from a managerial perspective. The impact of such field level constraints is visualized in the level of technology adoption. Since cane growers are a part of sugarcane development, area specific approach should be formulated to perk up the socioeconomic conditions, reinforce infrastructural facilities and credit facilities, and engage the numerous development organizations located in the area.

Methodical analysis of yield gap coupled with field level problem analysis must be done to have a direct impact on adoption of recommended technologies and assessment of the effect of related cane development programmes. Participating farmers of such an approach are to be objectively and effectively engaged in such attempts. By this way alone, effectual strategy formulation and its successive operation can be guaranteed.

16.5   INFORMATION AND COMMUNICATION TECHNOLOGY (ICT) INITIATIVES AT SUGARCANE BREEDING INSTITUTE

Information technology is blooming every day and has a very high presence in this era of communication explosion. ICT based module is a tool for communicators of all trades and a successful mechanism for change. Information technology is far more interactive and personalized that can provide services, specifically the information based on the needs and requirements of the end users. This type of facility makes a constructive impact on adoption and utilization of the innovative techniques in hi-technology agriculture. Information technology has a union of three strands of technologies, *viz.,* computers, microelectronics, and communication. It covers different things such as books, print, reprography, telephone network, broadcasting and computers. Transfer of new farm technology can be done through Satellite communication, Geographic Information System, Computer network, Video, Mobile, Internet, Radio and Reprography. The key to future expansion in use of ICT based technology transfer in farming lies not with the technology itself, but to a certain extent with the way the messages are treated to reach the intended clients. Historians, however testify that its not technology that changes the world, but the people who adopt and use technology make the changes. In this scenario, it is high time that the extension functionaries realize the brain waves of the users, develop suitable communication systems and sensitize the clients towards betterment.

The ICT initiatives at ICAR-SBI, Coimbatore include:

- databases on promising sugarcane varieties;
- interactive multimedia module on sugarcane;
- expert system on major pests affecting sugarcane crop
- 'caneinfo' website on sugarcane;
- short video films on specific topics;
- 'Cane Adviser' mobile app on sugarcane.

16.5.1 MULTIMEDIA

16.5.1.1 What is Multimedia?

The term multimedia describes a woven combination of text, sound, animation, graphic art, and video. When the end user, who is the viewer of a multimedia module has the facility to control what elements are delivered and when, it is referred as interactive multimedia. When a structure of linked elements is provided through which the actual user can navigate, interactive multimedia becomes hypermedia. Professionals who weave multimedia are known as multimedia developers (Tearle and Dillon, 2001).

The amalgamation of multimedia know-how into the communication milieu has the prospective to change an audience from passive recipients of information to dynamic participants in a media rich learning process.

16.5.1.2 Context of Multimedia Usage

Multimedia is proper whenever a human interface connects a person to electronic information of any type. Multimedia augments conventional text-only, computer interfaces, and yields measurable profit by gaining and holding consideration and interest. Multimedia enhances the retention of information. A properly woven multimedia can be intensely entertaining.

Studies on retention rate indicate that if stimulated with audio, retention rate is about 20%, that of audio visual is up to 30%, and in an interactive multimedia presentation with complete involvement, retention rate is as high as 60%.

16.5.1.3   Prerequisites for a Multimedia Module

Multimedia is always a team effort wherein artwork is done by graphic artists, video shoots by video producers, sound editing by audio technicians and programming by programmers. To create multimedia you need to have a real yearning to communicate, because multimedia is creating, essentially, an entirely new syntax for communications. One must have an interest in human psychology because you need to anticipate the brain waves of users. One should adopt a strategy that allows one to prototype and test the interactive design assumptions. The three basic items for preparing a multimedia module are software, hardware and creativity.

16.5.1.4   The Multimedia Production Team

High quality interactive multimedia is always the outcome of the efforts of a creative dedicated production team. Though several software applications are available, the production of fine quality high-end multimedia is generally the work of a team of specialists. Typically this team consists of the production manager, instructional designer, scriptwriter, content specialist(s), text editor, multimedia architect or (program authoring specialists), computer graphic artist, audio and video specialists and computer programmer. The quality multimedia production is a result of effective and synergistic teamwork.

16.5.1.5   Multimedia Authoring

Multimedia authoring consists of bringing together different elements of multimedia or the building blocks of multimedia such as text, graphics, sound, animation, video, etc., into a comprehensive presentation, and if and when needed, incorporation of interactive elements, such as navigation tools. The application software used for such integration of the elements into a presentation and incorporation of interactivity through navigational controls is called multimedia authoring software or simply authoring tool.

16.5.1.6 Selection of Multimedia Authoring Tool

Varied multimedia authoring tools are available in the market. The basic considerations are purpose of application, level of expertise available/amenable, time frame for the production, budget and level of cosmetics to be applied. The authoring tools generally used for education purpose includes Macromedia Director, Flash, Authorware, Asymetrix ToolBook, HyperCard, HyperStudio and SuperCard.

16.5.1.7 Multimedia on Sugarcane Production

Sugarcane Breeding Institute has led the way in developing a multimedia module on sugarcane production. It describes and illustrates the numerous steps implicated in scientific sugarcane cultivation and available as an information kiosk format for sugarcane growers, extension personnel and researchers interested in the practicalities of sugarcane production technology. The module envisages the use of diverse multimedia structure blocks such as text, audio, video, graphics and animation. The module contains exhaustive information from sett planting to harvest with appropriate image clippings allowing the learner to learn at his own rate of knots of learning (Rajula Shanthy and Thiagarajan, 2011).

16.5.1.8 Steps Involved in the Development of the Module

i.   Initially the storyboard for the module was finalized and detailed literature on sugarcane cultivation was collected.

ii. The available text matter was organized and generated as text files.

iii. Graphic elements on sugarcane production were collected and the same were scanned and generated as graphic files.

iv. Digital stills relevant to the text matter were generated and stored as graphic files.

v.   Relevant video files were created using the available CDs on sugarcane production technologies and suitable video clippings were also generated for the purpose.

vi. The module was integrated using Macromedia Flash as the authorware.

The module has three parts viz., sugarcane varieties, crop production (both for plant and ratoon crop) and crop protection technologies (pest, disease and nematodes). The module comes in the form of CD for nearly one-hour duration.

16.5.2 EXPERT SYSTEMS

Expert system is a recent advanced application in using computers based on the application oriented branch of artificial intelligence. So far, a great deal of expert systems had been developed and applied in office automation, science, and medicine including agriculture. Development of expert systems in agriculture has gained momentum due to the complexity of problems farmers face like yield losses, soil erosion, retreating market prices from international competition, mounting chemical pesticides expenditure and pest resistance and economic barriers hindering adoption of farming strategies. The application of expert system technology to agriculture seems natural in India. Extension professionals may not always be accessible for discussion or may not be able to optimize financially viable decisions. Expertise gained in the developed countries could also be disseminated to developing countries, though resistance to new technology would have to be overcome.

Definition: An "expert system" is an intelligent program that uses comprehension and inference procedures to unravel problems that are tricky enough to need significant human expertise for their clarification (Jay, 1997).

16.5.2.1 Components of Expert System

Even though no general standard for expert systems is available, most include: a knowledge base storing domain facts of the concerned topic and connected heuristics, an inference procedure or control structure for utilizing the knowledge base and a natural language user interface.

16.5.2.2 Expert Systems in Agriculture

Computer assisted farming through distributed information based expert system is a frontier area in transfer of technology. Generally, expert system is designed to answer questions typed at a keyboard on any diversified topic, for example, in pest control, the need to spray, selection of right chemical to spray, mixing and application, optimal machinery management practices, weather damage recovery such as freeze, frost or drought, etc., Popular expert systems developed for use in agriculture are Rice-Crop Doctor, AGREX, Farm Advisory System, Sericulture, Grain Marketing Advisor, COMAX, Gossym, Pomme, PLANT d/s, etc.

16.5.2.3 Work Done at Sugarcane Breeding Institute

Sugarcane Breeding Institute has developed an expert system that can diagnose major pests infesting sugarcane crop, provide details of the identified pest and suggest preventive/curative measures. SUGAR-EX illustrates the use of expert systems in agriculture and specifically in the area of sugarcane production through development of a prototype, taking into consideration the major pests limiting sugarcane yield (Rajula Shanthy and Mukunthan, 2009).

This prototype is a result of joint effort by Extension Scientists and Entomologists of the Institute. The subject matter expert knowledge on sugarcane pests was obtained from the previous publications of the Institute, reference books and personal experience. The pests included in the system for identification and suggesting preventive and curative measures are early shoot borer, internode borer, top borer, plassey borer, gurdaspur borer, stalk borer, root borer, white grub, scale insect, mealy bug, white fly, pyrilla, woolly aphid, blister mite, red mite and white mite.

16.5.2.4 The Logic Flow

The concise logic flow of the expert system is as follows:

i. Basic input: The basic input in the expert system can be given as:
   a. part of the sugarcane plant where damage symptoms are seen like stem/cane, leaf, root;
   b. depiction of the damage symptom;
   c. life stages of the pest – egg, larva, nymph, pupa, adult;
   d. description of the life stage.
ii. Taking into account these symptoms/description of the life stages, the pest is identified.

iii. The user then has the choice to either exit the window or further reset or proceed with diagnosis for other pests.

iv. After identification, the user can directly get hyperlinked to know more about the pest, get management measures as well as get a graphic view of the damage symptom/ life stages of the identified pest.

All these operations are done using a "mouse," which helps a user to get familiar with the system quickly and work on it easily. In this system, computer, graphics, digitized pictures and text are being used following a "Decision Tree," logic.

16.5.3 WEBSITE ON SUGARCANE FOR CANE DEVELOPMENT PERSONNEL/CANE GROWERS

Sugarcane Breeding Institute has developed an interactive and user-friendly website that provides a stage for sugarcane growers, cane development personnel, scientists and students to share information and knowledge on sugarcane. It is available at http://caneinfo.nic.in. The exclusive features of the website include ask a specialist, online fertilizer advisor service, discussion forum, multimedia gallery, newsfeeds, and streaming video programs on sugarcane.

16.5.4 MOBILE APP ON SUGARCANE

Mobile phones are invariably used by all categories of people, be it in developed or developing countries. With an intention of harnessing the potentialities of mobile technology for technology dissemination, a mobile app on sugarcane was developed. Initially, baseline surveys were conducted to understand the information needs of cane growers, the type of mobile phones they use, and the pattern of mobile phone usage through focus group discussions with cane growers and millers. Subsequently, a database of sugarcane farmers of over 1,50,000 covering 12 states of the country with details of mobile numbers and demographic profile to establish a network of sugarcane stakeholders through mobile usage was developed. The content for the mobile app covering all aspects of sugarcane cultivation right from sett plating to harvest was digitized and arranged as sugarcane varieties, crop production technologies and crop protection technologies (~220 pages) with relevant photographs (~650 nos.) and entire matter was translated from English to Hindi and Tamil so as to have the app in trilingual. Software for programming, integration, code review and configuration for the module was created and an android based mobile application software that would serve as an efficient information dissemination system on sugarcane was developed and made available in google playstore for free download.

The features of the app developed include static as well as dynamic platforms. The static display is on entire sugarcane agriculture including fertilizer schedule for all the cane growing states, both text and graphics which would serve as a digital compendium; The dynamic user-interface facets include login dialogue, downloader, tailor-made scheduler app with reminder messages for the individual registered users, query sender in text or graphic form, message sorting, facilities for short message service, etc. The app targets to provide farmers, cane development personnel, line department officials and rural residents with timely access to extension services such as advancements on scientific sugarcane production, advice on appropriate technology and availability of services.

16.5.5   IS THERE A SINGLE EXTENSION STRATEGY?

No solitary "magic bullet" can support demand-driven and/or decentralized extension providing better ways to address these changes and challenges. Rather, they are yielding another collage of responses which could probably include: specific client-focused extension, privatization of extension, farmer-to-farmer participatory extension, broadening the role of extension to include other related sectors, unified delivery of rural and agricultural development services, and demand driven extension.

Since the operational setting in Extension system is becoming increasingly complex, there is a need for extension professionals to:

- work under complex and fluid circumstances with little supervision;
- analyze growers' constraints efficiently;
- pay attention to and learn from growers;
- communicate successfully and toil with farmers and farmer groups;
- provide options, based on ideology of science and good farming practices.

Regardless of denigrations of linear technology transfer (trickle down) theory, there is still a need for access to reliable scientific information from the sugarcane research institutes, just as there is a need to provide for active participation by growers in research and development activities. Personal face to face contact, one-to-one swap over of information and advice, whether from farmer to farmer or from a change agent to the cane grower (and vice versa), will continue to be significant. New information and communication technologies will smooth the progress of training and information exchange. The future of public extension depends on our capability to construe trends and make use of information technology to deliver programs and teach crisis solving for the farmers. But such strategy will need to be supplemented by other extension strategies so as to enhance sugarcane productivity and ensure increased profits for the cane growers.

16.6 CONCLUSION

It is understood that no single theory, model or strategy is probably to be adequate by itself. If the likelihood of adoption is to be improved, communication channels must be utilized in a perfect sequence, progressing from mass media to interpersonal channels (Sill, 1958). Sugarcane production has an impact in rural areas and has served as a successful instrument for carrying progressive trends in the countryside. The development of sugarcane on a holistic sense in the years to come requires mix up of innovative extension approach with technological interventions appropriate for a given situation. It is merely the appropriateness of such blending that demands to formulate and implement a strategic plan for holistic sugarcane development.

KEYWORDS

- **sugarcane**
- **transfer of technology**
- **strategies**
- **ICT modules**
- **outreach programs**

REFERENCES

Chambers, R., & Ghildyal, B. P., (1985). Agricultural research for resource-poor farmers: *The Frmer-First-and Last-Model. Agricultural Administration 20*(1), 1–30.

Davidson, A. P., (2007). Participation, education and pluralism: towards a new extension ethic. *Journal of Development in Practice, 17*(1), 39–50.

Gupta, D. D., (2009). Sugarcane development – Technological interface between tradition and modernity. *Agrobios*, (India), Jodhpur, pp. 359.

Jay Liebowitz., (1997). *Worldwide Perspectives and Trends in Expert Systems.* AI Magazine, *18*(2), 115–119.

Nair, N. V., (2014). An overview of sugarcane agriculture in India and the future scenario. In: T. Rajula Shanthy et al., (Eds.). *Recent Technologies for Increase Sugarcane Productivity.* ISBN 9789380800257. PRDAG Print, Coimbatore, pp. 1–7.

Rajula Shanthy, T., (2011). Strategies for effective dissemination of appropriate technologies to sugarcane growers in India. *Sugar Tech., 13*(4), 354–359.

Rajula Shanthy, T., & Mukunthan, N., (2011). SUGAR-EX, an information and communication technology based decision making tool, *Sugar Tech., 11*(1), 69–72.

Rajula Shanthy, T., & Thiagarajan, R., (2011). Interactive multimedia instruction versus traditional training programmes: Analysis of their effectiveness and perception, *The Journal of Agricultural Education and Extension (Formerly European Journal of Agricultural Education and Extension), 17*(5), 459–472.

Rogers, E. M., (1983). The Innovation-Decision Process. In *Diffusion of Innovations.* Free Press, New York, pp. 163–206.

Sill, M., (1958). *Personal, Situational and Communicational Factors Associated with Farm Practice Adoption,* PhD thesis, University Park: Pennsylvania State University.

Solomon, S., (2011). The Indian Sugar Industry an Overview. *Sugar Tech., 13*(4), 255–265.

Tearle, Penni, & Dillon, P., (2001). The Development and Evaluation of a Multimedia Resource to Support ICT Training: Design Issues, Training Processes and User Experiences,' *Innovations in Education and Teaching International, 38*(1), 8–18.

Van den Ban, A. W., & Mkwawa, M. K., (1997). Towards a participatory and demand-driven Training and Visit system. *European Journal of Agricultural Education and Extension, 4*(1), 117–123.

INDEX

Printed and bound by CPI Group (UK) Ltd, Croydon, CR0 4YY

23/10/2024

01777703-0016